PROTEIN AGGREGATION IN BACTERIA

WILEY SERIES IN PROTEIN AND PEPTIDE SCIENCE

Vladimir N. Uversky, Series Editor

Metalloproteomics · Eugene Permyakov

Instrumental Analysis of Intrinsically Disordered Proteins: Assessing Structure and Conformation · Vladimir Uversky and Sonia Longhi

Protein Misfolding Diseases: Current and Emerging Principles and Therapies · Marina Ramirez-Alvarado, Jeffery W. Kelly, and Christopher M. Dobson

Calcium Binding Proteins · Eugene Permyakov and Robert H. Kretsinger

Protein Chaperones and Protection from Neurodegenerative Diseases · Stephan Witt

Transmembrane Dynamics of Lipids · Philippe Devaux and Andreas Herrmann

Flexible Viruses: Structural Disorder in Viral Proteins · Vladimir Uversky and Sonia Longhi

Protein and Peptide Folding, Misfolding, and Non-Folding · Reinhard Schweitzer-Stenner

Protein Oxidation and Aging · Tilman Grune, Betul Catalgol, and Tobias Jung

Protein Families: Relating Protein Sequence, Structure, and Function · Edited by Christine Orengo and Alex Bateman

Protein Aggregation in Bacteria: Functional and Structural Properties of Inclusion Bodies in Bacterial Cells · Edited by Silvia Maria Doglia and Marina Lotti

PROTEIN AGGREGATION IN BACTERIA

Functional and Structural Properties of Inclusion Bodies in Bacterial Cells

Edited by

SILVIA MARIA DOGLIA

MARINA LOTTI

WILEY

Published by John Wiley & Sons, Inc., Hoboken, New Jersey.
Published simultaneously in Canada.

Library of Congress Cataloging-in-Publication Data

Protein aggregation in bacteria : functional and structural properties of inclusion bodies in bacterial cells / edited by Silvia Maria Doglia, Marina Lotti.
p. ; cm. – (Wiley series in protein and peptide science)
Includes bibliographical references and index.
ISBN 978-1-118-44852-6 (cloth)
I. Doglia, Silvia Maria, editor of compilation. II. Lotti, Marina, editor of compilation. III. Series: Wiley series in protein and peptide science.
[DNLM: 1. Inclusion Bodies–physiology. 2. Bacterial Proteins–chemistry. 3. Bacterial Proteins–physiology. 4. Bacterial Proteins–therapeutic use. 5. Inclusion Bodies–chemistry. 6. Industrial Microbiology–methods. QW 51]
QP552.C34
572′.69–dc23

2013049564

Printed in the United States of America

10 9 8 7 6 5 4 3 2 1

CONTENTS

CONTRIBUTORS

Diletta Ami, Department of Biotechnology and Biosciences and Department of Physics "G. Occhialini," University of Milano-Bicocca, Milano, Italy

Marina Caldara, Functional Biology Group, Molecular Physiology of Plants and Micro-organisms, KU Leuven, Belgium

Olivia Cano-Garrido, Institut de Biotecnologia i de Biomedicina and Department de Genètica i de Microbiologia, Universitat Autònoma de Barcelona, Bellaterra, Barcelona, Spain, and CIBER en Bioingeniería, Biomateriales y Nanomedicina, Bellaterra, Barcelona, Spain

Yongli Chen, Department of Natural Sciences, Hawaii Pacific University, Kaneohe, Hawaii

José Luis Corchero, CIBER en Bioingeniería, Biomateriales y Nanomedicina, Bellaterra, 08193 Barcelona, Spain, and Institut de Biotecnologia i de Biomedicina and Department de Genètica i de Microbiologia, Universitat Autònoma de Barcelona, Bellaterra, Barcelona, Spain

Rafael Cubarsí, Departament de Matemàtica Aplicada IV, Universitat Politècnica de Catalunya, Barcelona, Spain

Ario de Marco, Department of Biomedical Sciences and Engineering, University of Nova Gorica, Vipava, Slovenia, and Therapeutic Antibody Platform, Institut Curie, Paris, France

Silvia Maria Doglia, Department of Biotechnology and Biosciences and Department of Physics "G. Occhialini," University of Milano-Bicocca, Milano, Italy

Vanessa Franssens, Functional Biology Group, Molecular Physiology of Plants and Micro-organisms, KU Leuven, Belgium

Elena García-Fruitós, CIBER en Bioingeniería, Biomateriales y Nanomedicina, Bellaterra, Barcelona, Spain, and Institut de Biotecnologia i de Biomedicina and Department de Genètica i de Microbiologia, Universitat Autònoma de Barcelona, Bellaterra, Barcelona, Spain

Pierre Goloubinoff, Department of Plant Molecular Biology, Faculty of Biology and Medicine, University of Lausanne, Lausanne, Switzerland

Marina Lotti, Department of Biotechnology and Biosciences, University of Milano-Bicocca, Milano, Italy

Rayees U.H. Mattoo, Department of Plant Molecular Biology, Faculty of Biology and Medicine, University of Lausanne, Lausanne, Switzerland

Antonino Natalello, Department of Biotechnology and Biosciences, University of Milano-Bicocca, Milano, Italy

Bernd Nidetzky, Institute of Biotechnology and Biochemical Engineering, Graz University of Technology, Graz, Austria

Spela Peternel, EN-FIST Centre of Excellence, Ljubljana, Slovenia

Loredano Pollegioni, Department of Biotechnology and Life Sciences, Insubria University, Varese, Italy, and "The Protein Factory" Center, Milano, Italy

Imma Ratera, Department of Molecular Nanoscience and Organic-Materials, Institut de Ciència de Materials de Barcelona, Bellaterra, Barcelona, Spain, and CIBER en Bioingeniería, Biomateriales y Nanomedicina, Bellaterra, Barcelona, Spain

Escarlata Rodríguez-Carmona, CIBER en Bioingeniería, Biomateriales y Nanomedicina, Bellaterra, Barcelona, Spain, and Institut de Biotecnologia i de Biomedicina and Department de Genètica i de Microbiologia, Universitat Autònoma de Barcelona, Bellaterra, Barcelona, Spain

Joaquin Seras-Franzoso, Institut de Biotecnologia i de Biomedicina and Department de Genètica i de Microbiologia, Universitat Autònoma de Barcelona, Bellaterra, Barcelona, Spain, and CIBER en Bioingeniería, Biomateriales y Nanomedicina, Bellaterra, Barcelona, Spain

Vladimir N. Uversky, Department of Molecular Medicine and USF Health Byrd Alzheimer's Research Institute, College of Medicine, University of South Florida, Tampa, Florida, and Institute for Biological Instrumentation, Russian Academy of Sciences, Pushchino, Moscow Region, Russia

Esther Vazquez, Institut de Biotecnologia i de Biomedicina and Department de Genètica i de Microbiologia, Universitat Autònoma de Barcelona, Bellaterra, Barcelona, Spain, and CIBER en Bioingeniería, Biomateriales y Nanomedicina, Bellaterra, 08193 Barcelona, Spain

Jaume Veciana, Department of Molecular Nanoscience and Organic-Materials, Institut de Ciència de Materials de Barcelona, Bellaterra, Barcelona, Spain, and CIBER en Bioingeniería, Biomateriales y Nanomedicina, Bellaterra, Barcelona, Spain

Salvador Ventura, Departament de Bioquímica i Biologia Molecular, Facultat de Biociències, Universitat Autònoma de Barcelona, Bellaterra, Spain, and Institut de Biotecnologia i de Biomedicina, Universitat Autònoma de Barcelona, Bellaterra, Spain

Anna Villar-Piqué, Departament de Bioquímica i Biologia Molecular, Facultat de Biociències, Universitat Autònoma de Barcelona, Bellaterra, Spain

Antonio Villaverde, Institut de Biotecnologia i de Biomedicina and Department de Genètica i de Microbiologia, Universitat Autònoma de Barcelona, Bellaterra, Barcelona, Spain, and CIBER en Bioingeniería, Biomateriales y Nanomedicina, Bellaterra, Barcelona, Spain

Marielle Wälti, Department of Chemistry and Applied Biosciences, ETH Zurich, Zurich, Switzerland

Lei Wang, Department of Natural Sciences, Hawaii Pacific University, Kaneohe, Hawaii

Christian Wasmer, Department of Biological Chemistry and Molecular Pharmacology, Harvard Medical School, Boston, Massachusetts

Joris Winderickx, Functional Biology Group, Molecular Physiology of Plants and Micro-organisms, KU Leuven, Belgium

PREFACE

Only a few years ago, nobody would have written a book devoted to inclusion bodies, a topic that was considered of relevance only for the impact of protein aggregation on the effectiveness of the biotechnological production of recombinant proteins. What brings bacterial protein aggregates into the spotlight today is a combination of novel scientific achievements and of a more curiosity-driven approach to the subject.

From the point of view of fundamental knowledge, a wealth of information has been produced in recent years about protein misfolding and aggregation. This knowledge integrates results on the thermodynamics and kinetics of protein folding and on the action of "cell quality control systems," chaperones and folding helpers, in keeping proteins in their native and soluble conformation. The biological role of aggregation was also investigated and the features of the protein sequences driving their propensity to aggregate were studied by new bioinformatics tools.

More specifically, growing attention was addressed to the mechanistic and basic issues leading to the deposition of aggregated structures of recombinant proteins in bacterial cells. One major breakthrough in the field was the finding that aggregates might be formed by (fully or at least partially) properly folded recombinant proteins, a result suggested by the presence of native-like secondary structures that were revealed unexpectedly by structural studies on inclusion bodies containing a number of different polypeptides. This important advancement was further substantiated by the detection of biological activity (enzyme activity, fluorescence, binding ability) associated with inclusion bodies. These results first had an impact on protein production, since a large body of experimental evidence indicated that the presence of native-like structure aggregates, obtained by proper modulation of the expression condition, allows for easy protein refolding.

After the initial reports on their structural properties, several additional features of inclusion bodies attracted the interest of researchers working in different fields of protein science. Most intriguing were the amyloid-like properties of bacterial protein aggregates, which suggested the possible use of bacterial cells as a model system for the study of amyloid aggregation and toxicity in a cellular environment. Actually, the amyloid-like behavior of inclusion bodies was confirmed when the presence of fibrils was detected within aggregates extracted from bacteria. Thanks to this experimental evidence, bacterial cells are today key model systems for the study in vivo of amyloid aggregation and of the effects induced by anti-amyloidogenic compounds.

In addition, several applications are blooming in different research fields. For example, the biological activity and the ability of inclusion bodies to penetrate eukaryotic cells, suggested their use as nanoparticles in medicine for the controlled release of drugs. Applications to tissue engineering were also achieved successfully for the construction of protein scaffolds in regenerative medicine.

Such an integrated development in knowledge and in the possibilities of application reflects in the structure of this book, which covers a broad spectrum of issues related both to fundamental knowledge and to application. Thermodynamics, as well as the mechanistic and cellular aspects of protein folding, are treated primarily with reference to bacteria but also with some attention to simple eukaryotes. Some of the chapters also deal with applications, which means not only implementation of the protocols for the production of recombinant proteins but also the direct use of aggregated proteins as immobilized biocatalysts and as natural nanostructured surfaces. Finally, we should emphasize the potential of using "easy to handle" protein aggregates to study the evolution of pathogenic aggregation processes (amyloidosis) and to screen for drugs.

We warmly acknowledge the effort of all the authors to provide a comprehensive view of different but interconnected issues related to protein aggregation in unicellular organisms.

Silvia Maria Doglia
Marina Lotti
State University of Milano-Bicocca, Italy

INTRODUCTION TO THE WILEY SERIES IN PROTEIN AND PEPTIDE SCIENCE

Proteins and peptides are the major functional components of the living cell. They are involved in all aspects of the maintenance of life. Their structural and functional repertoires are endless. They may act alone or in conjunction with other proteins, peptides, nucleic acids, membranes, small molecules, and ions during various stages of life. Dysfunction of proteins and peptides may result in the development of various pathological conditions and diseases. Therefore, the protein/peptide structure–function relationship is a key scientific problem lying at the junction point of modern biochemistry, biophysics, genetics, physiology, molecular and cellular biology, proteomics, and medicine.

The *Wiley Series in Protein and Peptide Science* is designed to supply a complementary perspective from current publications by focusing each volume on a specific protein- or peptide-associated question and endowing it with the broadest possible context and outlook. The volumes in this series should be considered required reading for biochemists, biophysicists, molecular biologists, geneticists, cell biologists, and physiologists as well as those specialists in drug design and development, proteomics, and molecular medicine with an interest in proteins and peptides. I hope that each reader will find in the volumes in this book series interesting and useful information.

First and foremost I would like to acknowledge the assistance of Anita Lekhwani of John Wiley & Sons, Inc. throughout this project. She has guided me through countless difficulties in the preparation of this book series, and her enthusiasm, input, suggestions, and efforts were indispensable in bringing the *Wiley Series in Protein and Peptide Science* into existence. I would like to take this opportunity to thank everybody whose contribution in one way or another has helped and supported this project. Finally, a special thank you goes to my wife, sons, and mother for their constant support, invaluable assistance, and continuous encouragement.

Vladimir N. Uversky

1

FUNDAMENTALS OF PROTEIN FOLDING

VLADIMIR N. UVERSKY

1.1 FOLDING–MISFOLDING–NONFOLDING CROSSROADS

Data accumulated to date suggest that a protein can exist in at least three global forms: functional and folded, functional and intrinsically disordered (nonfolded), and nonfunctional and misfolded or aggregated. Therefore, in *in vitro* experiments or in the cellular environment, a polypeptide chain faces a choice among three potential routes: nonfolding, folding, and misfolding, with the last two representing competitive routes to higher structural order (see Figure 1.1). For a single-chain protein, folding, nonfolding, and misfolding pathways represent a choice of each individual molecule, whereas unproductive protein aggregation, which frequently follows protein misfolding and is often associated with the pathogenesis of several diseases, is a fate of the ensemble of molecules.

Multiple factors, originating from the peculiarities of the protein amino acid sequence and/or features of the protein environment, might affect the choice among nonfolding, folding, and misfolding. The primary selection between low structural complexity (nonfolding) and increased complexity (folding) is

Protein Aggregation in Bacteria: Functional and Structural Properties of Inclusion Bodies in Bacterial Cells, First Edition. Edited by Silvia Maria Doglia and Marina Lotti.

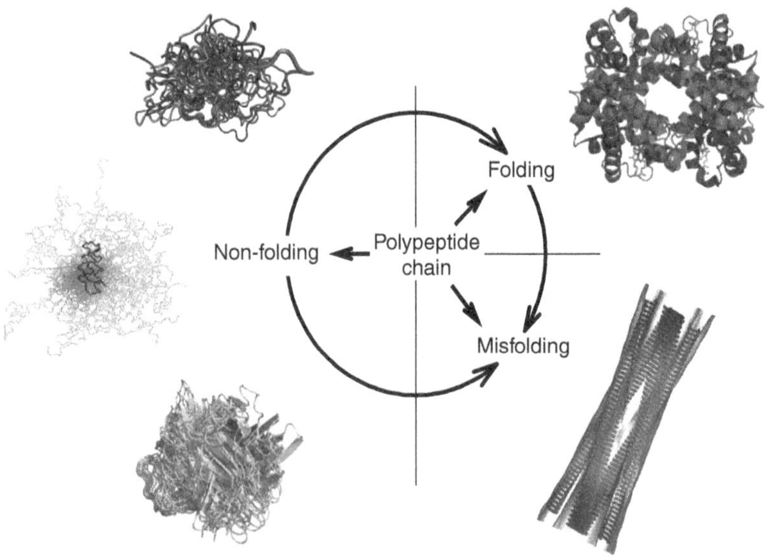

FIGURE 1.1 Fate of a newly synthesized polypeptide chain in a cell. (*See insert for color representation of the figure.*)

determined only by the amino acid composition. An abnormally highly charged polypeptide with low overall hydrophobicity will not fold, giving rise to a natively unfolded protein, whereas usually, under identical conditions a polypeptide chain will choose the folding path. On the other hand, some changes in the amino acid sequence (point mutations) may favor the misfolding pathway for both natively unfolded and natively folded proteins.

Importantly, for a given polypeptide chain a chosen fate is not a final one and a choice may be modulated further by environmental pressure (Figure 1.1) (Uversky, 2003b). For example, intrinsically disordered proteins (IDPs) may be forced to fold or misfold via the modification of their environment (i.e., addition of natural binding partners, changes in properties of solvent, etc.), whereas a destabilizing environment may push a natively folded protein to the misfolding route. Alternatively, the presence of chaperones may reverse the misfolding route and effectively dissolve small aggregates (Ben-Zvi and Goloubinoff, 2001). Another important point is that to some extent, the misfolding of natively unfolded proteins resembles the process of protein folding [i.e., it represents a pathway from a simple, flexible, and disordered conformation (e.g., an unfolded polypeptide chain), via a somehow more ordered partially folded intermediate, to a complex and rigid structure (e.g., an amyloid fibril)]. On the other hand, misfolding a rigid globular protein involves a step of transient disordering and formation of a partially unfolded intermediate, which is followed by the subsequent increase in the order originated from the formation of specific protein aggregates.

1.2 PROTEIN FOLDING

The ability of proteins to adopt their functional highly structured states in the intracellular environment during or after biosynthesis on the ribosome is one of the most remarkable evolutionary achievements in biology. In recent years our understanding of the mechanisms of the protein self-organization process has increased dramatically. This understanding has been achieved as a result of the application of a variety of novel experimental and theoretical approaches to this complex task (Bryngelson et al., 1995; Dill and Chan, 1997; Dobson et al., 1998; Dobson and Karplus, 1999; Dinner et al., 2000; Fersht and Daggett, 2002; Uversky, 2003b; Turoverov et al., 2010). In this chapter we cover several protein folding–related problems, such as a protein-folding code; major models of protein folding; polymer aspects underlying protein folding and structural peculiarities of the unfolded state and folding intermediates; and peculiarities of protein amino acid compositions favoring formation of equilibrium partially folded intermediate(s).

1.2.1 Protein-Folding Code

In protein biosynthesis, the information encoded in the DNA/mRNA nucleotide sequence is read step by step, and the corresponding amino acids are gathered into the polypeptide chain one after another. Therefore, the one-dimensional information encoded in the DNA nucleotide sequence is sequentially transformed into the one-dimensional information included in the protein amino acid sequence. As the interactions between remote amino acid residues play a crucial role in protein folding, this process obviously deviates from the linear information transduction. Only some amino acid residues are crucial for protein folding. Therefore, proteins with very low sequence homology can have similar structures, whereas a single amino acid replacement can significantly affect the rate of protein folding or in extreme cases can halt correct protein folding completely (Turoverov et al., 2010).

Many proteins have rigid globular structures in aqueous solutions and are functional only in this state. The native state of these proteins is a unique conformation, which is entropically unfavorable since it has significant restrictions on conformational freedom. On the other hand, the unfolded state of a polypeptide chain is entropically favorable, representing a dynamic ensemble of a large number of conformations originating from the main-chain rotational isomerization around ϕ and ψ angles. Therefore, the possibility of a given polypeptide chain folding into a compact state is determined by its ability to form numerous intramolecular contacts that differ in their physical nature, to compensate for the free-energy increase due to the decrease in the entropy component (Finkelstein and Ptitsyn, 2002). Although a native state of a globular protein has a clearly defined and unique three-dimensional structure, the folding and ordering degree can vary greatly for the various parts of a given protein. In x-ray data this is seen from the *B*-factor values, which characterize the mobility of separate atoms (Berman et al., 2000). For example, the atoms in the active center of an enzyme typically have high *B*-factors. Furthermore, some globular proteins are shown to have disordered, highly dynamic fragments (i.e., termini, loops, etc.) which could not be detected by x-ray analysis, thus corresponding to regions of missing electron density.

The first direct evidence that all the information necessary for a given polypeptide chain to fold into a unique tertiary structure is encoded in its amino acid sequence was obtained by Anfinsen's group (Anfinsen, 1973). The reduced and urea-denatured ribonuclease A was shown to restore its native structure and functional state completely after removal of the denaturant and the reducing agent. Later, the capability to regain the native structure *in vitro* was demonstrated for a variety of proteins. In essence, protein folding can be regarded as a second part of the genetic code, as the protein amino acid sequence contains information about its functional three-dimensional structure.

The folding of a typical globular protein occurs on a millisecond-to-second time scale. However, for a small protein consisting of 100 amino acid residues, a simple search for a native state (which meets the requirements of the free-energy minimum) among all the alternative conformations would take a billion years. This contradiction represents the essence of Levinthal's paradox (Levinthal, 1968), which is resolved by the fact that amino acid sequences bear information related not only to their native structures but also to the pathways of their formation.

Some globular proteins fold into a unique globular structure only after ligand binding. A typical example of such proteins is a globular actin, which denatures after the removal of ligands (Ca^{2+} and ATP) (Kuznetsova et al., 1999; Altschuler and Willison, 2008). It is very likely that such proteins could not attain ordered structured spontaneously and should therefore be classified as IDPs.

1.2.2 Protein-Folding Models

Several models of protein folding have been developed (Turoverov et al., 2010). The *nucleation and growth model* is based on the assumption that protein folding is similar to the crystallization process and that the limiting step in the folding process is nucleus formation (Radford, 2000). This model describes the folding of small single-domain proteins that follow the "all-or-none" principle.

The *sequential protein-folding model*, also known as the *framework* or *hierarchic model*, was proposed in 1973 by O.B. Ptitsyn (1973). It suggests that folding starts with the backbone forming secondary structure elements, which then interact to form a more advanced folding intermediate; the specific packing of the side chains concludes the process. Each stage of the folding process stabilizes the major structural elements formed in the preceding state, suggesting the existence of several folding intermediates. Therefore, long before the appearance of convincing experimental data, Ptitsyn put forward the idea of a partially folded conformation that serves as a universal folding intermediate. Eight years later, such a folding intermediate was found in a test tube (Dolgikh et al., 1981) and termed the *molten globule state* (Ohgushi and Wada, 1983). Other partially folded intermediates (e.g., pre-molten globule and highly ordered molten globule) were found later (Uversky, 2003b).

According to the current view, protein folding is realized via different pathways that are determined by the protein's energy landscape (Radford, 2000; Jahn and Radford, 2005). This landscape describes the dependence of the free energy on all the coordinates determining the protein conformation. The number of conformational states accessible by a polypeptide chain is reduced while approaching the native state. Therefore, this energetic surface is often called the *energy funnel* (Figure 1.2). The unfolded polypeptide

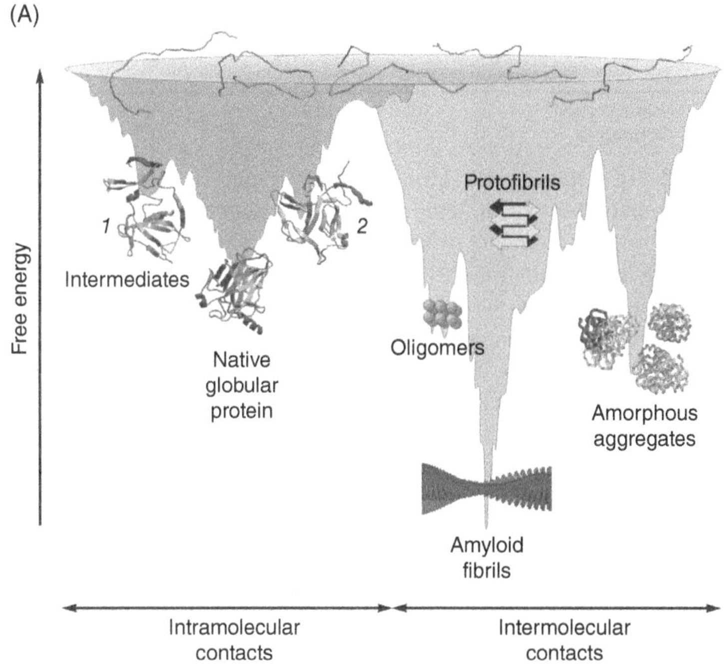

FIGURE 1.2 The energy landscape model, illustrating the formation of native globular and intrinsically disordered proteins, supramolecular complexes, amorphous aggregates, and amyloid fibrils. (A) Globular proteins. In the globular protein folding, the increase in the free energy associated with the folding-induced entropy decrease is compensated by the formation of specific intramolecular contacts. Local free-energy minima at the energy landscape correspond to the formation of partially folded intermediates. Intermolecular contacts of partially folded protein molecules can result in the formation of oligomers, amorphous aggregates, or amyloid fibrils. (*See insert for color representation of the figure.*)

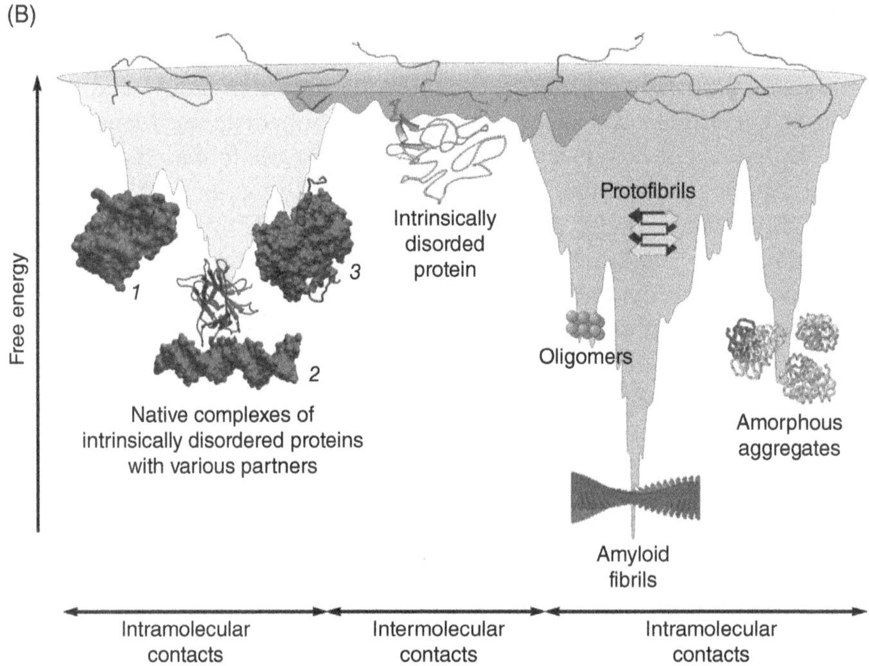

FIGURE 1.2 (*Continued*) (B) Intrinsically disordered proteins. Many native proteins with distinctive biological functions lack compact globular structure in aqueous solutions. Disordered segments of these proteins can gain ordered structure at the interaction with specific binding partners in a case if the free energy of such complexes is lower than the free energies of the intrinsically disordered protein and its partner. The propensity of native completely or partially disordered proteins to interact with various partners determines their biological functions in recognition of various binding partners (ligands, nucleic acids, and other proteins), in regulation of almost all cellular processes, and in signal transduction. In contrast to the folded globular proteins, which have to unfold to become amyloidogenic, disordered proteins seem always to be ready for such intermolecular interactions. 1, 2, and 3 represent native complexes of intrinsically disordered proteins with various partners. This figure is based on the energy funnel model developed for globular proteins (Schultz, 2000; Jahn and Radford, 2005). [Adapted from Turoverov et al. (2010).] (*See insert for color representation of the figure.*)

chain's free energy represents a large "hilly plateau" describing the dynamic ensemble of a large number of conformations. Hills on the plateau correspond to the forbidden conformations, and the plateau is separated from the entrance to the folding funnel by high energetic barrier(s), corresponding to the transitional state(s) (Finkelstein and Ptitsyn, 2002). This barrier is of great importance for proper protein functioning, as its existence guarantees the structural identity of all the native protein molecules. The ability of native globular proteins to form crystals is the major proof of this hypothesis.

In principle, the transition from the unfolded state to the uniquely folded native state can be realized via different pathways. Frequently, it seems that the experimental data prove the validity of this hypothesis, but in fact the rates of the various folding–unfolding stages might change dramatically, depending on the nature of the denaturing conditions (or the concentration of a given denaturant). As a result, not all of the folding or unfolding steps can be registered experimentally, suggesting that the character of the unfolding pathway depends on the choice of the denaturant (or denaturant concentration). At the same time, our experimental data on several proteins, including actin, showed that the number and the order of appearance of intermediate states is not dictated by denaturing agents (Kuznetsova et al., 2005; Povarova et al., 2007).

The energy landscape model not only elucidates the mechanisms of the globular protein folding, but also explains the nature of the IDPs, describes the formation of their supramolecular complexes, and delineates the formation of potentially pathogenic oligomers, amorphous aggregates, and amyloid-like fibrils (Turoverov et al., 2010).

1.2.3 Polymer Aspects of Protein Folding

It is generally accepted that the protein molecule has a unique primary sequence, which governs its three-dimensional structure and ensures proper biological activity. In this respect, each protein represents a unique case. That is why an understanding of the effect of sequence variations on biological performance represents a difficult challenge. It is also thought that natural polypeptides have originated as random copolymers of amino acids and were evolutionarily adjusted or "edited" (based on the principle of natural selection) to acquire and refine their various unique three-dimensional structures and functional properties (Ptitsyn and Volkenstein, 1986; Lau and Dill, 1990; Ptitsyn, 1995). Since protein molecules are remarkably unique, a serious question arises as to the existence of any general features in protein self-organization. To understand the common physicochemical principles underlying the protein-folding process, it is important to delineate the common polymer roots and their impact on the protein structures. The traditional way of performing such an analysis is a determination of the correlation between different physical characteristics of a polymer (e.g., its molecular density) and its length.

The molecular dimensions of 180 proteins in a variety of conformational states have been analyzed to establish a potential correlation between the hydrodynamic dimensions and the length of the polypeptide chain (Tcherkasskaya and Uversky, 2001; Uversky, 2002a, 2002c, 2003b; Tcherkasskaya et al., 2003; Turoverov et al., 2010). The protein categories analyzed included native globular proteins with nearly spherical shapes, equilibrium molten globules (MGs), compact denatured [or pre-molten globules (pre-MGs)], denaturant-unfolded proteins without cross-links in the presence of strong denaturants (8 M urea or 6 M GdmHCl), and extended intrinsically disordered proteins (native coils and native pre-MGs). Figure 1.3 represents the results of this analysis and shows that in all the cases a correlation between the apparent molecular density [determined as $\rho = M/(4\pi R_S^3/3)$, where M is a molecular mass and R_S is a hydrodynamic radius of a given protein] and molecular mass is observed. This gave rise to a set of standard equations $R_S = K_h M^{\varepsilon}$ for a number of conformational states of a polypeptide chain (Uversky, 2002a, 2002c, 2003b; Tcherkasskaya et al., 2003):

$$\log R_S^{\text{globular}} = -(0.204 \pm 0.023) + (0.357 \pm 0.005) \cdot \log(M) \quad (1)$$

$$\log R_S^{\text{MG}} = -(0.053 \pm 0.094) + (0.334 \pm 0.021) \cdot \log(M) \quad (2)$$

$$\log R_S^{\text{pre MG}} = -(0.21 \pm 0.18) + (0.392 \pm 0.041) \cdot \log(M) \quad (3)$$

$$\log R_S^{\text{native-coil}} = -(0.551 \pm 0.032) + (0.493 \pm 0.008) \cdot \log(M) \quad (5)$$

$$\log R_S^{\text{native pre-MG}} = -(0.239 \pm 0.055) + (0.403 \pm 0.012) \cdot \log(M) \quad (6)$$

$$\log R_S^{\text{8 M urea}} = -(0.649 \pm 0.016) + (0.521 \pm 0.004) \cdot \log(M) \quad (7)$$

$$\log R_S^{\text{6 M GdmHCl}} = -(0.723 \pm 0.033) + (0.543 \pm 0.007) \cdot \log(M) \quad (8)$$

This analysis showed that for a given conformational state, parameters K_h and ε were invariable over a wide range of chain lengths. Furthermore, statistical analysis has revealed that the relative errors of the recovered approximations exhibited random distribution over a wide range of chain lengths and did not generally exceed 10% (Uversky, 2002a,c, 2003b; Tcherkasskaya et al., 2003). This meant that the effective protein dimensions in a variety of conformational states can be predicted based on the chain length, with an accuracy of 10%. Thus, regardless of the differences in the amino acid sequences and biological functions, protein molecules behave as polymer homologs in a number of conformational states.

It is important to remember that the most unambiguous characteristic of the conformation of a polymer molecule remains the molecular density. For example, the density of a globule is expected to be independent of chain length, whereas the density of a partially collapsed or swelled macromolecules depends on both the chain length (and therefore on its molecular weight M)

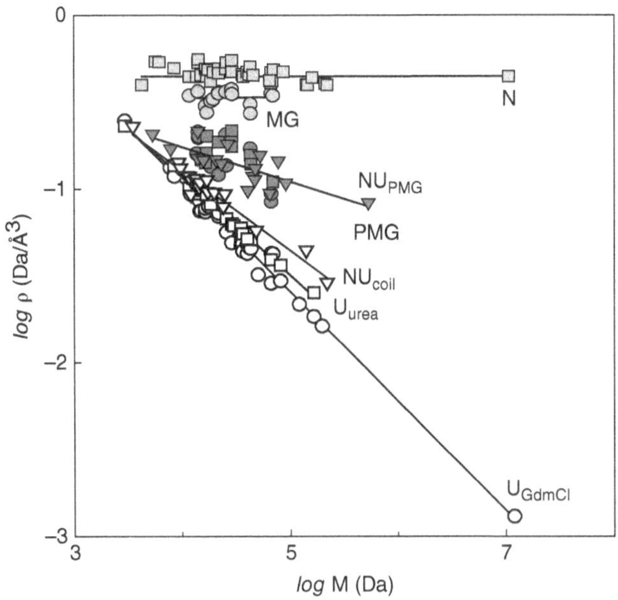

FIGURE 1.3 Variation of the density of protein molecules, ρ, with protein molecular weight, M, for a number of thermodynamically stable conformational states: native, molten globule; pre-molten globules (intermediates accumulated during the unfolding by urea or GdmCl are shown by circles; proteins with intact disulfide bridges in 8 M urea or 6 M GdmHCl are shown as squares; native pre-molten globules are shown as reversed triangles); native coils; proteins without cross-links or with reduced cross-links unfolded in 8 M urea; proteins without cross-links or with reduced cross-links unfolded in 6 M GdmHCl. The proteins used for this analysis are from Tcherkasskaya et al. (2003), Uversky (2002a,c), and Uversky (2003b). The solid lines represent the best fit of the data to the standard function $R_S = K_h M^{\varepsilon}$.

and on the nonspecific interactions of the monomer units with the solvent (Grossberg and Khohlov, 1989).

1.2.4 Different Conformations Seen in Protein Folding

Unfolded State: General Information on the Unfolded State of Proteins. Obviously, to solve the problem of protein folding it is necessary first to characterize the unfolded state, which represents the starting point of the folding reaction. It is known that the unfolded state represents an ensemble of rapidly interchanging conformations, some of which are extended and some of which are more compact. It is possible that when stabilizing interactions occur, they induce a more populated ensemble of chain conformations, and if such structures exist in the unfolded state, they would probably guide the folding process and function as folding-initiation sites (Baldwin, 1986). In this respect, it is important to mention that theoretical studies have shown that small preferences for native-like interactions in the unfolded state will substantially increase the probability of reaching the native state.

Returning to the polymer roots, under conditions known as *ideal* or *Θ-conditions* (i.e., when the attractions of the macromolecular segments are balanced by those with the solvent), the density of macromolecules is expected to follow $M^{-0.5}$, thereby $R_S \sim lN^{0.5}$, with l being a statistical chain length (Tanford, 1961, 1968; Grossberg and Khohlov, 1989). Here the polymer is assumed to be in a random coil conformation, and its conformational behavior can be described using Gaussian statistics (Grossberg and Khohlov, 1989). Further, in a good solvent, the macromolecular coil is expanded due to the prevalence of the repulsive interactions between polymer segments, and the molecular dimensions change more significantly with increasing chain length, $R_S \sim (l^2B)^{0.2} N^{0.6}$.

Residual Structure in Proteins Unfolded by Strong Denaturants Note that the "fully" unfolded states induced by GdmHCl or urea provide ε = 0.54 and 0.52, respectively (Tanford, 1961, 1968; Grossberg and Khohlov, 1989). Given that these ε values are less than 0.6, it appears that under these conditions the unfolded polypeptide chains exhibit features of macromolecular coils in Θ-solvents. This observation is in good agreement with the results of earlier studies (Tanford, 1968; Smith et al., 1996), where, in particular, it has been shown that the size of unfolded proteins in 6 M GdmCl is in reasonable agreement with the random coil model (Tanford, 1968). Recently, this conclusion was supported further by an examination of the correlation between the denatured-state radii of gyration, R_g, of 26 proteins and their polypeptide lengths, ranging from 16 to 549 residues (Kohn et al., 2004). This analysis revealed that the dimensions of most chemically denatured proteins scale with polypeptide length by means of the power-law relationship with a best-fit exponent, 0.598 ± 0.028, coinciding closely with the 0.588 predicted for an excluded-volume random coil, suggesting that the mean dimensions of chemically denatured proteins are effectively indistinguishable from the mean dimensions of a random coil ensemble (Kohn et al., 2004).

However, it has been also pointed out that the inclusion of "knots" of collapsed structure into the random coil model would not have a great influence on the hydrodynamic dimensions of a coil (Fitzkee and Rose, 2004). In fact, analysis of model systems where 33 proteins of known structure were used to generate disordered conformers by varying backbone torsion angles at random for approximately 8% of the residues, with the remaining approximately 92%

of the residues being remained fixed in their native conformation, revealed that despite this extreme degree of imposed internal structure, the ensembles analyzed were shown to have end-to-end distances and mean radii of gyration that agree well with random coil expectations (Fitzkee and Rose, 2004). Furthermore, the values of the hydrodynamic dimensions that Tanford measured for the unfolded proteins (Tanford, 1968) correspond better to a model where 20% of the residues are located in the collapsed structures (Miller and Goebel, 1968). In agreement with these observations, more recent Monte Carlo analysis of the effects of two types of local structures, α-helix and polyproline II (PPII) helix, on the dimensions of random coil polyalanine chains viewed as a model of highly denatured proteins revealed that although Flory's power-law scaling, long regarded as a signature of random coil behavior, holds for chains containing up to 90% α-helix or PPII, the absolute magnitude of the chain dimensions is sensitive to helix content (Wang et al., 2007). Furthermore, this study showed that the increase in α-helix content was accompanied by detectable chain contraction, reaching a minimum radius at about 70% α-helix, after which the chain dimensions expand rapidly. The presence of about 20% α-helix generated an unfolded state with the experimentally observed radii of gyration. Importantly, at an α-helix content of about 87%, polypeptide chains were also characterized by R_g values similar to those measured in experiments. On the other hand, R_g increased monotonically with increasing PPII content, always being more expanded than the dimensions observed experimentally. These results suggest that PPII is unlikely to be the sole dominant preferred conformation for unfolded proteins (Wang et al., 2007).

A great many protein-folding studies have revealed the presence of an assured residual structure even under the most severe denaturing conditions, such as high concentrations of strong denaturants. For example, considerable residual structure involving both α-helical and β-structural elements has been detected in the staphylococcal nuclease (SNase) (James et al., 1992; Alexandrescu et al., 1994; Wang and Shortle, 1995, 1996, 1997; Gillespie and Shortle, 1997a, b; Shortle and Ackerman, 2001). A stretch of continuous nonpolar residues ranging between Ile 95 and Tyr 103, forming β-strand 3, was shown to be involved in the formation of a compact cluster in the unfolded state of the tryptophan synthetase α-subunit (Saab-Rincon et al., 1993, 1996). Similarly, a contiguous stretch of nonpolar residues comprising Val 54, Val 56, Trp 58, and Leu 59 was found to form a cluster in a urea-unfolded fragment of the protein 434 (Neri et al., 1992a, b, c). Analysis of the urea-unfolded human fibroblast growth factor 1 (hFGF-1), a protein that consists of 12 antiparallel β-strands arranged into a β-trefoil architecture, by direct measurement of the hydrogen/deuterium (H/D)-exchange rates revealed that 38 residues whose heteronuclear single quantum coherence cross-peaks can be observed after exchange show higher protection level 2 than those predicted for the same residues in a random coil conformation, suggesting the existence of residual structure(s) (Wang and Yu, 2011).

Investigations of the SH3 have identified turn-like structure in regions that fold into β-strands (Zhang and Forman-Kay, 1995, 1997). The existence of defined residual structure has been observed in unfolded states of barstar (Nolting et al., 1997) and barnase (Arcus et al., 1995). The WW-domain retains a native-like core in high concentrations of GdmCl and urea (Koepf et al., 1999). The unfolded states of BPTI have been reported to be relatively compact in 6 M GdmCl under reducing conditions (Amir et al., 1992; Gottfried and Haas, 1992). Human carbonic anhydrase II, HCA II, has also been shown to possess a compact unfolded state (Martensson et al., 1993; Svensson et al., 1995; Hammarstrom and Carlsson, 2000). Unfolded apomyoglobin (8 M urea at pH 2) displays distinct regions with dramatically different backbone mobility, assuming the existence of residual structure (Schwarzinger et al., 2002). Extensive clusters of hydrophobic structure exist in the unfolded state of lysozyme even under strongly denaturing conditions (Klein-Seetharaman et al., 2002). The residual structure in fully unfolded photoactive yellow protein (PYP) was shown to be affected by isomerization of its *p*-coumaric acid (pCA) chromophore, suggesting that the residual structure in the fully unfolded state of PYP can be controlled directly by pCA photoexcitation (Lee et al., 2010). The *Escherichia coli* outer membrane protein X (OmpX) contains two polypeptide segments that present nonrandom residual structure in 8 M aqueous urea, whereas the remainder of the protein is in a flexibly disordered conformation (Tafer et al., 2004). Residual structure was also found in the chemically unfolded state of the N-terminal domain of a multidomain protein enzyme I from *Streptomyces coelicolor* (Romero-Beviar et al., 2010).

NMR analysis of urea-induced unfolding of the molten globule state of α-lactalbumin revealed a remarkably stable part of the protein in the core of the helical domain, which comprises interactions in both the N- and C-terminal parts of the protein (Schulman et al., 1997). More recent comparison of the unfolded states of three homologous proteins with very similar folds—the reduced unfolded states of hen lysozyme and the calcium-binding proteins bovine and human α-lactalbumins—by heteronuclear NMR spectroscopy showed the presence of noticeable residual structure manifested as significant deviations from random coil predictions (Wirmer et al., 2006). Furthermore, although these homologous proteins possessed very similar structures in their native states, their unfolded states differed noticeably from each other. For example, the residual structure was found in the α- and the β-domains in bovine α-lactalbumin and lysozyme, and only in the α-domain of human α-lactalbumin (Wirmer et al., 2006).

Using a combination of NMR experiments and molecular dynamics simulations it was shown that although the denatured state of chymotrypsin inhibitor 2 is highly unfolded, it contains some residual native helical structure along with hydrophobic clustering in the center of the chain (Kazmirski et al., 2001). Residual dipolar couplings (RDCs) for HN–N and HA–CA bond vectors measured by NMR spectroscopy for the 70-residue protein eglin C revealed the existence of a native-like global structure in the urea unfolded state (Ohnishi et al., 2004). Based on ^{19}F NMR analysis it has been concluded that a residual structure in unfolded intestinal fatty acid–binding protein with incorporated fluorinated aromatic amino acids consists of amino acids that are neighbors in the native state (Ropson et al., 2006). In yeast alcohol dehydrogenase (YADH) with the cysteine residues covalently modified by *N*-(1-pyrenyl) maleimide (PM) residual structure was detected even in the presence of 5 M GdnHCl using the excimer fluorescence of PM-YADH (Santra et al., 2006). By a combination of circular dichroism (CD) and small-angle x-ray scattering (SAXS), some residual structure was found in the unfolded state of the HIV-1 protease (Kogo et al., 2009).

In the unfolded state of the ultrafast folder "Trp-cage" miniprotein TC5b, residual structure due to hydrophobic collapse, with strong interresidue contacts between side chains that are relatively distant from one another in the native state, was detected in photochemically induced dynamic nuclear polarization NMR pulse-labeling experiments (Mok et al., 2007). When a general photochemical modification with methylene carbene was used to evaluate the accessible surface area (ASA) of the polypeptide chain in different conformational states of *Bacillus licheniformis* β-lactamase (BL-betaL), the unfolded state in 7 M urea was labeled just 60% more than the native state (Ureta et al., 2007). Since the retrieved value was well below the increment of ASA expected from theoretical estimates, it has been concluded that the unfolded BL-betaL possessed noticeable residual organization (Ureta et al., 2007).

Comparative analysis of the unfolded states of the ribosomal protein S16 isolated from a mesophilic ($S16_{meso}$) and a hyperthermophilic ($S16_{thermo}$) bacterium by a variety of biophysical methods revealed that the denatured-state ensemble of $S16_{thermo}$ is more compact relative to $S16_{meso}$, suggesting that the extreme temperature tolerance of a hyperthermophilic protein is coupled to residual structure in its unfolded state (Wallgren et al., 2008). Similarly, detailed comparison of unfolded states of ribonucleases H from organisms that grow at different temperatures revealed that a tolerance to higher temperatures correlates well with the residual structure in the unfolded state of the thermophilic proteins (Ratcliff and Marqusee, 2010).

Using a combination of residual dipolar coupling (RDC) and paramagnetic relaxation enhancement (PRE) data as constraints in ensemble structure calculations, it was shown that only a relatively small number of conformers is necessary to fully reproduce the experimental RDCs, PREs, and average radius of gyration of the urea-unfolded ubiquitin (Huang and Grzesiek, 2010). This conformational ensemble was characterized by a characteristic residual structure corresponding to an intact native first β-hairpin and α-helix combined with the nonnative α-helical conformations in the C-terminal half of the protein (Huang and Grzesiek, 2010). Therefore, by combining modern RDC, PRE, and computational methods, statistically significant detection of subconformations in the unfolded ensemble at population levels of a few percent is achievable (Huang and Grzesiek, 2010).

All these facts show unambiguously that the existence of profound residual structure might be a general characteristic of an unfolded polypeptide chain under aggressively denaturing conditions (Dill and Shortle, 1991; Shortle, 1996a, b; Baldwin and Zimm, 2000; Plaxco and Gross, 2001). This residual structure may be important in the folding process. Therefore, unfolded states of proteins exhibit behavior that is not random coil in nature, which is not surprising considering the complexity of polypeptides. In fact, it has been pointed out that a total lack of intraresidue interactions would be unexpected in the unfolded state, because certain (e.g., hydrophobic) side chains have high affinity for each other in a folded protein (Hammarstrom and Carlsson, 2000). In addition, some secondary structure within unfolded protein could be expected, due to the preferential distribution of ϕ and ψ angles (Ramachandran and Sasisekharan, 1968; Smith et al., 1996; Shortle, 2002); and some residual hydrophobic interactions can also be present (Hammarstrom and Carlsson, 2000; Shimizu and Chan, 2002). All this restricts considerably the conformational space of the unfolded polypeptide chain. Thus, it seems most likely that the polypeptide chains under "strong denaturing conditions" are still below the critical point (bad solvent conditions) and can easily be transformed to the compact state. For example, a small fluctuation of temperature from 25 to 30°C was shown to encourage the cooperative collapse of the fully unfolded proteins in 6 M GdmCl (Ptitsyn, 1995). Thus, globular proteins are never random coils without positional correlations, and biological polypeptide chains represent the macromolecular coils below a critical point even under harsh denaturing conditions.

Nonglobular Pre-molten Globule State Now let us consider a situation when the thermodynamic quality of the solvent worsens. In this case the binary interactions between monomers become mainly attractive (Tcherkasskaya et al., 2003). As a result, the polymer chain collapses partially, leading to an increase in the molecular density and bringing many-body interactions into the scene. It has been found that many proteins under the appropriate conditions can form a specific compact partially folded conformation, a pre-molten globule state (Palleros et al., 1993; Uversky and Ptitsyn, 1994; Ptitsyn et al., 1995; Uversky and Ptitsyn, 1996a; Zhang et al., 1996; Karnoup and Uversky, 1997; Uversky et al., 1998a; Bushmarina et al., 2001; Tcherkasskaya and Uversky, 2001; Kuznetsova et al., 2002; Georlette et al., 2003a, b; Tcherkasskaya et al., 2003). Major structural characteristics of this intermediate are summarized below. This conformation is characterized by considerable secondary structure, although much less than that of the molten globule. The pre-molten globule state is considerably less compact than the molten globule state, but it is still more compact than the corresponding random coil. Pre-molten globules can interact with the hydrophobic fluorescent probe ANS, although more weakly than in the molten globule state. This means that at least some hydrophobic clusters are already formed in the pre-molten globule state, although there is no globular structure.

Finally, the pre-molten globule and the molten globule are separated by an all-or-none phase transition, reflecting the fact that these partially folded intermediates represent discrete phase states (Uversky et al., 1992; Uversky and Ptitsyn, 1996b). Importantly, several structural elements of pre-molten globules, these squeezed coils, may occupy native-like positions (Tcherkasskaya and Uversky, 2001). We assume that this compact denatured state might represent a general intermediate in protein folding. Obviously, the existence of such a state substantially reduces any search through the conformational space, ensuring rapid folding. Given that this state might comprise a specific native-like core with burial of hydrophobic residues, the transition from this state to a molten globule state or native state would not require significant energy changes and could occur quite easily. It may well be the case that the protein folding *in vivo* involves mostly the molten globule and the denatured compact states. On this occasion it would require only slight changes to make the coil "condense onto itself" and form a globule.

Analysis of hydrodynamic data reveals that the molecular dimensions of pre-molten globules follow the chain length as $R_S = 0.6M^{0.40}$ (see above). This indicates the bad solvent conditions. Thus, one can conclude that this conformation exhibits behavior that is typical for squeezed macromolecular coils. Therefore, any small variations in the protein environment (i.e., changes in the thermodynamic quality of the solvent, or changes induced by proton transfer, interactions with a ligand, fluctuations of temperature, etc.) can trigger the transition of the compact protein molecule to the more rigid molten globule or native states (Grossberg and Khohlov, 1989). Importantly, Figure 1.3 shows that stabilization of the pre-molten globule state might be achieved by the incorporation of disulfide bridges into the protein sequence.

In fact, the hydrodynamic behavior of unfolded proteins with intact disulfide bridges was shown to be close to that of pre-molten globules. Finally, the pre-exponential term $K_h = 0.6$ retrieved for the pre-molten globule state is significantly larger than those retrieved for the fully unfolded species, which probably indicates developing multiple body interactions.

Globular Molten Globule and Ordered States The theory of *coil–globule transition* predicts that the overall dimension of a polymer globule, R_S, is anticipated to change with the chain length, N, as $R \sim (C/B)^{1/3}N^{1/3}$. Here B and C are the second and third virial coefficients, which characterize the pair collisions and three-body interactions of the monomer units of the polymer chain (Grossberg and Khohlov, 1989). The density of the globules is expected to show no change with increasing chain length, owing to $\rho \sim N/R^3 \sim (-B/C)$. These results are in excellent agreement with data obtained for the ordered and molten globules of proteins (Figure 1.3, see also Tcherkasskaya and Uversky, 2001; Uversky, 2002a,c, 2003b; Tcherkasskaya et al., 2003). In particular it has been found that for ordered and molten globule proteins, the parameter K_h has a value of 0.75 and 0.9, respectively, whereas ε equals 0.33 for both conformational states. Given that the overall dimensions of globules are determined by the balance of pair collisions and three-body interactions, the difference in preexponential terms K_h observed for the native and molten globules probably reflects the larger probability of three-body interactions in the molten globule state. This could be because of the compact but flexible nature of the molten globules.

Amino Acid Composition Favoring Equilibrium Partially Folded Intermediate(s) A variety of different physicochemical forces play a role in stabilizing the unique three-dimensional structure of a protein. Both the strength and specificity of many of these forces are strongly dependent on environmental conditions in such a way that changes in the environment can reduce or even eliminate some of the conformational interactions, while the remaining interactions are unchanged or even intensified. Under some environmental conditions, the native protein structure can be transformed into new conformations with properties intermediate between those of the native and completely unfolded states. Thus, the ability of a protein to adopt different stable partially folded conformations should be considered as an intrinsic property of a polypeptide chain. Since all the necessary and sufficient information to fold into the native, biologically active conformation is thought to be present in a protein amino acid sequence (Anfinsen et al., 1961), the capability of a given protein to adopt equilibrium partially folded conformation(s) may also be encoded in specific features of its amino acid sequence.

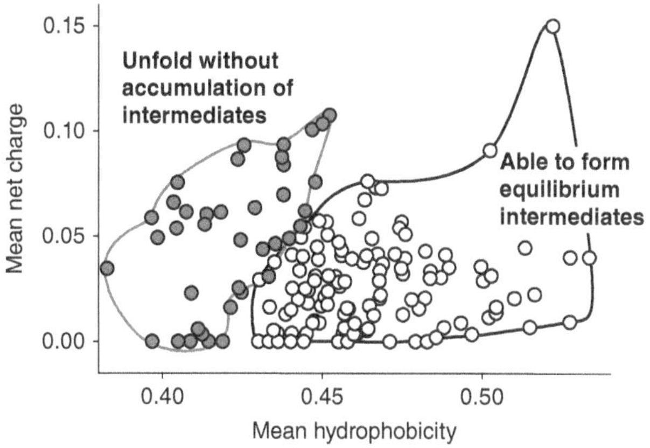

FIGURE 1.4 Comparison of mean net charge vs. mean hydrophobicity for the set of 115 proteins able to form equilibrium intermediates (open symbols and black lines) and the set of 39 proteins shown to unfold without accumulation of partially folded conformations (gray symbols and lines). [Modified from data presented by Uversky (2002b).]

Interestingly, it has been shown that not all proteins (even homologous ones) have an identical response to changes in their environment. For example, hen egg white lysozyme represents a textbook illustration for the two-state model of denaturant-induced unfolding (Tanford, 1968), whereas accumulation of classical molten globules under different experimental conditions was described for its homolog, α-lactalbumin (Dolgikh et al., 1981, 1985; Kuwajima, 1996). Analysis of literature data on equilibrium unfolding of globular proteins induced by changes in pH, temperature, or strong denaturants (urea or guanidinium chloride) revealed that unfolding in 115 proteins is accompanied by accumulation of equilibrium intermediate states of one sort or another. Another set comprises 39 proteins, which were shown to unfold according to a simple two-state model; that is, no equilibrium intermediate of any kind was formed during their unfolding (Uversky, 2002b).

In an attempt to understand which factors may be responsible for such tremendous differences in the formation of equilibrium partially folded intermediates, the general sequence features of proteins from both groups have been analyzed using a simple method comparing global sequence charge and hydrophobicity (Uversky, 2002b). Figure 1.4 represents the results of this analysis as a plot of mean hydrophobicity versus mean net charge (i.e., as a distribution of groups within the charge-hydrophobicity phase space). Figure 1.4 shows that the reliable separation of both groups of proteins takes place in such coordinates. These data imply that the competency of a protein to form equilibrium intermediate(s) may be predetermined by the bulk content of hydrophobic and charged amino acid residues (Uversky, 2002b).

In other words, this competency may be encoded in the charge/hydrophobicity ratio of its polypeptide chain, not its sequence. This may mean that partially folded conformations are stabilized mostly by nonspecific side-chain/side-chain interactions of hydrophobic amino acid residues (Uversky, 2002b). Interestingly, proteins that do not have equilibrium intermediates are less hydrophobic and have, in general, a larger net charge than those competent to form discrete intermediate states. This may indicate that such proteins are less strengthened by hydrophobic interactions and more disturbed by electrostatic repulsion. Thus, smaller environmental changes may be required to overcome the marginal stabilization energies, leading to immediate and complete unfolding of the protein.

Concluding Remarks on Protein Folding Regardless of the differences in primary amino acid sequences, protein molecules behave as polymer homologs in a number of conformational states, allowing speculation as to the volume interactions being a driving force in a formation of equilibrium structures. For example, both native and molten globules exhibit key features of polymer globules, where fluctuations of the molecular density are expected to be much less than the molecular density itself. Protein molecules in the pre-molten globule state possess properties of squeezed coils. Furthermore, even high concentrations of strong denaturants are more likely to constitute bad solvents for protein chains. Thus, globular proteins are never random coils without positional correlations, and biological polypeptide chains represent the macromolecular coils below a critical point, even under harsh denaturing conditions.

1.3 NONFOLDING

1.3.1 Intrinsically Disordered Proteins and Their Abundance

It is recognized now that a considerable number of biologically active proteins are not completely rigid but possess some amount of disorder under physiological conditions. These proteins are known as *intrinsically disordered proteins* (IDPs), among other names. Structurally, IDPs are highly diverse, and some compact IDPs possess noticeable secondary structure and behave as native molten globules, whereas other IDPs are extended and possess little residual structure (i.e., these IDPs behave as native coils or native pre-molten globules) (Uversky, 2011a; Uversky and Dunker, 2010).

IDPs and intrinsically disordered protein regions (IDPRs) constitute a unique tribe of the protein kingdom (Uversky, 2003b, 2011a, Turoverov et al., 2010;

Tompa, 2011), and the number of IDPs and IDPRs is amplifying rapidly. For example, the first list of extended IDPs (also known as natively unfolded proteins, i.e., proteins that do not possess almost any specific structure under physiological conditions), published in 2000 (Uversky et al., 2000), contained 91 members (the full-length proteins and their domains with a chain length of more than 50 amino acid residues). Now this set has almost doubled.

The commonness of intrinsic disorder was estimated further by predicting disorder for whole genomes, including both known and putative protein sequences. First predictions of such type have been published for 31 genomes that span the three kingdoms: archaea, bacteria, and eukaryotes. The percentage of sequences in each genome with segments predicted to have ≥ 40 consecutive disordered residues was used to gain an overview of proteomic disorder. For so many consecutive predictions of disorder, the false-positive error rate was estimated from ordered proteins to be less than 0.5% of the segments of 40 and less than 6% of the fully ordered proteins (Dunker et al., 2000, 2001). The eukaryotes exhibited more disorder by this measure than either the prokaryotes or the archaea, with *C. elegans*, *A. thaliana*, *S. cerevisiae*, and *D. melanogaster* predicted to have 52 to 67% of their proteins with such long predicted regions of disorder, while bacteria and archaea were predicted to have 16 to 45% and 26 to 51% of their proteins with such long disorder regions, respectively (Dunker et al., 2000; Oldfield et al., 2005a). The increased amount of disorder in the eukaryota may be related to the increase in cellular signaling in the eukaryota (Dunker et al., 2000, 2001; Oldfield et al., 2005a). Recently, this analysis was extended to include entire proteomes of 3484 species from the three main kingdoms of life and in viruses (Xue et al., 2012). This analysis revealed that the process of evolution is characterized by the unique patterns of changes in the protein intrinsic disorder content. In fact, viruses were characterized by the widest spread of the proteome disorder content (the number of disordered residues ranges from 7.3% in human coronavirus NL63 to 77.3% in *avian carcinoma virus*) (Xue et al., 2012). For several organisms, a clear correlation was seen between their disorder contents and habitats. In multicellular eukaryotes, there was a weak correlation between the organism complexity (evaluated as a number of different cell types) and the overall disorder content. For both prokaryotes and eukaryotes, the disorder content was generally independent of proteome size. However, it showed a sharp increase associated with the transition from prokaryotic to eukaryotic cells (see Figure 1.5). This suggests that the increased disorder content in eukaryotic proteomes might be used by nature to deal with the increased cell complexity due to the appearance of the various cellular compartments (Xue et al., 2012).

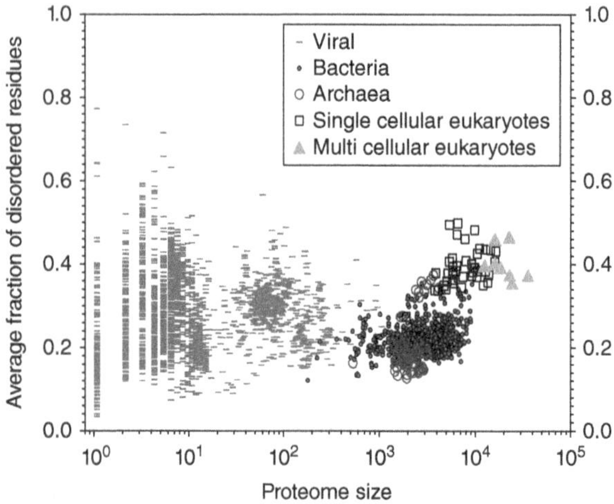

FIGURE 1.5 Correlation between the intrinsic disorder content and proteome size for 3484 species from viruses, archaea, bacteria, and eukaryotes. Each symbol indicates a species. There are a total of five groups of species: viruses (red horizontal marks), bacteria (small green circles), archaea (open blue circles), unicellular eukaryotes (open brown squares), and multicellar eukaryotes (pink triangles). The proteome size is the number of proteins in the proteome of that species and is shown on a logarithmic scale. The average fraction of disordered residues is calculated by averaging the fraction of disordered residues of each sequence over all the sequences of that species. Disorder prediction is evaluated by PONDR-VSL2B. [Adapted from Xue et al. (2012).] (*See insert for color representation of the figure.*)

1.3.2 Some Functional Advantages of IDPs

For several reasons, the intriguing family of IDPs has attracted an excessive attention of researchers over the past decade and a half. First, the existence of IDPs, with their lack of ordered rigid structure under "physiological" conditions *in vitro*, disputes the validity of one of the cornerstones in protein biology, chemistry, and physics (i.e., the structure–function paradigm). Note that this paradigm, formulated more than 100 years ago as a lock-and-key concept for explaining the amazing specificity of the enzymatic hydrolysis of glucosides (Fischer, 1894), claims that a specific function of a protein is determined by its unique and rigid three-dimensional structure. IDPs, being disordered (at least in part), are involved in countless biological activities (see below) and effectively resist those evolutionary pressures that normally favor stable globular folds. For example, α-synuclein (one of the best-characterized IDPs) isolated from different organisms possesses a high degree of sequence conservation; for example, the mouse and rat α-synucleins are identical throughout the first 93 residues, and the human and canary proteins differ from them by only two residues (Clayton and George, 1998). To overcome the conflict in the existence of functionally active unfolded polypeptides with the structure–function paradigm, it has been suggested that the lack of rigid globular structure under physiological conditions might represent a considerable functional advantage for IDPs, as their large plasticity allows them to interact efficiently with several different targets (Wright and Dyson, 1999; Dunker et al., 2001, 2002a,b; Tompa, 2002, 2011; Uversky, 2002a,c, 2011a; Uversky, 2003b; Dyson and Wright, 2005; Uversky and Dunker, 2010). Furthermore, a disorder-to-order transition induced in IDPs during the binding of specific targets *in vivo* might represent a simple mechanism for the regulation of numerous cellular processes, including the regulation of transcription and translation and cell-cycle control (reviewed by (Wright and Dyson, 1999; Dunker et al., 2001, 2002a,b; Tompa, 2002; 2011; Uversky, 2002a,c, 2003b, 2011a; Dyson and Wright, 2005; Uversky and Dunker, 2010).

Therefore, in addition to being highly abundant in nature (Dunker et al., 2000; Romero et al., 1998; Uversky et al., 2000; Ward et al., 2004; Xue et al., 2012), IDPs/IDPRs possess numerous intriguing properties (Uversky and Dunker, 2010), are intimately involved in various cellular processes (Wright and Dyson, 1999, 2009; Dunker et al., 2001, 2002a,b, 2005; Iakoucheva et al., 2002; Tompa, 2002; Uversky, 2002a,c; Dyson and Wright, 2005; Uversky et al., 2005; Vucetic et al., 2007; Xie et al., 2007a,b; Kim et al., 2008; Oldfield et al., 2008; Liu et al., 2009), and are commonly found to be related to the pathogenesis of various diseases (Iakoucheva et al., 2002; Cheng et al., 2006; Xie et al., 2007a; Uversky, 2008, 2009; Uversky et al., 2008, 2009; Midic et al., 2009a, b). The common theme of protein disorder–based functionality is recognition, and IDPs/IDPRs are frequently involved in complex protein–protein, protein–nucleic acid, and protein–small molecule interactions. Some of these interactions can induce a disorder-to-order transition in the entire IDP or in its part (Wright and Dyson, 1999, 2009; Uversky et al., 2000; Dunker et al., 2001; Tompa, 2002; Uversky, 2002a,b, 2011a,b; 2013; Dyson and Wright, 2005; Oldfield et al., 2005b; Mohan et al., 2006; Vacic et al., 2007a; Dosztanyi et al., 2009; Meszaros et al., 2009). Furthermore, intrinsic disorder opens a unique capability for one protein to be involved in interaction with several unrelated binding partners and to gain different bound structures (Oldfield et al., 2008; Hsu et al., 2012). Some IDPs/IDPRs can form highly stable complexes; others are involved in signaling interactions, where they undergo constant "bound–unbound" transitions, thus acting as dynamic and sensitive "on–off" switches. These proteins typically return to their intrinsically disordered state after the completion of a particular function. Many of the IDPs/IDPRs can gain different conformations, depending on the environmental peculiarities (Oldfield et al., 2005b; Hsu et al., 2012). All this constitutes an important arsenal of the unique physiological properties of IDPs/IDPRs, which determines their ability to exert different functions in different cellular contests according to a specific conformational state (Uversky and

Dunker, 2010). The folding-at-binding principle is believed to help IDPs or IDPRs to obtain maximal specificity in a protein–protein interaction without very high affinity (Liu et al., 2009). This combination of high specificity with low affinity defines the broad utilization of intrinsic disorder in regulatory interactions, where turning a signal off is as important as turning it on (Dunker et al., 2001). Although some partial folding during the IDP/IDPR-based interactions is a widespread phenomenon, with a significant fraction (about one-third) of the interacting residues in IDPs/IDPRs adopting α-helix and β-strand structures (Mohan et al., 2006; Vacic et al., 2007a), there are still many other IDPs/IDPRs that are involved in the formation of "fuzzy complexes," where an IDP/IDPR keeps a certain amount of disorder in its bound conformations (Tompa and Fuxreiter, 2008; Hazy and Tompa, 2009; Meszaros et al., 2011; Uversky, 2011a,b).

Often, the interacting regions in IDPs are observed as loosely structured fragments in their unbound forms. These disorder-based binding sites are known as *molecular recognition elements* or *features* (MoREs or MoRFs) (Oldfield et al., 2005b; Mohan et al., 2006) preformed structural elements (Fuxreiter et al., 2004), or pre-structured motifs (PreSMos) (Chi et al., 2007). Although the existence of such loosely structured regions suggests that IDPs can adopt their bound structure(s) at a free-energy cost that is not too high, it is important to remember that increasing the stability of the bound conformation does not necessarily enhance the binding affinity (Wright and Dyson, 2009). Another important feature of the disorder-based interactions is their increased speed due to the greater capture radius and the ability to search spatially through interaction space (called the "fly-casting" mechanism) (Shoemaker et al., 2000), or due to the fewer encounter complexes required for the binding event with IDPs/IDPRs (Huang and Liu, 2009). Linking all these considerations with the recent report showing that IDP affinities are tuned mostly by association rates (Prakash, 2011) suggests that the degree of pre-adoption of binding conformations in IDPs has to be limited, but not unfavorable.

1.3.3 Function-Induced Folding of IDPs

An important feature of IDPs is that they are able to undergo a disorder-to-order transition (i.e., partial or complete folding) during or prior to their biological function (Wright and Dyson, 1999, 2009; Dunker et al., 2001, 2002a,b, 2005; Tompa, 2002, 2011; Uversky, 2002a,b, 2003b, 2011a,b; Dyson and Wright, 2005; Uversky et al., 2005; Tompa and Fuxreiter, 2008; Uversky and Dunker, 2010). In other words, IDPs are likely to be stabilized *in vivo* by function-related binding to specific targets and ligands (such as a variety of small molecules, substrates, cofactors, other proteins, nucleic acids, membranes, etc.). The functional importance of being disordered has been analyzed intensively, and it has been established that increased intrinsic plasticity represents an important prerequisite for effective molecular recognition. The diapason of biological functions of the IDPs/IDPRs is extremely wide, including cell-cycle control, transcriptional and translational regulation, modulation of activity and/or assembly of other proteins, and even regulation of nerve cell function [reviewed in Dunker et al., 2001; 2002a, b; 2005; Dyson and Wright, 2002, 2005; Tompa, 2002; Uversky and Dunker, 2010; Uversky, 2011a; 2013]. It has been suggested that the persistence of IDPs throughout evolution may be due to the definite advantages of disorder-to-order transitions accompanying functional performance of flexible structures in comparison with action of rigid proteins (Dunker et al., 1998, 2001, 2005; Wright and Dyson, 1999; Dyson and Wright, 2002; Uversky et al., 2005; Uversky and Dunker, 2010). Some of these potential advantages of intrinsic lack of structure and function-related folding include:

1. The ability of binding to several different targets, known as one-to-many signaling
2. The possibility of high specificity coupled with low affinity
3. The precise control and simple regulation of the binding thermodynamic
4. The capability to overcome steric restrictions, enabling essentially larger interaction surfaces in the complex than could be obtained for the rigid partners
5. The increased rates of specific macromolecular association
6. The reduced lifetime of the intrinsically disordered proteins in the cell, possibly representing a mechanism of rapid turnover of the important regulatory molecules

Finally, a few words should be added about the extent of function-related folding of IDPs. It has been shown that the range of conformational changes induced in natively unfolded proteins as a result of their interaction with natural partners is very wide (Uversky, 2002a,c, 2011b). In fact, the examples of all possible conformational transitions have been described, including function-induced transitions of coil to pre-molten globule, coil to molten globule, coil to ordered conformation, pre-molten globule to molten globule, pre-molten globule to rigid structure, and molten globule to ordered and rigid form (Uversky, 2002a,c). This means that the structure–function paradigm, which emphasizes that ordered three-dimensional structures represent the indispensable prerequisite to the effective protein functioning, should be redefined to include IDPs (Wright and Dyson, 1999; Dunker et al., 2001; Uversky and Dunker, 2010). According to this redefined paradigm, native proteins (or their functional regions) can exist in any of the known conformational states, ordered, molten globule, pre-molten globule, and coil. Function can arise from any of these conformations and transitions between them. Thus, not just the ordered state but any of the known polypeptide conformations can be the native state of a protein.

It was suggested that the structure-forming effect of natural partners might be explained by their influence on the hydrophobicity and/or net charge of the natively unfolded polypeptide. Obviously, these parameters could be changed in such a way that they will approach values typical of ordered native proteins forcing a polypetide to fold (Uversky et al., 2000); that is, the formation of protein–ligand complexes may increase a "critical mass" of the folding unit, thus initiating a chain of folding reactions.

There is, however, an alternative view of the problem of existence and the functionality of IDPs. It is possible that these proteins are folded in the cell as "normal" proteins, due to the fact that the concentration of their ligands is always high enough *in vivo* to fold them. Thus, evolution never sees them as any different from normal proteins. In this view, the property of being intrinsically disordered is merely an artifact resulting from the proteins being studied after being purified to homogeneity. Evolutionarily, IDPs persist not because there is an advantage to being intrinsically disordered, but because there is no disadvantage as long as there are enough ligands around *in vivo* to fold the proteins. The stability of the folded, ligand-bound form may only be a few kilocalories per mole, but this would be enough to keep the vast majority of molecules folded and functional (even most normal, folded proteins are only marginally stable).

1.3.4 IDPs and Human Diseases

Because of the fact that ID proteins play crucial roles in numerous biological processes, it was not too surprising to find that many of them are involved in human diseases. For example, a number of human diseases originate from the deposition of stable, ordered, filamentous protein aggregates, commonly referred to as *amyloid fibrils*. In each of these pathological states, a specific protein or protein fragment changes from its natural soluble form into insoluble fibrils, which accumulate in a variety of organs and tissues (Kelly, 1998; Bellotti et al., 1999; Dobson, 1999; Uversky et al., 1999b,c; Rochet and Lansbury, 2000; Uversky and Fink, 2004). Approximately 20 different proteins are known so far to be involved in these diseases. These proteins are unrelated in terms of sequence or starting structure. Several ID proteins are found in this list of 20, being associated with the development of several neurodegenerative diseases (Uversky and Fink, 2004, 2005). An incomplete list of disorders associated with ID proteins includes Alzheimer's disease [deposition of amyloid-β, tau-protein, α-synuclein fragment NAC (Glenner and Wong, 1984; Masters et al., 1985; Lee et al., 1991; Ueda et al., 1993); Niemann–Pick disease type C, subacute sclerosing panencephalitis, argyrophilic grain disease, myotonic dystrophy, and motor neuron disease with neurofibrillary tangles (accumulation of tau-protein in the form of neurofibrillary tangles (Lee et al., 1991)]; Down's syndrome [nonfilamentous amyloid-β deposits (Wisniewski et al., 1985)]; Parkinson's disease, dementia with Lewy body, diffuse Lewy body disease, Lewy body variant of Alzheimer's disease, multiple-system atrophy and Hallervorden–Spatz disease (deposition of α-synuclein in the form of a Lewy body, or Lewy neuritis (Dev et al., 2003)]; prion diseases [deposition of PrP^{SC} (Prusiner, 2001)]; and a family of polyQ diseases, a group of neurodegenerative disorders caused by expansion of GAC trinucleotide repeats coding for polyQ in the gene products (Zoghbi and Orr, 1999). Furthermore, most mutations in rigid globular proteins associated with accelerated fibrillation and protein deposition diseases have been shown to destabilize the native structure, increasing the steady-state concentration of partially folded (disordered) conformers (Kelly, 1998; Bellotti et al., 1999; Dobson, 1999; Rochet and Lansbury, 2000; Uversky et al., 1999b,c; Uversky and Fink, 2004).

The maladies listed above have been called *conformational diseases*, as they are characterized by the conformational changes, misfolding, and aggregation of an underlying protein. However, there is another side to this coin: protein functionality. In fact, many of the proteins associated with the conformational disorders are also involved in recognition, regulation, and cell signaling. For example, functions ascribed to α-synuclein, a protein involved in several neurodegenerative disorders, include binding fatty acids and metal ions; regulation of certain enzymes, transporters, and neurotransmitter vesicles; and regulation of neuronal survival (reviewed by Dev et al., 2003). Overall, about 50 proteins and ligands have been shown to physically interact and/or co-localize with this protein. Furthermore, α-synuclein has amazing structural plasticity and adopts a series of different monomeric, oligomeric, and insoluble conformations (reviewed by Uversky, 2003a). The choice between these conformations is determined by the peculiarities of the protein environment, assuming that α-synuclein has an exceptional ability to fold in a template-dependent manner. Based on these observations, we hypothesize that the development of the conformational diseases may originate from misidentification, misregulation, and missignaling, accompanied by misfolding. In other words, mutations and/or changes in the environment may result in protein confusion, for which its ID becomes lost, thus reducing its capability to recognize proper binding partners and leading to the formation of nonfunctional and deadly aggregates.

Recent analysis of polyglytamine diseases gives support to this hypothesis (Okazawa, 2003). *Polyglytamine diseases* are a specific group of hereditary neurodegeneration caused by expansion of CAG triplet repeats in an exon of disease genes, which leads to the production of a disease protein containing an expanded polyglutamine (polyQ) stretch. Nine neurodegenerative disorders, including Kennedy's disease, Huntington's diseases, spinocerebellar atrophy-1, -2, -3, -6, -7, -17, and dentatorubral pallidoluysian atrophy are known to belong to this class of diseases (Cummings and Zoghbi, 2000; Gusella and MacDonald, 2000; Fischbeck, 2001; Orr, 2001). In most polyQ diseases, expansion to over 40 repeats leads to the onset (Fischbeck, 2001). It has been emphasized that such molecular processes as unfolded protein response, protein transport, and synaptic transmission and transcription are implicated in the pathology of polyQ diseases (Okazawa, 2003). Importantly, more than 20 transcription-related factors have been reported to interact with pathological polyQ proteins. Furthermore, these interactions were shown to repress the transcription, leading finally to neuronal dysfunction and death (reviewed by Okazawa, 2003). These results suggest that polyQ diseases represent a type of transcriptional disorder (Okazawa, 2003), supporting our misidentification hypothesis for at least some conformational disorders.

Generally, three computational/bioinformatics approaches were elaborated to estimate the abundance of IDPs under various pathological conditions. The first approach is based on the assembly of specific data sets of proteins associated with a given disease and the computational analysis of these data sets using a number of disorder predictors (Iakoucheva et al., 2002; Cheng et al., 2006; Uversky et al., 2006; Mohan et al., 2008; Uversky, 2008, 2009). In essence, this is an analysis of individual proteins extended to a set of independent proteins. A second approach utilized diseasome, a network of genetic diseases in which the related proteins are interlinked within one disease and between different diseases (Midic et al., 2008). A third approach is based on evaluation of the association between a particular protein function (including the disease-specific functional keywords) and the level of intrinsic disorder in a set of proteins known to carry out this function (Vucetic et al., 2007; Xie et al., 2007a, b). These three approaches are described briefly below and the results of their use are presented in a subsequent section.

The data set analysis approach was used for the first time in 2002, when it was found that 79% of cancer-associated and 66% of cell-signaling proteins contain predicted regions of disorder of 30 residues or longer (Iakoucheva et al., 2002). In contrast, only 13% of a set of proteins with well-defined ordered structures contained such long regions of predicted disorder. For this study, cancer-associated proteins were defined as those human proteins in Swiss-Prot containing the keyword *oncogene* (this included anti- and proto-oncogenes) or containing the word *tumor* in the description field. In experimental studies, the presence of disorder has been observed directly in several cancer-associated proteins, including p53 (Lee et al., 2000), $p57^{kip2}$ (Adkins and Lumb, 2002), Bcl-X_L and Bcl-2 (Chang et al., 1997), c-Fos (Campbell et al., 2000), and most recently, a thyroid cancer–associated protein, TC-1 (Sunde et al., 2004). Following a similar analytical model, a data set of 487 proteins related to cardiovascular disease (CVD) was collected and analyzed (Cheng et al., 2006). On average, CVD-related proteins were found to be highly disordered. The percentage of proteins with 30 or more consecutive disordered residues was 61% for CVD-associated proteins. Many proteins were predicted to be wholly disordered, with 101 proteins from the CVD data set predicted to have a total of almost 200 specific disorder-based binding motifs (thus, about two binding sites per protein), α-MoRFs (Cheng et al., 2006). Finally, data set analysis revealed that in addition to being abundant in cancer- and CVD-related proteins, intrinsic disorder is commonly found in such maladies as neurodegenerative diseases and diabetes (Uversky et al., 2008; Uversky, 2009).

The human diseasome systematically links the human disease phenome (which includes 1284 human genetic diseases, 867 of which had at least one link to other diseases, and 516 diseases formed a giant component) with the human disease genome (which contains 1777 disease genes, of which 1377 were shown to be connected to other disease genes, and 903 genes belonged to a giant cluster) (Goh et al., 2007). The abundance of intrinsic disorder in the human diseasome was evaluated using a set of computational tools such as PONDR VSL2, CDF analysis, CH plots, and α-MoRF prediction (Midic et al., 2008). These analyses uncovered an unfoldome associated with human genetic diseases and revealed that intrinsic disorder is common in proteins associated with many human genetic diseases. It was also shown that different disease classes vary in the IDP content of their associated proteins and that α-MoRFs are common in the diseasome, and their abundance correlates with the intrinsic disorder level. Finally, some disease classes were shown to have a significant fraction of genes affected by alternative splicing, and the alternatively spliced regions in the corresponding proteins are predicted to be highly disordered and in some diseases contain a significant number of MoRFs (Midic et al., 2008).

The studies on correlation of ID with various functional key words (Vucetic et al., 2007; Xie et al., 2007a, b) revealed that many diseases were strongly correlated with proteins predicted to be disordered. Contrary to this, no disease-associated proteins were found to be strongly correlated with an absence of disorder (Xie et al., 2007a). Among disease-related Swiss-Prot keywords strongly associated with ID were *oncoproteins, malaria, trypanosomiasis, human immunodeficiency virus* (HIV) and *acquired immunodeficiency syndrome* (AIDS), *deafness, obesity, cardiovascular disease, diabetes mellitus, albinism*, and *prion* (Xie et al., 2007a). In agreement with this bioinformatics analysis, at least one illustrative, experimentally validated example of functional disorder or order was found for the vast majority of functional keywords related to diseases (Xie et al., 2007a).

Summarizing, intrinsic disorder is highly abundant among proteins associated with various human diseases. Since ID proteins are very common in various diseases, the "disorder in disorders" (D^2) concept was introduced to summarize work in this area (Uversky et al., 2008), and concepts of the disease-related unfoldome and unfoldomics were developed (Uversky et al., 2009).

1.3.5 How Does an Amino Acid Sequence Encode Intrinsic Disorder?

As has already been pointed out, the correct folding of a protein into its rigid biologically active conformation is determined by its amino acid sequence (Anfinsen et al., 1961). This gave rise to the reasonable assumption that the absence of rigid structure in natively unfolded proteins may also somehow be encoded in the specific features of their amino acid sequences. All the functional and structural peculiarities of IDPs and IDPRs are encoded in their amino acid sequences. In fact, it was recognized long ago that there are significant differences between the ordered proteins/domains and IDPs and IDPRs at the level of their amino acid sequences (Dunker et al., 1998, 2001; Uversky et al., 2000).

In early studies on finding the relationship between sequence and disorder, Dunker and co-authors developed several neuronal network predictors and applied them to large data sets (Dunker et al., 1998, 2000, 2001; Romero et al., 1998, 2001). The results of these analyses were more than impressive, as it was established that the disordered regions shared at least some common sequence features between many proteins and that more than 15,000 proteins in the Swiss-Prot database were identified as having long IDPRs (Romero et al., 1998).

Interestingly, the disordered proteins with the highest scores were shown to have low sequence complexity, assuming that the sequences of IDPs may essentially be degenerated. However, it was later established that distributions of the complexity values for ordered and disordered sequences overlapped, suggesting that low sequence complexity did not represent the only characteristic feature of intrinsically disordered proteins (Romero et al., 2001). Overall, the sequence of the intrinsically disordered proteins is characterized by amino acid compositional bias and the existence of highly predictable flexibility (Dunker et al., 1998; Garner et al., 1998). In fact, in comparison with ordered proteins, IDPs and IDPRs are characterized by noticeable biases in their amino acid compositions (Dunker et al., 1998, 2001; Uversky et al., 2000; Radivojac et al., 2007; Vacic et al., 2007b; Uversky and Dunker, 2010), containing less of

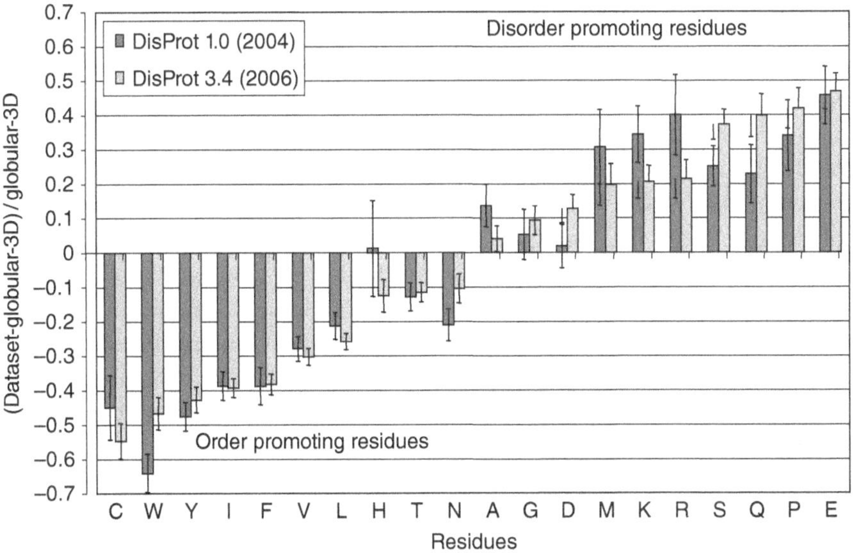

FIGURE 1.6 Peculiarities of amino acid composition of IDPs. An order/disorder composition profile is shown for comparisons of amino acid compositions of ordered protein with each of two databases of disordered protein. The ordinates are (% amino acid in disordered data set − % amino acid in ordered data set)/(% amino acid in ordered data set). Names of each database indicate how the disordered regions were identified. Negative values indicate that the disordered database has less than order; positive indicates more than order.

the "order-promoting" residues (cysteine, tryptophan, isoleucine, tyrosine, phenylalanine, leucine, histidine, valine, asparagines, and methionine, which are mostly hydrophobic residues commonly found within the hydrophobic cores of foldable proteins) and more of the "disorder-promoting" residues (lysine, glutamine, serine, glutamic acid, and proline, which are mostly polar and charged residues typically located at the surface of foldable proteins) (see Figure 1.6).

Also, in one of the early studies on these subjects, comparison of the overall hydrophacy and the net charge of ordered proteins and extended IDPs might be used to predict whether a given amino acid sequence encodes a native (folded) or an intrinsically disordered protein (Uversky et al., 2000). In fact, this survey shows that the natively unfolded proteins are specifically localized within a unique region of the charge–hydrophacy phase space, being completely isolated from the native globular proteins. Obviously, this allows estimation of the "boundary" mean hydrophacy value, below which a polypeptide chain with a given mean net charge will most probably be unfolded (Uversky et al., 2000).

1.3.6 Polymer Aspects of Nonfolding

It is interesting to apply the same formalism of polymer physics as that represented in the first part of the review to the two classes of extended IDPs (native coils and native pre-molten globules). This type of analysis is available because a considerable number of data have been accumulated on the hydrodynamic properties of these proteins. Obviously, the behavior of any polypeptide chain in a solution is determined by interaction of the polymer with the solvent. The fact that natively unfolded proteins, with their depleted hydrophobicity, are noncompact under physiological conditions indicates that "salted water" (typical "physiological" buffer contains 100 to 150 mM NaCl) does not represent for them a poor solvent. In other words, these conditions do not force polymer segments to interact specifically with each other and thus do not force them to be excluded effectively from the solvent. On the other hand, it has already been noted that even high concentrations of strong denaturants do not represent a good solvent for a polypeptide chain encoding for a typical globular protein, and a globular protein was assumed never to be a random coil.

In this view, the data on the hydrodynamic analysis of extended IDPs are extremely important. Such an analysis reveals that the molecular dimensions of extended IDPs follow the chain length as $R_S = 0.28\,M^{0.49}$ or $R_S = 0.6\,M^{0.40}$ for the native coils and native pre-molten globules, respectively (see Figure 1.3). Figure 1.3 clearly reflects the fact that native coils belong to the class of relatively extended unfolded conformations. Importantly, these coils show the highest K_h values and the smallest ε values between different unfolded conformations of a polypeptide chain [cf. equations (5), (7), and (8)]. This means that native coils under physiological conditions are in considerably worsened solvent conditions than those of globular proteins in urea or GdmHCl solutions (the lowest ε value), which gives rise to the increased probability of multiple body interactions (the highest K_h value). Finally, Figure 1.3 shows that native pre-molten globules follow exactly the same dependence as that of pre-molten globules of normal globular proteins. Thus, these proteins may exhibit the structural features of a squeezed polymer coil.

1.4 MISFOLDING

1.4.1 Molecular Mechanisms of Protein Misfolding

As has already been noted, the sequences of proteins have evolved in such a way that their unique native states can be found very efficiently even in the complex environment inside a living cell. However, under some conditions, proteins fail to fold properly or to remain folded correctly; this misfolding can lead to the development of different pathological conditions. A number of human diseases, including the amyloidoses and several neurodegenerative disorders, such as Alzheimer's disease, Parkinson's disease, and transmissible spongiform encephalopathies, originate from the deposition of the stable, ordered, and filamentous protein aggregates commonly known as *amyloid fibrils*. In each of these pathological states, a specific protein or protein fragment changes from its natural soluble form into insoluble fibrils, which accumulate in a variety of organs and tissues (Kelly, 1998; Bellotti et al., 1999; Dobson and Karplus, 1999; Uversky et al., 1999b,c; Rochet and Lansbury, 2000). Approximately 20 different proteins are known to be involved in the amyloidoses (extracellular deposits), which are unrelated structurally or at the level of primary structure. In addition, a number of diseases also originate from the deposition of fibrillar proteins, but within cells (i.e., intracellular deposits).

Importantly, prior to fibrillation, amyloidogenic polypeptides may be rich in β-sheet, α-helix, β-helix, or contain both α-helices and β-sheets. They may be well folded or IDPs. Despite these differences, the fibrils from different pathologies display many common properties, including a core cross-β-sheet structure in which continuous β-sheets are formed with β-strands running perpendicular to the long axis of the fibrils (Sunde et al., 1997). Amyloid fibrils have been shown to form *in vitro* from disease-associated as well as disease-unrelated proteins and peptides. Furthermore, it is clear that the ability to form fibrils is a generic property of the polypeptide chain; that is, many proteins, perhaps all, are able to form amyloid fibrils under appropriate conditions (Dobson and Karplus, 1999). If so, this would dramatically extend the structural diversity of polypeptide chains able to fibrillate.

Since all fibrils independent of the original structure of the given amyloidogenic proteins have a common cross-β structure, considerable conformational rearrangements have to occur prior to fibrillation. Such changes cannot happen in a native protein, due to its stable and rigid tertiary structure. Thus, protein destabilization favoring partial unfolding and culminating in the formation of a partially unfolded conformation is required. Presumably, such a partially unfolded conformation favors reciprocal and specific intermolecular interactions, including electrostatic attraction, hydrogen bonding, and hydrophobic contacts, which are necessary for oligomerization and fibrillation (Fink, 1998; Kelly, 1998; Bellotti et al., 1999; Dobson and Karplus, 1999; Lansbury, 1999; Uversky et al., 1999b,c; Rochet and Lansbury, 2000; Dobson, 2001a, b; Zerovnik, 2002).

Obviously, this model does take into account a class of IDPs, as they are devoid of rigid tertiary structure in their native state. The primary step in the fibrillation of these proteins represents partial folding, that is, stabilization of a partially folded conformation (Schweers et al., 1994; Teplow, 1998; Kayed et al.,

1999; Uversky et al., 2001a,b; Pavlov et al., 2002). Thus, the general hypothesis of fibrillogenesis states: Structural transformation of a polypeptide chain into a partially folded conformation represents an important prerequisite for successful protein fibrillation.

A question then arises as to the nature of amyloidogenic intermediate(s). It has already been mentioned that the conformational space of globular proteins involves four general conformations: the native state, unfolded states, and two classes of partially folded intermediates. The latter are compact, relatively well-folded intermediates with substantial native-like secondary structure but little tertiary structure, often referred to as molten globules; and significantly less compact, relatively unfolded intermediates, with substantially less secondary structure, often known as pre-molten globules. Potentially, either of these latter conformations may play a role as the crucial amyloidogenic species. Detailed structural analysis of the early fibrillation events in several proteins has demonstrated that the amyloidogenic conformation is only slightly folded and shares many structural properties with the pre-molten globule state.

1.4.2 Fibrillogenesis of Globular Proteins: Requirement for Partial Unfolding

Data have been reported indicating that the first critical step in protein fibrillogenesis is partial unfolding of the protein. Due to structural fluctuations (conformational breathing) the structure of a globular protein under physiological conditions represents a mixture of tightly folded and multiple partially unfolded conformations, with a great prevalence of the former. Most mutations associated with accelerated fibrillation and protein deposition diseases have been shown to destabilize the native structure, increasing the steady-state concentration of partially folded conformers (Colon et al., 1996; Kelly, 1996, 1998; Wetzel, 1997; Bellotti et al., 1999; Canet et al., 1999; Dobson and Karplus, 1999; Lashuel et al., 1999; Uversky et al., 1999b,c; Rochet and Lansbury, 2000; Heegaard et al., 2001; Saraiva, 2001). Conversely, it has been shown that the amyloidogenicity of a protein can be reduced significantly by stabilization of the native structure, for example, via specific binding of ligands (Miroy et al., 1996; Chiti et al., 2001; Nielsen et al., 2001).

The fibrillation-provoking destabilization of a rigid protein is achieved *in vitro* at low or high pH, high temperatures, and low to moderate concentrations of strong denaturants, organic solvents, and so on. This fact is well illustrated in recent studies on the fibrillogenesis of transthyretin (TTR), also known as prealbumin, which is a homotetramer composed of 127 amino acid subunits characterized by 2,2,2 molecular symmetry. TTR is found in human plasma (0.1 to 0.4 mg/mL) and cerebral spinal fluid (0.017 mg/mL), with the plasma form being the amyloidogenic precursor. Wild-type TTR amyloidogenesis may cause senile systemic amyloidosis, characterized by deposition and pathology in the heart after age 60. Early-onset amyloid formation (as early as the second decade) by one of more than 80 single-site TTR variants provokes a number of diseases collectively termed *familial amyloid polyneuropathy* (Saraiva et al., 1984). TTR can be converted into amyloid *in vitro* by acid-mediated dissociation of the homotetramer into monomers. The pH required for disassembly also results in tertiary structural changes within the monomeric subunits, finally leading to the enhanced fibrillation (Lai et al., 1996). Recently, in a quest to understand the relationship between the tertiary structural changes and amyloidogenicity, a monomeric mutant with native-like structure and stability has been designed, which was nonamyloidogenic unless partially unfolded (Jiang et al., 2001).

Light chain, or AL, amyloidosis is a pathological condition arising from systemic extracellular deposition of monoclonal immunoglobulin light chain variable domains in the form of insoluble amyloid fibrils, especially in the kidneys (Buxbaum, 1992). Structural and fibrillation properties of one of the amyloidogenic light chain variable domains, SMA, have been analyzed under a variety of conditions (Khurana et al., 2001). The results of biophysical analysis revealed that a decrease in pH resulted in the accumulation of two partially folded intermediates. A relatively native-like intermediate, I_N, was observed between pH 4 and 6, and was characterized by little loss of secondary structure, combined with significant changes in tertiary structure and enhanced ANS binding. At a pH below 3, a relatively unfolded but compact intermediate, I_U, with decreased tertiary and secondary structure, was observed. The I_U intermediate readily forms amyloid fibrils, whereas I_N preferentially leads to amorphous aggregates (Khurana et al., 2001). Comparable data have recently been reported for another light chain variable domain, LEN (Souillac et al., 2002a, b).

α-Lactalbumin, or α-LA, is a small acidic protein with a single Ca^{2+}-binding site. It is very attractive for studies of partially folded conformations, since it adopts the classic molten globule conformation at acidic pH, moderate guanidinium–chloride concentrations, or elevated temperatures (apo form) (Permyakov and Berliner, 2000). α-LA is comprised of a large α-helix domain and a small β-sheet domain connected by a calcium-binding loop and four disulfide bridges (Permyakov and Berliner, 2000). α-LA forms amyloid fibrils at low pH. *S*-Carboxymethyl-α-lactalbumin, a disordered form of the protein with three of the disulfide bridges reduced, was even more susceptible to fibrillation. *S*-Carboxymethyl-α-lactalbumin exhibits the properties of a pre-molten globule, and its fibrillation is orders of magnitude faster than when starting with the molten globule conformation (Goers et al., 2002). Other partially folded conformations induced in α-LA at neutral pH, either by removal of Ca^{2+} or by binding of Zn^{2+} to the Ca^{2+}-protein, did not fibrillate, although Zn^{2+}-loaded α-lactalbumin precipitated out of solution as amorphous aggregates. Based on these data it was concluded that the transformation from native state to a substantially unfolded conformation is required for a successful fibril formation, whereas less unfolded species may form amorphous aggregates (Goers et al., 2002).

Finally, fibrillation of bovine β-lactoglobulin represents another illustrative example of the importance of being unfolded (Hamada and Dobson, 2002). It has been shown that this protein will form fibrils in urea solutions; the process is denaturant concentration-dependent, showing the highest efficiency in the vicinity of 5 M urea, which corresponds roughly to the C_m value. Importantly, it was shown that molten globule-like intermediate was not accumulated during the urea-induced unfolding of β-lactoglobulin (Hamada and Dobson, 2002). Thus, it has been concluded that amyloid fibril formation by bovine β-lactoglobulin is promoted under conditions where significant accumulation of unfolded protein occurs, but is inhibited under conditions where higher denaturant concentrations destabilize intermolecular interactions (Hamada and Dobson, 2002). Generally speaking, it has been concluded that amyloid formation *in vitro* can be achieved by destabilizing the native state of the protein under conditions in which noncovalent interactions still remain favorable (Ramirez-Alvarado et al., 2000).

1.4.3 Fibrillogenesis of IDPs: Requirement for Partial Folding

We now consider details of the fibrillogenesis of intrinsically disordered proteins, which constitute a significant portion of known amyloidogenic proteins (Uversky, 2008, 2009). It seems to be reasonable to assume that such proteins are well suited for amyloidogenesis, as they lack significant secondary and tertiary structure as well as many specific intrachain interactions. In the absence of such conformational constraints, they would be expected to be substantially more conformationally flexible and thus able to polymerize more readily than tightly packed globular proteins. Substantial evidence suggests that the earliest stage of fibrillation of these proteins is their partial refolding.

α-Synuclein α-Synuclein is a small (14 kDa), soluble, intracellular, highly conserved protein that is abundant in various regions of the brain. This protein has been estimated to account for as much as 1% of the total protein in soluble cytosolic brain fractions. Structurally, purified α-synuclein is a typical extended IDP (Uversky et al., 2001b), which is unstructured under conditions of neutral pH but does not represent a random coil. It has some residual structure [at least a region with a preference for helical conformation (Eliezer et al., 2001)], leading to partial compaction (Uversky et al., 2001b; Uversky, 2003a).

Either a decrease in pH, or an increase in temperature, transformed α-synuclein into a partially folded conformation. This partially folded conformation resembles the pre-molten globule state, an intermediate, preceding the molten globule in the refolding of globular proteins (Uversky and Ptitsyn, 1994, 1996a; Uversky, 2003b).

Misfolding, dysfunction, and aggregation of α-synuclein are associated with several diseases, known collectively as *synucleinopathies*: a group of neurodegenerative disorders characterized by fibrillary aggregates of α-synuclein protein in the cytoplasm of selective populations of neurons and glia (Goedert, 1999; Spillantini and Goedert, 2000; Galvin et al., 2001; Trojanowski and Lee, 2003). Clinically, synucleinopathies are characterized by a chronic and progressive decline in motor, cognitive, behavioral, and autonomic functions, depending on the distribution of the lesions. Because of clinical overlap, differential diagnosis is sometimes very difficult (Marti et al., 2003). The neuropathological spectrum of synucleinopathies has been discussed intensively (Goedert, 1999, 2001; Duda et al., 2000; Spillantini and Goedert, 2000; Dickson, 2001; Galvin et al., 2001; Dev et al., 2003; Jellinger, 2003; Marti et al., 2003; Mitra et al., 2003; Trojanowski and Lee, 2003; Norris et al., 2004), and the potential mechanisms linking the α-synuclein aggregation with the development of several of these diseases are the major focus of numerous studies.

Several observations implicate α-synuclein in the pathogenesis of Parkinson's disease (PD). For example, a direct role for α-synuclein in the neurodegenerative processes in PD is demonstrated by genetic evidence and autosomal dominant early-onset PD is associated with three different missense mutations in the α-synuclein gene, corresponding to A30P, E46K, and A53T substitutions in α-synuclein (Polymeropoulos et al., 1997; Kruger et al., 1998; Zarranz et al., 2004) or with the hyperexpression of the wild-type α-synuclein protein, due to gene triplication (Farrer et al., 2004; Singleton et al., 2003, 2004). Antibodies to α-synuclein detect this protein in LBs and LNs, the hallmark lesions of PD. A substantial portion of fibrillar material in these specific inclusions was shown to be comprised of α-synuclein, and the insoluble α-synuclein filaments were recovered from purified LBs (Spillantini et al., 1997, 1998).

The fibrillogenesis of this protein is probably the most precisely studied and the best understood among other amyloidogenic members of the family of natively unfolded proteins. In particular, accumulated data strongly suggest that the formation of a partially folded intermediate (possessing the major characteristics of the pre-molten globule) represents the critical first step of α-synuclein fibrillogenesis. This partially folded intermediate can be stabilized by numerous factors, including high temperatures, low pH (Uversky et al., 2001b), the presence of low concentrations of organic solvents (Munishkina et al., 2003) and TMAO (Uversky et al., 2001c), the presence of different metal ions (Uversky et al., 2001e), various salts (Munishkina et al., 2004), several common pesticides and herbicides (Uversky et al., 2001d, 2002a; Manning-Bog et al., 2002), heparin and other glycosoaminoglycans (Cohlberg et al., 2002), some polycations (Goers et al., 2003), or as a result of a spontaneous oligomerization both *in vitro* and *in vivo* (Uversky et al., 2001a). Furthermore, the addition of various alcohols was shown to increase the content of ordered secondary structure in α-synuclein (Munishkina et al., 2003). Interestingly, the structural transformations induced by high solvent concentrations were dependent on the type of alcohol, with simple alcohols inducing a β-sheet-enriched conformation and fluorinated alcohols promoting α-helix-rich species (Munishkina et al., 2003). Interestingly, both α-helical and β-structural species were shown to be initially monomeric but underwent association over longer times, and β-sheet-rich conformations were strongly prone to form amorphous aggregates (Munishkina et al., 2003). Oligomeric α-helical globular species potentially possessing rigid tertiary structure were induced in α-synuclein by high concentrations of TMAO (Uversky et al., 2001c).

Importantly, under conditions stabilizing the pre-molten globule-like conformation, α-synuclein was shown to undergo significantly enhanced fibrillation. In contrast, fibril formation was considerably slowed or inhibited under conditions favoring the formation of more folded conformations, or by stabilization of the fully unfolded form (e.g., by oxidation of its methionines) (Uversky et al., 2002b).

Amylin In addition to insulin, pancreatic islet β-cells produce a polypeptide called *amylin* or *islet amyloid polypeptide*, IAAP (Cooper et al., 1987). Amylin has several functions associated with the normal regulation of energy metabolism. Dysfunction of amylin due to mutation and/or amyloid fibril formation has been associated with the development of non-insulin-dependent diabetes mellitus (NIDDM), also known as type 2 diabetes (Higham et al., 2000; Jaikaran and Clark, 2001; Jaikaran et al., 2001). Type 2 diabetes is characterized by chronic insulin resistance and progressive decline in pancreatic β-cell function. One of the most common pathological features of type 2 diabetes is the deposition of amyloid fibrils in the islets of Langerhans of the pancreas (Schneider et al., 1980; Clark et al., 1988; Rocken et al., 1992; Kahn et al., 1999). Human IAPP or amylin, is the major protein component of these amyloid deposits (Cooper et al., 1987; Westermark et al., 1987). Amylin is an unstructured peptide hormone of 37 amino acid residues. The natively unfolded nature of this peptide was established using far-ultraviolet (UV) circular dichroism (CD) spectroscopy (Goldsbury et al., 2000; Jaikaran et al., 2001; Yoon and Welsh, 2005) and electron paramagnetic resonance spectroscopy (Jayasinghe and Langen, 2004). For example, eight spin-labeled derivatives of IAPP were analyzed using electron paramagnetic resonance spectroscopy.

In solution, all eight derivatives gave rise to electron paramagnetic resonance spectra with sharp lines indicative of rapid motion on a sub-nanosecond time scale, which are consistent with a rapidly tumbling and highly dynamic peptide (Jayasinghe and Langen, 2004). Human amylin and its 8–37 fragment were shown to form fibrils under physiological conditions. The process of polymerization is relatively fast (lag times were 100 and 50 min for full-length amylin and its 8–37 fragment, respectively) and results in the appearance of typical amyloid fibrils (Goldsbury et al., 2000). Interestingly, both peptides showed the formation of a partially folded (pre-molten globule-like) intermediate early in the fibrillation process. It takes about 90 min for full-length amylin to form such an intermediate, whereas this period was almost half as long for the truncated peptide, showing excellent agreement with the fibrillation lag times (Goldsbury et al., 2000).

Amyloid β-Protein (Aβ) Alzheimer's disease (AD) is the most prevalent age-dependent dementia. AD is characterized pathologically by the accumulation of extracellular amyloid deposits in the cerebral neuropil and vasculature and of intracellular neurofibrillary tangles. Amyloid deposits contain the amyloid β-protein (Aβ), which is a 40- to 42-residue peptide that is produced by endoproteolytic cleavage of the amyloid β-protein precursor (APP). Many lines of evidence support the crucial role of Aβ in AD. Fibrillation of Aβ is associated with development of the cascade of neuropathogenetic events, ending with the appearance of cognitive and behavioral features typical of AD. Aβ appears to be unfolded at the beginning of the fibrillation under physiological conditions. NMR studies have shown that monomers of Aβ(1–40) or Aβ(1–42) possess no α-helical or β-sheet structure (Zagorski et al., 1999); they exist predominately as "random" extended chains. Partial refolding to the pre-molten globule-like conformation has been detected at the earliest stages of Aβ amyloidosis (Kirkitadze et al., 2001).

Tau-Protein Tau, a microtubule assembly protein isolated from brain microtubules, represents a family of isoforms which migrate as close bands of 55 to 62 kDa in SDS gel electrophoresis. Heterogeneity is explained in part by alternative mRNA splicing, leading to the appearance of one, two, three, or four repeats in the C-terminal region (Himmler, 1989; Himmler et al., 1989). Post-translational phosphorylation of tau is an additional source of microheterogeneity (Kenessey and Yen, 1993). *In vitro*, tau binds to microtubules, promotes microtubule assembly, and affects the dynamic instability of individual microtubules (Cleveland et al., 1977a, b; Drechsel et al., 1992; Brandt and Lee, 1993a, b). In situ, tau is highly enriched in the axons (Binder et al., 1985). In living cells and brain tissue, tau protein has been estimated as comprising 0.025 to 0.25% of total protein (Drubin et al., 1985;

Khatoon et al., 1992). On the basis of its *in vitro* activity and its distribution, it is believed that tau regulates the organization of neuronal microtubules. Interest in tau increased dramatically with the discovery of its aggregation in neuronal cells in the progress of Alzheimer's disease and various other neurodegenerative disorders, especially frontotemporal dementia (Delacourte and Buee, 1997; Crowther and Goedert, 2000). In these cases, specific tau-containing neurofibrillary tangles (paired helical filaments) are formed (Delacourte and Buee, 1997). Filaments isolated from end-stage AD are particularly well characterized and consist of all six full-length tau isoforms extensively phosphorylated and organized into twisted paired helical filaments (PHFs) and nontwisted straight filaments (SFs) (Lee et al., 1991). Hyperphosphorylation was shown to be a common characteristic of pathological tau (Vulliet et al., 1992). Hyperphosphorylated tau isolated from patients with AD was shown to be unable to bind to microtubules and promote microtubule assembly. However, both of these activities were restored after enzymatic dephosphorylation of tau protein (Lu and Wood, 1993; Alonso et al., 1994, 1996; Iqbal et al., 1994).

During brain development, tau is phosphorylated at many residues, including sites phosphorylated with GSK-3β, cdk 5, and MAPK (Watanabe et al., 1993). *In vitro*, tau can be phosphorylated on multiple sites by several kinases (for a review, see Billingsley and Kincaid, 1997). Most of the *in vitro* phosphorylation sites of tau are located within the microtubule interacting region (repeat domain) and sequences flanking the repeat domain. Many of these sites are also phosphorylated in PHF-tau (Morishima-Kawashima et al., 1995a,b). In fact, 10 major phosphorylation sites have been identified in tau isolated from PHFs from patients with AD (Morishima-Kawashima et al., 1995a,b). All of these sites are located in regions flanking tau's repeat domain and constitute recognition sites for several AD diagnostic antibodies, which may point to an important role for these phosphorylation sites for AD pathogenesis. Hyperphosphorylation was shown to be accompanied by transformation from the unfolded state of tau into a partially folded conformation (Hagestedt et al., 1989; Uversky et al., 1998b), accelerating dramatically the self-assembly of this protein into paired helical filaments *in vitro* (Alonso et al., 1996). To analyze the potential role of tau hyperphosphorylation in tauopathies, mutated tau proteins have been produced in which all 10 serine/threonine residues known to be highly phosphorylated in PHF-tau were substituted for negatively charged residues, thus producing a model for a defined and permanent hyperphosphorylation-like state of tau protein (Eidenmuller et al., 2000). It has been demonstrated that, like hyperphosphorylation, glutamate substitutions induce compact structure elements and SDS-resistant conformational domains in tau protein, as well as lead to the dramatic acceleration of its fibrillation (Eidenmuller et al., 2000).

Prior to aggregation, tau protein was shown to be in a mostly random coil structure. This conclusion followed from the conformational analysis of this protein by circular dichroism, Fourier transform infrared, x-ray scattering, and biochemical assays (von Bergen et al., 2006). Analysis of the primary structure reveals a very low content of hydrophobic amino acids and a high content of charged residues, which was sufficient to explain the lack of folding (von Bergen et al., 2006). Analysis of the hydrodynamic radii confirms a mostly random coil structure of various tau isoforms and tau domains. However, the protein was further unfolded by high concentrations of strong denaturant GdmCl, indicating the presence of some residual structure in this protein. This conclusion was supported by a FRET-based approach where the distances between different domains of tau were determined. The combined data show that tau is mostly disordered and flexible but tends to assume a hairpin-like overall fold which may be important in the transition to a pathological aggregate (von Bergen et al., 2006).

Intriguingly, purified recombinant tau isoforms do not detectably aggregate over days of incubation under physiological conditions. However, aggregation and fibrillization of tau protein can be greatly accelerated under nearly-physiological conditions *in vitro* by the addition of anionic surfactants (Chirita et al., 2003). Based on the detailed analysis of tau fibrillation in the presence of anionic inducers using a set of spectroscopic techniques (circular dichroism spectroscopy and reactivity with thioflavin S and 8-anilino-1-naphthalene-sulfonic acid fluorescent probes), it has been established that the inducer stabilized a monomeric partially folded species with the structural characteristics of a pre-molten globule state (Chirita et al., 2005). The stabilization of this intermediate was sufficient to trigger the fibrilliation of full-length tau protein (Chirita et al., 2005).

Prothymosin α Prothymosin α is a very acidic protein, containing about 50% aspartic and glutamic acid, no aromatic or cysteine residues, and very few large hydrophobic aliphatic amino acids (Gast et al., 1995). Because of these features, prothymosin α adopts a random coil-like conformation with no regular secondary structure at neutral pH (Gast et al., 1995; Uversky et al., 1999a). However, at acidic pH, prothymosin α folds into a partially folded pre-molten globule-like conformation (Uversky et al., 1999a). Interestingly, it has recently been shown that at low pH [below pH 3, i.e., under conditions favoring the formation of the partially folded conformation (Uversky et al., 1999a)], prothymosin α is capable of relatively fast formation (lag time of about 100 min) of regular elongated fibrils with a flat ribbon-like structure 4_5 nm in height and 12 to 13 nm in width (Pavlov et al., 2002).

1.4.4 Conformational Prerequisites for Amyloidogenesis

Based on a comparison of far-UV CD spectra measured for "pure" amyloidogenic conformations of 11 proteins (SH3-domain: cytocrome c_{552}, monellin, methionine aminopeptidase, SMA, α-lactalbumin, phosphoglycerate kinase, PGK, amylin, prothymosin α, Aβ, and α-synuclein) with those retrieved for the four basic protein conformations—native, molten globule, pre-molten globule, and unfolded states—it has been concluded that all amyloidogenic conformations can be grouped with the set of data for the pre-molten globule proteins (Uversky, 2003a; Uversky and Fink, 2004). This is an extremely important observation which demonstrates the importance of this conformation for protein fibrillogenesis.

The central model of protein misfolding is illustrated by Figure 1.7, which shows that the process of fibrillation (which is used as an example of a more general phenomenon—aggregation) can be divided into three major steps: (1) structural transformation of a native soluble protein (rigid globular or flexible IDP) into the "sticky" amyloidogenic precursor intermediate, (2) nucleation, and (3) fibril elongation/growth. Thus, transformation into the partially folded conformation is a critical early stage of fibrillogenesis and precedes the appearance of any aggregated material (Uversky, 2003a).

1.4.5 Multiple Pathways of Protein Misfolding

Although the data presented above were mostly devoted to a consideration of protein fibrillation, the process of amyloid fibril formation does not represent the only misfolding route. In fact, contrary to the process of productive protein folding leading to the appearance of a rigid conformation with the specific function, the end products of misfolding may have a different appearance. The morphology of these end products depends on the particular experimental conditions, and misfolded product may appear as soluble oligomers, amorphous aggregates, or amyloid-like fibrils. Any of these three species could be cytotoxic, thus giving rise to the development of pathological conditions. Potentially, the reason for such a morphological difference is connected with the diversity of the partially folded intermediates favoring protein self-association. In fact, multiple environmental factors, such as point mutations, a decrease in pH, an increase in temperature, or the presence of small organic molecules or metal ions or other charged molecules, might induce structural rearrangements within a protein molecule, shifting equilibrium toward the partially folded conformation(s). As different factors may stabilize slightly different partially folded intermediates, the formation of morphologically different aggregates is expected. This idea is illustrated by Figure 1.6, which represents a model of α-synuclein misfolding. It has been shown that

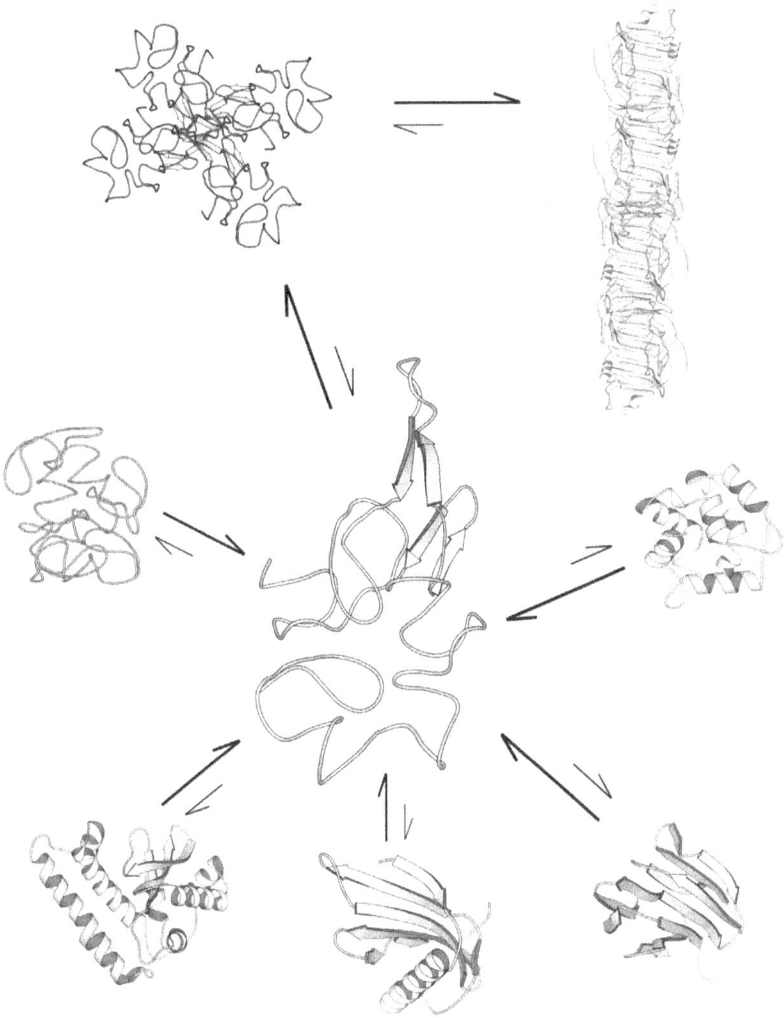

FIGURE 1.7 General model for protein misfolding (fibril formation is considered as an example). Three general stages of the process include the structural transformation of a native soluble protein (rigid or natively unfolded) into the "sticky" amyloidogenic precursor intermediate, nucleation, and fibril elongation/growth. Additional conformational changes may occur between the aggregation-competent intermediate and the fibrils. Structures at the top represent natively folded proteins with different structures. A common amyloidogenic intermediate for different rigid and natively unfolded proteins is shown for convenience only. Experimental data support an idea that amyloidogenic intermediates probably fall in the class of pre-molten globules. However, this still leaves a lot of room for structural diversity. The oligomeric intermediate (representing the nucleus or a soluble aggregate) is shown as a tetramer for convenience only, and could be much larger.

aggregation of this protein depends dramatically on the experimental conditions and might lead to the appearance of one of the three misfolded forms discussed above (soluble oligomers, amorphous aggregates, or amyloid fibrils) or mixture thereof.

1.4.6 Polymer Aspects of Protein Misfolding

The behavior of a given polymer in a given solution is determined by the peculiarities of polymer segment–solvent interactions. For example, the major reason for the appearance of globular conformation (in our particular case, we are talking about the correctly folded form of a normal globular protein) in a poor solvent (water) is that this conformation effectively excludes a portion of segments from the unfavorable contacts with the solvent and forms the shielding interface between the polymer interior and solvent. In turn, the stability of globular conformation also depends on the peculiarities of interactions between protein globule and solvent. Obviously, many factors may affect the efficiency of coil–globule transition (i.e., the efficiency and direction of the process of protein folding) as well as change the efficiency of the shield (interface between the polymer and solvent) and thus may modulate the stability of a native protein molecule. Basically, point amino acid substitutions, and changes in pH, temperature, and numerous other environmental circumstances, may considerably affect the mode of polymer–solvent interactions. Thus, protein misfolding (aggregation) may originate from the changes in the relative quality of solvent which appear either due to specific changes in protein amino acid composition or because of solvent composition modifications.

REFERENCES

Adkins JN, Lumb KJ (2002) Intrinsic structural disorder and sequence features of the cell cycle inhibitor p57Kip2. Proteins 46:1–7.

Alexandrescu AT, Abeygunawardana C, Shortle D (1994) Structure and dynamics of a denatured 131-residue fragment of staphylococcal nuclease: a heteronuclear NMR study. Biochemistry 33:1063–1072.

Alonso AC, Zaidi T, Grundke-Iqbal I, Iqbal K (1994) Role of abnormally phosphorylated tau in the breakdown of microtubules in Alzheimer disease. Proc Natl Acad Sci U S A 91:5562–5566.

Alonso AC, Grundke-Iqbal I, Iqbal K (1996) Alzheimer's disease hyperphosphorylated tau sequesters normal tau into tangles of filaments and disassembles microtubules. Nat Med 2:783–787.

Altschuler GM, Willison KR (2008) Development of free-energy-based models for chaperonin containing TCP-1 mediated folding of actin. J R Soc Interface 5:1391–1408.

Amir D, Krausz S, Haas E (1992) Detection of local structures in reduced unfolded bovine pancreatic trypsin inhibitor. Proteins 13:162–173.

Anfinsen CB (1973) Principles that govern the folding of protein chains. Science 181:223–230.

Anfinsen CB, Haber E, Sela M, White FH, Jr. (1961) The kinetics of formation of native ribonuclease during oxidation of the reduced polypeptide chain. Proc Natl Acad Sci U S A47:1309–1314.

Arcus VL, Vuilleumier S, Freund SM, Bycroft M, Fersht AR (1995) A comparison of the pH, urea, and temperature-denatured states of barnase by heteronuclear NMR: implications for the initiation of protein folding. J Mol Biol 254:305–321.

Baldwin RL (1986) Seeding protein folding. Trends Biochem Sci 11:6–9.

Baldwin RL, Zimm BH (2000) Are denatured proteins ever random coils? Proc Natl Acad Sci U S A 97:12391–12392.

Bellotti V, Mangione P, Stoppini M (1999) Biological activity and pathological implications of misfolded proteins. Cell Mol Life Sci 55:977–991.

Ben-Zvi AP, Goloubinoff P (2001) Review: mechanisms of disaggregation and refolding of stable protein aggregates by molecular chaperones. J Struct Biol 135:84–93.

Berman HM, Westbrook J, Feng Z, Gilliland G, Bhat TN, Weissig H, Shindyalov IN, Bourne PE (2000) The Protein Data Bank. Nucleic Acids Res 28:235–242.

Billingsley ML, Kincaid RL (1997) Regulated phosphorylation and dephosphorylation of tau protein: effects on microtubule interaction, intracellular trafficking and neurodegeneration. Biochem J 323 (Pt 3):577–591.

Binder LI, Frankfurter A, Rebhun LI (1985) The distribution of tau in the mammalian central nervous system. J Cell Biol 101:1371–1378.

Brandt R, Lee G (1993a) The balance between tau protein's microtubule growth and nucleation activities: implications for the formation of axonal microtubules. J Neurochem 61:997–1005.

Brandt R, Lee G (1993b) Functional organization of microtubule-associated protein tau: identification of regions which affect microtubule growth, nucleation, and bundle formation in vitro. J Biol Chem 268:3414–3419.

Bryngelson JD, Onuchic JN, Socci ND, Wolynes PG (1995) Funnels, pathways, and the energy landscape of protein folding: a synthesis. Proteins 21:167–195.

Bushmarina NA, Kuznetsova IM, Biktashev AG, Turoverov KK, Uversky VN (2001) Partially folded conformations in the folding pathway of bovine carbonic anhydrase II: a fluorescence spectroscopic analysis. ChemBioChem 2:813–821.

Buxbaum J (1992) Mechanisms of disease: monoclonal immunoglobulin deposition—amyloidosis, light chain deposition disease, and light and heavy chain deposition disease. Hematol Oncol Clin North Am 6:323–346.

Campbell KM, Terrell AR, Laybourn PJ, Lumb KJ (2000) Intrinsic structural disorder of the C-terminal activation domain from the bZIP transcription factor Fos. Biochemistry 39:2708–2713.

Canet D, Sunde M, Last AM, Miranker A, Spencer A, Robinson CV, Dobson CM (1999) Mechanistic studies of the folding of human lysozyme and the origin of amyloidogenic behavior in its disease-related variants. Biochemistry 38:6419–6427.

Chang BS, Minn AJ, Muchmore SW, Fesik SW, Thompson CB (1997) Identification of a novel regulatory domain in Bcl-X(L) and Bcl-2. EMBO J 16:968–977.

Cheng Y, LeGall T, Oldfield CJ, Dunker AK, Uversky VN (2006) Abundance of intrinsic disorder in protein associated with cardiovascular disease. Biochemistry 45:10448–10460.

Chi SW, Kim DH, Lee SH, Chang I, Han KH (2007) Pre-structured motifs in the natively unstructured preS1 surface antigen of hepatitis B virus. Protein Sci 16:2108–2117.

Chirita CN, Necula M, Kuret J (2003) Anionic micelles and vesicles induce tau fibrillization in vitro. J Biol Chem 278:25644–25650.

Chirita CN, Congdon EE, Yin H, Kuret J (2005) Triggers of full-length tau aggregation: a role for partially folded intermediates. Biochemistry 44:5862–5872.

Chiti F, Taddei N, Stefani M, Dobson CM, Ramponi G (2001) Reduction of the amyloidogenicity of a protein by specific binding of ligands to the native conformation. Protein Sci 10:879–886.

Clark A, Wells CA, Buley ID, Cruickshank JK, Vanhegan RI, Matthews DR, Cooper GJ, Holman RR, Turner RC (1988) Islet amyloid, increased A-cells, reduced B-cells and exocrine fibrosis: quantitative changes in the pancreas in type 2 diabetes. Diabetes Res 9:151–159.

Clayton DF, George JM (1998) The synucleins: a family of proteins involved in synaptic function, plasticity, neurodegeneration and disease. Trends Neurosci 21:249–254.

Cleveland DW, Hwo SY, Kirschner MW (1977a) Physical and chemical properties of purified tau factor and the role of tau in microtubule assembly. J Mol Biol 116:227–247.

Cleveland DW, Hwo SY, Kirschner MW (1977b) Purification of tau, a microtubule-associated protein that induces assembly of microtubules from purified tubulin. J Mol Biol 116:207–225.

Cohlberg JA, Li J, Uversky VN, Fink AL (2002) Heparin and other glycosaminoglycans stimulate the formation of amyloid fibrils from alpha-synuclein in vitro. Biochemistry 41:1502–1511.

Colon W, Lai Z, McCutchen SL, Miroy GJ, Strang C, Kelly JW (1996) FAP mutations destabilize transthyretin facilitating conformational changes required for amyloid formation. Ciba Found Symp 199:228–238; discussion 39–42.

Cooper GJ, Willis AC, Clark A, Turner RC, Sim RB, Reid KB (1987) Purification and characterization of a peptide from amyloid-rich pancreases of type 2 diabetic patients. Proc Natl Acad Sci U S A 84:8628–8632.

Crowther RA, Goedert M (2000) Abnormal tau-containing filaments in neurodegenerative diseases. J Struct Biol 130:271–279.

Cummings CJ, Zoghbi HY (2000) Fourteen and counting: unraveling trinucleotide repeat diseases. Hum Mol Genet 9:909–916.

Delacourte A, Buee L (1997) Normal and pathological tau proteins as factors for microtubule assembly. Int Rev Cytol 171:167–224.

Dev KK, Hofele K, Barbieri S, Buchman VL, van der Putten H (2003) Part II: Alpha-synuclein and its molecular pathophysiological role in neurodegenerative disease. Neuropharmacology 45:14–44.

Dickson DW (2001) Alpha-synuclein and the Lewy body disorders. Curr Opin Neurol 14:423–432.

Dill KA, Chan HS (1997) From Levinthal to pathways to funnels. Nat Struct Biol 4:10–19.

Dill KA, Shortle D (1991) Denatured states of proteins. Annu Rev Biochem 60:795–825.

Dinner AR, Sali A, Smith LJ, Dobson CM, Karplus M (2000) Understanding protein folding via free-energy surfaces from theory and experiment. Trends Biochem Sci 25:331–339.

Dobson CM (1999) Protein misfolding, evolution and disease. Trends Biochem Sci 24:329–332.

Dobson CM (2001a) Protein folding and its links with human disease. Biochem Soc Symp: 1–26.

Dobson CM (2001b) The structural basis of protein folding and its links with human disease. Philos Trans R Soc Lond B 356:133–145.

Dobson CM, Karplus M (1999) The fundamentals of protein folding: bringing together theory and experiment. Curr Opin Struct Biol 9:92–101.

Dobson CM, Sali A, Karplus M (1998) Protein folding: a perspective from theory and experiment. Angew Chem Int Ed 37:868–893.

Dolgikh DA, Gilmanshin RI, Brazhnikov EV, Bychkova VE, Semisotnov GV, Venyaminov S, Ptitsyn OB (1981) Alpha-Lactalbumin: compact state with fluctuating tertiary structure? FEBS Lett 136:311–315.

Dolgikh DA, Abaturov LV, Bolotina IA, Brazhnikov EV, Bychkova VE, Gilmanshin RI, Lebedev Yu O, Semisotnov GV, Tiktopulo EI, Ptitsyn OB, et al. (1985) Compact state of a protein molecule with pronounced small-scale mobility: bovine alpha-lactalbumin. Eur Biophys J 13:109–121.

Dosztanyi Z, Meszaros B, Simon I (2009) ANCHOR: Web server for predicting protein binding regions in disordered proteins. Bioinformatics 25:2745–2746.

Drechsel DN, Hyman AA, Cobb MH, Kirschner MW (1992) Modulation of the dynamic instability of tubulin assembly by the microtubule-associated protein tau. Mol Biol Cell 3:1141–1154.

Drubin DG, Feinstein SC, Shooter EM, Kirschner MW (1985) Nerve growth factor–induced neurite outgrowth in PC12 cells involves the coordinate induction of microtubule assembly and assembly-promoting factors. J Cell Biol 101:1799–1807.

Duda JE, Lee VM, Trojanowski JQ (2000) Neuropathology of synuclein aggregates. J Neurosci Res 61:121–127.

Dunker AK, Garner E, Guilliot S, Romero P, Albrecht K, Hart J, Obradovic Z, Kissinger C, Villafranca JE (1998) Protein disorder and the evolution of molecular recognition: theory, predictions and observations. Pac Symp Biocomput: 473–484.

Dunker AK, Obradovic Z, Romero P, Garner EC, Brown CJ (2000) Intrinsic protein disorder in complete genomes. Genome Inform Ser Workshop Genome Inform 11:161–171.

Dunker AK, Lawson JD, Brown CJ, Williams RM, Romero P, Oh JS, Oldfield CJ, Campen AM, Ratliff CM, Hipps KW, Ausio J, Nissen MS, Reeves R, Kang C, Kissinger CR, Bailey RW, Griswold MD, Chiu W, Garner EC, Obradovic Z (2001) Intrinsically disordered protein. J Mol Graph Model 19:26–59.

Dunker AK, Brown CJ, Lawson JD, Iakoucheva LM, Obradovic Z (2002a) Intrinsic disorder and protein function. Biochemistry 41:6573–6582.

Dunker AK, Brown CJ, Obradovic Z (2002b) Identification and functions of usefully disordered proteins. Adv Protein Chem 62:25–49.

Dunker AK, Cortese MS, Romero P, Iakoucheva LM, Uversky VN (2005) Flexible nets: the roles of intrinsic disorder in protein interaction networks. FEBS J 272:5129–5148.

Dyson HJ, Wright PE (2002) Coupling of folding and binding for unstructured proteins. Curr Opin Struct Biol 12:54–60.

Dyson HJ, Wright PE (2005) Intrinsically unstructured proteins and their functions. Nat Rev Mol Cell Biol 6:197–208.

Eidenmuller J, Fath T, Hellwig A, Reed J, Sontag E, Brandt R (2000) Structural and functional implications of tau hyperphosphorylation: information from phosphorylation-mimicking mutated tau proteins. Biochemistry 39:13166–13175.

Eliezer D, Kutluay E, Bussell R, Jr., Browne G (2001) Conformational properties of alpha-synuclein in its free and lipid-associated states. J Mol Biol 307: 1061–1073.

Farrer M, Kachergus J, Forno L, Lincoln S, Wang DS, Hulihan M, Maraganore D, Gwinn-Hardy K, Wszolek Z, Dickson D, Langston JW (2004) Comparison of kindreds with parkinsonism and alpha-synuclein genomic multiplications. Ann Neurol 55:174–179.

Fersht AR, Daggett V (2002) Protein folding and unfolding at atomic resolution. Cell 108:573–582.

Fink AL (1998) Protein aggregation: folding aggregates, inclusion bodies and amyloid. Fold Des 3:R9–R23.

Finkelstein AV, Ptitsyn OB (2002) Protein Physics: A Course of Lectures. San Diego, CA: Academic Press.

Fischbeck KH (2001) Polyglutamine expansion neurodegenerative disease. Brain Res Bull 56:161–163.

Fischer E (1894) Einfluss der Configuration auf die Wirkung der Enzyme. Ber Dt Chem Ges 27:2993.

Fitzkee NC, Rose GD (2004) Reassessing random-coil statistics in unfolded proteins. Proc Natl Acad Sci U S A 101:12497–12502.

Fuxreiter M, Simon I, Friedrich P, Tompa P (2004) Preformed structural elements feature in partner recognition by intrinsically unstructured proteins. J Mol Biol 338:1015–1026.

Galvin JE, Lee VM, Trojanowski JQ (2001) Synucleinopathies: clinical and pathological implications. Arch Neurol 58:186–190.

Garner E, Cannon P, Romero P, Obradovic Z, Dunker AK (1998) Predicting disordered regions from amino acid sequence: common themes despite differing structural characterization. Genome Inform Ser Workshop Genome Inform 9:201–213.

Gast K, Damaschun H, Eckert K, Schulze-Forster K, Maurer HR, Muller-Frohne M, Zirwer D, Czarnecki J, Damaschun G (1995) Prothymosin alpha: a biologically active protein with random coil conformation. Biochemistry 34:13211–13218.

Georlette D, Blaise V, Dohmen C, Bouillenne F, Damien B, Depiereux E, Gerday C, Uversky VN, Feller G (2003a) Cofactor binding modulates the conformational stabilities and unfolding patterns of NAD(+)-dependent DNA ligases from *Escherichia coli* and *Thermus scotoductus*. J Biol Chem 278:49945–49953.

Georlette D, Damien B, Blaise V, Depiereux E, Uversky VN, Gerday C, Feller G (2003b) Structural and functional adaptations to extreme temperatures in psychrophilic, mesophilic, and thermophilic DNA ligases. J Biol Chem 278: 37015–37023.

Gillespie JR, Shortle D (1997a) Characterization of long-range structure in the denatured state of staphylococcal nuclease: II. Distance restraints from paramagnetic relaxation and calculation of an ensemble of structures. J Mol Biol 268: 170–184.

Gillespie JR, Shortle D (1997b) Characterization of long-range structure in the denatured state of staphylococcal nuclease: I. Paramagnetic relaxation enhancement by nitroxide spin labels. J Mol Biol 268:158–169.

Glenner GG, Wong CW (1984) Alzheimer's disease and Down's syndrome: sharing of a unique cerebrovascular amyloid fibril protein. Biochem Biophys Res Commun 122:1131–1135.

Goedert M (1999) Filamentous nerve cell inclusions in neurodegenerative diseases: tauopathies and alpha-synucleinopathies. Philos Trans R Soc Lond B 354: 1101–1118.

Goedert M (2001) Parkinson's disease and other alpha-synucleinopathies. Clin Chem Lab Med 39:308–312.

Goers J, Permyakov SE, Permyakov EA, Uversky VN, Fink AL (2002) Conformational prerequisites for alpha-lactalbumin fibrillation. Biochemistry 41:12546–12551.

Goers J, Uversky VN, Fink AL (2003) Polycation-induced oligomerization and accelerated fibrillation of human alpha-synuclein in vitro. Protein Sci 12: 702–707.

Goh KI, Cusick ME, Valle D, Childs B, Vidal M, Barabasi AL (2007) The human disease network. Proc Natl Acad Sci U S A 104:8685–8690.

Goldsbury C, Goldie K, Pellaud J, Seelig J, Frey P, Muller SA, Kistler J, Cooper GJ, Aebi U (2000) Amyloid fibril formation from full-length and fragments of amylin. J Struct Biol 130:352–362.

Gottfried DS, Haas E (1992) Nonlocal interactions stabilize compact folding intermediates in reduced unfolded bovine pancreatic trypsin inhibitor. Biochemistry 31:12353–12362.

Grossberg AY, Khohlov AR (1989) Statistical Physics of Macromolecules. Moscow: Nauka.

Gusella JF, MacDonald ME (2000) Molecular genetics: unmasking polyglutamine triggers in neurodegenerative disease. Nat Rev Neurosci 1:109–115.

Hagestedt T, Lichtenberg B, Wille H, Mandelkow EM, Mandelkow E (1989) Tau protein becomes long and stiff upon phosphorylation: correlation between paracrystalline structure and degree of phosphorylation. J Cell Biol 109:1643–1651.

Hamada D, Dobson CM (2002) A kinetic study of beta-lactoglobulin amyloid fibril formation promoted by urea. Protein Sci 11:2417–2426.

Hammarstrom P, Carlsson U (2000) Is the unfolded state the Rosetta Stone of the protein folding problem? Biochem Biophys Res Commun 276:393–398.

Hazy E, Tompa P (2009) Limitations of induced folding in molecular recognition by intrinsically disordered proteins. ChemPhysChem 10:1415–1419.

Heegaard NH, Sen JW, Kaarsholm NC, Nissen MH (2001) Conformational intermediate of the amyloidogenic protein beta 2-microglobulin at neutral pH. J Biol Chem 276:32657–32662.

Higham CE, Jaikaran ET, Fraser PE, Gross M, Clark A (2000) Preparation of synthetic human islet amyloid polypeptide (IAPP) in a stable conformation to enable study of conversion to amyloid-like fibrils. FEBS Lett 470:55–60.

Himmler A (1989) Structure of the bovine tau gene: alternatively spliced transcripts generate a protein family. Mol Cell Biol 9:1389–1396.

Himmler A, Drechsel D, Kirschner MW, Martin DW, Jr. (1989) Tau consists of a set of proteins with repeated C-terminal microtubule-binding domains and variable N-terminal domains. Mol Cell Biol 9:1381–1388.

Hsu WL, Oldfield C, Meng J, Huang F, Xue B, Uversky VN, Romero P, Dunker AK (2012) Intrinsic protein disorder and protein–protein interactions. Pac Symp Biocomput:116–127.

Huang JR, Grzesiek S (2010) Ensemble calculations of unstructured proteins constrained by RDC and PRE data: a case study of urea-denatured ubiquitin. J Am Chem Soc 132:694–705.

Huang Y, Liu Z (2009) Kinetic advantage of intrinsically disordered proteins in coupled folding-binding process: a critical assessment of the "fly-casting" mechanism. J Mol Biol 393:1143–1159.

Iakoucheva LM, Brown CJ, Lawson JD, Obradovic Z, Dunker AK (2002) Intrinsic disorder in cell-signaling and cancer-associated proteins. J Mol Biol 323: 573–584.

Iqbal K, Zaidi T, Bancher C, Grundke-Iqbal I (1994) Alzheimer paired helical filaments: restoration of the biological activity by dephosphorylation. FEBS Lett 349:104–108.

Jahn TR, Radford SE (2005) The yin and yang of protein folding. FEBS J 272: 5962–5970.

Jaikaran ET, Clark A (2001) Islet amyloid and type 2 diabetes: from molecular misfolding to islet pathophysiology. Biochim Biophys Acta 1537:179–203.

Jaikaran ET, Higham CE, Serpell LC, Zurdo J, Gross M, Clark A, Fraser PE (2001) Identification of a novel human islet amyloid polypeptide beta-sheet domain and factors influencing fibrillogenesis. J Mol Biol 308:515–525.

James E, Wu PG, Stites W, Brand L (1992) Compact denatured state of a staphylococcal nuclease mutant by guanidinium as determined by resonance energy transfer. Biochemistry 31:10217–10225.

Jayasinghe SA, Langen R (2004) Identifying structural features of fibrillar islet amyloid polypeptide using site-directed spin labeling. J Biol Chem 279: 48420–48425.

Jellinger KA (2003) Neuropathological spectrum of synucleinopathies. Mov Disord 18 Suppl 6:S2–S12.

Jiang X, Smith CS, Petrassi HM, Hammarstrom P, White JT, Sacchettini JC, Kelly JW (2001) An engineered transthyretin monomer that is nonamyloidogenic, unless it is partially denatured. Biochemistry 40:11442–11452.

Kahn SE, Andrikopoulos S, Verchere CB (1999) Islet amyloid: a long-recognized but underappreciated pathological feature of type 2 diabetes. Diabetes 48:241–253.

Karnoup AS, Uversky VN (1997) Sequential compaction of a random copolymer of hydrophilic and hydrophobic amino acid residues. Macromolecules 30: 7427–434.

Kayed R, Bernhagen J, Greenfield N, Sweimeh K, Brunner H, Voelter W, Kapurniotu A (1999) Conformational transitions of islet amyloid polypeptide (IAPP) in amyloid formation in vitro. J Mol Biol 287:781–796.

Kazmirski SL, Wong KB, Freund SM, Tan YJ, Fersht AR, Daggett V (2001) Protein folding from a highly disordered denatured state: the folding pathway of chymotrypsin inhibitor 2 at atomic resolution. Proc Natl Acad Sci U S A 98: 4349–4354.

Kelly JW (1996) Alternative conformations of amyloidogenic proteins govern their behavior. Curr Opin Struct Biol 6:11–17.

Kelly JW (1998) The alternative conformations of amyloidogenic proteins and their multi-step assembly pathways. Curr Opin Struct Biol 8:101–106.

Kenessey A, Yen SH (1993) The extent of phosphorylation of fetal tau is comparable to that of PHF-tau from Alzheimer paired helical filaments. Brain Res 629: 40–46.

Khatoon S, Grundke-Iqbal I, Iqbal K, (1992) Brain levels of microtubule-associated protein tau are elevated in Alzheimer's disease: a radioimmuno-slot-blot assay for nanograms of the protein. J Neurochem 59:750–753.

Khurana R, Gillespie JR, Talapatra A, Minert LJ, Ionescu-Zanetti C, Millett I, Fink AL (2001) Partially folded intermediates as critical precursors of light chain amyloid fibrils and amorphous aggregates. Biochemistry 40:3525–3535.

Kim PM, Sboner A, Xia Y, Gerstein M (2008) The role of disorder in interaction networks: a structural analysis. Mol Syst Biol 4:179.

Kirkitadze MD, Condron MM, Teplow DB (2001) Identification and characterization of key kinetic intermediates in amyloid beta-protein fibrillogenesis. J Mol Biol 312:1103–1119.

Klein-Seetharaman J, Oikawa M, Grimshaw SB, Wirmer J, Duchardt E, Ueda T, Imoto T, Smith LJ, Dobson CM, Schwalbe H (2002) Long-range interactions within a nonnative protein. Science 295:1719–1722.

Koepf EK, Petrassi HM, Sudol M, Kelly JW (1999) WW: An isolated three-stranded antiparallel beta-sheet domain that unfolds and refolds reversibly; evidence for a structured hydrophobic cluster in urea and GdnHCl and a disordered thermal unfolded state. Protein Sci 8:841–853.

Kogo H, Takeuchi K, Inoue H, Kihara H, Kojima M, Takahashi K (2009) Urea-dependent unfolding of HIV-1 protease studied by circular dichroism and small-angle x-ray scattering. Biochim Biophys Acta 1794:70–74.

Kohn JE, Millett IS, Jacob J, Zagrovic B, Dillon TM, Cingel N, Dothager RS, Seifert S, Thiyagarajan P, Sosnick TR, Hasan MZ, Pande VS, Ruczinski I, Doniach S, Plaxco KW (2004) Random-coil behavior and the dimensions of chemically unfolded proteins. Proc Natl Acad Sci U S A 101:12491–12496.

Kruger R, Kuhn W, Muller T, Woitalla D, Graeber M, Kosel S, Przuntek H, Epplen JT, Schols L, Riess O (1998) Ala30Pro mutation in the gene encoding alpha-synuclein in Parkinson's disease. Nat Genet 18:106–108.

Kuwajima K (1996) The molten globule state of alpha-lactalbumin. FASEB J 10: 102–109.

Kuznetsova IM, Biktashev AG, Khaitlina SY, Vassilenko KS, Turoverov KK, Uversky VN (1999) Effect of self-association on the structural organization of partially folded proteins: inactivated actin. Biophys J 77:2788–2800.

Kuznetsova IM, Stepanenko OV, Turoverov KK, Zhu L, Zhou JM, Fink AL, Uversky VN (2002) Unraveling multistate unfolding of rabbit muscle creatine kinase. Biochim Biophys Acta 1596:138–155.

Kuznetsova IM, Forge V, Turoverov KK (2005) [Structural dynamics, stability, and folding of proteins.] Tsitologiia 47:943–952.

Lai Z, Colon W, Kelly JW (1996) The acid-mediated denaturation pathway of transthyretin yields a conformational intermediate that can self-assemble into amyloid. Biochemistry 35:6470–6482.

Lansbury PT, Jr. (1999) Evolution of amyloid: what normal protein folding may tell us about fibrillogenesis and disease. Proc Natl Acad Sci U S A 96:3342–3344.

Lashuel HA, Wurth C, Woo L, Kelly JW (1999) The most pathogenic transthyretin variant, L55P, forms amyloid fibrils under acidic conditions and protofilaments under physiological conditions. Biochemistry 38:13560–13573.

Lau KF, Dill KA (1990) Theory for protein mutability and biogenesis. Proc Natl Acad Sci U S A 87:638–642.

Lee BC, Kumauchi M, Hoff WD (2010) Modulating native-like residual structure in the fully denatured state of photoactive yellow protein affects its refolding. J Biol Chem 285:12579–12586.

Lee H, Mok KH, Muhandiram R, Park KH, Suk JE, Kim DH, Chang J, Sung YC, Choi KY, Han KH (2000) Local structural elements in the mostly unstructured transcriptional activation domain of human p53. J Biol Chem 275:29426–29432.

Lee VM, Balin BJ, Otvos L, Jr., Trojanowski JQ (1991) A68: a major subunit of paired helical filaments and derivatized forms of normal tau. Science 251:675–678.

Levinthal C (1968) Are there pathways for protein folding? Chim Phys 65:44–45.

Liu J, Faeder JR, Camacho CJ (2009) Toward a quantitative theory of intrinsically disordered proteins and their function. Proc Natl Acad Sci U S A 106:19819–19823.

Lu Q, Wood JG (1993) Functional studies of Alzheimer's disease tau protein. J Neurosci 13:508–515.

Manning-Bog AB, McCormack AL, Li J, Uversky VN, Fink AL, Di Monte DA (2002) The herbicide paraquat causes up-regulation and aggregation of alpha-synuclein in mice: paraquat and alpha-synuclein. J Biol Chem 277:1641–1644.

Martensson LG, Jonsson BH, Freskgard PO, Kihlgren A, Svensson M, Carlsson U (1993) Characterization of folding intermediates of human carbonic anhydrase: II. Probing substructure by chemical labeling of SH groups introduced by site-directed mutagenesis. Biochemistry 32:224–231.

Marti MJ, Tolosa E, Campdelacreu J (2003) Clinical overview of the synucleinopathies. Mov Disord 18 Suppl 6:S21–S17.

Masters CL, Multhaup G, Simms G, Pottgiesser J, Martins RN, Beyreuther K (1985) Neuronal origin of a cerebral amyloid: neurofibrillary tangles of Alzheimer's disease contain the same protein as the amyloid of plaque cores and blood vessels. EMBO J 4:2757–2763.

Meszaros B, Simon I, Dosztanyi Z (2009) Prediction of protein binding regions in disordered proteins. PLoS Comput Biol 5:e1000376.

Meszaros B, Simon I, Dosztanyi Z (2011) The expanding view of protein–protein interactions: complexes involving intrinsically disordered proteins. Phys Biol 8: 035003.

Midic U, Oldfield CJ, Dunker AK, Obradovic Z, Uversky VN (2008) Protein disorder in the human diseasome: unfoldomics of human genetic diseases. PLoS Computational Biology 10:S12.

Midic U, Oldfield CJ, Dunker AK, Obradovic Z, Uversky VN (2009a) Unfoldomics of human genetic diseases: illustrative examples of ordered and intrinsically disordered members of the human diseasome. Protein Pept Lett 16:1533–1547.

Midic U, Oldfield CJ, Dunker AK, Obradovic Z, Uversky VN (2009b) Protein disorder in the human diseasome: unfoldomics of human genetic diseases. BMC Genomics 10 Suppl 1:S12.

Miller WG, Goebel CV (1968) Dimensions of protein random coils. Biochemistry 7:3925–935.

Miroy GJ, Lai Z, Lashuel HA, Peterson SA, Strang C, Kelly JW (1996) Inhibiting transthyretin amyloid fibril formation via protein stabilization. Proc Natl Acad Sci U S A 93:15051–15056.

Mitra K, Gangopadhaya PK, Das SK (2003) Parkinsonism plus syndrome: a review. Neurol India 51:183–188.

Mohan A, Oldfield CJ, Radivojac P, Vacic V, Cortese MS, Dunker AK, Uversky VN (2006) Analysis of molecular recognition features (MoRFs). J Mol Biol 362: 1043–1059.

Mohan A, Sullivan WJ, Jr., Radivojac P, Dunker AK, Uversky VN (2008) Intrinsic disorder in pathogenic and non-pathogenic microbes: discovering and analyzing the unfoldomes of early-branching eukaryotes. Mol Biosyst 4:328–340.

Mok KH, Kuhn LT, Goez M, Day IJ, Lin JC, Andersen NH, Hore PJ (2007) A pre-existing hydrophobic collapse in the unfolded state of an ultrafast folding protein. Nature 447:106–109.

Morishima-Kawashima M, Hasegawa M, Takio K, Suzuki M, Yoshida H, Titani K, Ihara Y (1995a) Proline-directed and non-proline-directed phosphorylation of PHF-tau. J Biol Chem 270:823–829.

Morishima-Kawashima M, Hasegawa M, Takio K, Suzuki M, Yoshida H, Watanabe A, Titani K, Ihara Y (1995b) Hyperphosphorylation of tau in PHF. Neurobiol Aging 16:365–371; discussion 71–80.

Munishkina LA, Phelan C, Uversky VN, Fink AL (2003) Conformational behavior and aggregation of alpha-synuclein in organic solvents: modeling the effects of membranes. Biochemistry 42:2720–2730.

Munishkina LA, Henriques J, Uversky VN, Fink AL (2004) Role of protein–water interactions and electrostatics in alpha-synuclein fibril formation. Biochemistry 43:3289–3300.

Neri D, Billeter M, Wider G, Wuthrich K (1992a) NMR determination of residual structure in a urea-denatured protein, the 434-repressor. Science 257:1559–1563.

Neri D, Wider G, Wuthrich K (1992b) Complete ^{15}N and ^{1}H NMR assignments for the amino-terminal domain of the phage 434 repressor in the urea-unfolded form. Proc Natl Acad Sci U S A 89:4397–4401.

Neri D, Wider G, Wuthrich K (1992c) ^{1}H, ^{15}N and ^{13}C NMR assignments of the 434 repressor fragments 1–63 and 44–63 unfolded in 7 M urea. FEBS Lett 303: 129–135.

Nielsen L, Khurana R, Coats A, Frokjaer S, Brange J, Vyas S, Uversky VN, Fink AL (2001) Effect of environmental factors on the kinetics of insulin fibril formation: elucidation of the molecular mechanism. Biochemistry 40:6036–6046.

Nolting B, Golbik R, Soler-Gonzalez AS, Fersht AR (1997) Circular dichroism of denatured barstar suggests residual structure. Biochemistry 36:9899–9905.

Norris EH, Giasson BI, Lee VM (2004) Alpha-synuclein: normal function and role in neurodegenerative diseases. Curr Top Dev Biol 60:17–54.

Ohgushi M, Wada A (1983) "Molten-globule state": a compact form of globular proteins with mobile side-chains. FEBS Lett 164:21–24.

Ohnishi S, Lee AL, Edgell MH, Shortle D (2004) Direct demonstration of structural similarity between native and denatured eglin C. Biochemistry 43:4064–4070.

Okazawa H (2003) Polyglutamine diseases: a transcription disorder? Cell Mol Life Sci 60:1427–1439.

Oldfield CJ, Cheng Y, Cortese MS, Brown CJ, Uversky VN, Dunker AK (2005a) Comparing and combining predictors of mostly disordered proteins. Biochemistry 44:1989–2000.

Oldfield CJ, Cheng Y, Cortese MS, Romero P, Uversky VN, Dunker AK (2005b) Coupled folding and binding with alpha-helix-forming molecular recognition elements. Biochemistry 44:12454–12470.

Oldfield CJ, Meng J, Yang JY, Yang MQ, Uversky VN, Dunker AK (2008) Flexible nets: disorder and induced fit in the associations of p53 and 14-3-3 with their partners. BMC Genomics 9 Suppl 1:S1.

Orr HT (2001) Beyond the Qs in the polyglutamine diseases. Genes Dev 15: 925–932.

Palleros DR, Shi L, Reid KL, Fink AL (1993) Three-state denaturation of DnaK induced by guanidine hydrochloride: evidence for an expandable intermediate. Biochemistry 32:4314–4321.

Pavlov NA, Cherny DI, Heim G, Jovin TM, Subramaniam V (2002) Amyloid fibrils from the mammalian protein prothymosin alpha. FEBS Lett 517:37–40.

Permyakov EA, Berliner LJ (2000) alpha-Lactalbumin: structure and function. FEBS Lett 473:269–274.

Plaxco KW, Gross M (2001) Unfolded, yes, but random? Never! Nat Struct Biol 8:659–660.

Polymeropoulos MH, Lavedan C, Leroy E, Ide SE, Dehejia A, Dutra A, Pike B, Root H, Rubenstein J, Boyer R, Stenroos ES, Chandrasekharappa S, Athanassiadou A, Papapetropoulos T, Johnson WG, Lazzarini AM, Duvoisin RC, Di Iorio G, Golbe LI, Nussbaum RL (1997) Mutation in the alpha-synuclein gene identified in families with Parkinson's disease. Science 276:2045–2047.

Povarova OI, Kuznetsova IM, Turoverov KK (2007) Different disturbances—one pathway of protein unfolding: actin folding–unfolding and misfolding. Cell Biol Int 31:405–412.

Prakash MK (2011) Insights on the role of (dis)order from protein–protein interaction linear free-energy relationships. J Am Chem Soc 133:9976–9979.

Prusiner SB (2001) Shattuck lecture: neurodegenerative diseases and prions. N Engl J Med 344:1516–1526.

Ptitsyn OB (1973) [Stages in the mechanism of self-organization of protein molecules]. Dokl Akad Nauk SSSR 210:1213–1215.

Ptitsyn OB (1995) Molten globule and protein folding. Adv Protein Chem 47: 83–229.

Ptitsyn OB, Volkenstein MV (1986) Protein structure and neutral theory of evolution. J Biomol Struct Dyn 4:137–156.

Ptitsyn OB, Bychkova VE, Uversky VN (1995) Kinetic and equilibrium folding intermediates. Philos Trans R Soc Lond B 348:35–41.

Radford SE (2000) Protein folding: progress made and promises ahead. Trends Biochem Sci 25:611–618.

Radivojac P, Iakoucheva LM, Oldfield CJ, Obradovic Z, Uversky VN, Dunker AK (2007) Intrinsic disorder and functional proteomics. Biophys J 92:1439–1456.

Ramachandran GN, Sasisekharan V (1968) Conformation of polypeptides and proteins. Adv Protein Chem 23:283–438.

Ramirez-Alvarado M, Merkel JS, Regan L (2000) A systematic exploration of the influence of the protein stability on amyloid fibril formation in vitro. Proc Natl Acad Sci U S A 97:8979–8984.

Ratcliff K, Marqusee S (2010) Identification of residual structure in the unfolded state of ribonuclease H1 from the moderately thermophilic *Chlorobium tepidum*: comparison with thermophilic and mesophilic homologues. Biochemistry 49:5167–5175.

Rochet JC, Lansbury PT, Jr. (2000) Amyloid fibrillogenesis: themes and variations. Curr Opin Struct Biol 10:60–68.

Rocken C, Linke RP, Saeger W (1992) Immunohistology of islet amyloid polypeptide in diabetes mellitus: semi-quantitative studies in a post-mortem series. Virchows Arch A Pathol Anat Histopathol 421:339–344.

Romero P, Obradovic Z, Kissinger CR, Villafranca JE, Garner E, Guilliot S, Dunker AK (1998) Thousands of proteins likely to have long disordered regions. Pac Symp Biocomput: 437–448.

Romero P, Obradovic Z, Li X, Garner EC, Brown CJ, Dunker AK (2001) Sequence complexity of disordered protein. Proteins 42:38–48.

Romero-Beviar M, Martinez-Rodriguez S, Prieto J, Goormaghtigh E, Ariz U, Martinez-Chantar Mde L, Gomez J, Neira JL (2010) The N-terminal domain of the enzyme I is a monomeric well-folded protein with a low conformational stability and residual structure in the unfolded state. Protein Eng Des Sel 23:729–742.

Ropson IJ, Boyer JA, Dalessio PM (2006) A residual structure in unfolded intestinal fatty acid binding protein consists of amino acids that are neighbors in the native state. Biochemistry 45:2608–2617.

Saab-Rincon G, Froebe CL, Matthews CR (1993) Urea-induced unfolding of the alpha subunit of tryptophan synthase: one-dimensional proton NMR evidence for residual structure near histidine-92 at high denaturant concentration. Biochemistry 32:13981–13990.

Saab-Rincon G, Gualfetti PJ, Matthews CR (1996) Mutagenic and thermodynamic analyses of residual structure in the alpha subunit of tryptophan synthase. Biochemistry 35:1988–1994.

Santra MK, Dasgupta D, Panda D (2006) Pyrene excimer fluorescence of yeast alcohol dehydrogenase: a sensitive probe to investigate ligand binding and unfolding pathway of the enzyme. Photochem Photobiol 82:480–486.

Saraiva MJ (2001) Transthyretin amyloidosis: a tale of weak interactions. FEBS Lett 498:201–203.

Saraiva MJ, Birken S, Costa PP, Goodman DS (1984) Amyloid fibril protein in familial amyloidotic polyneuropathy, Portuguese type: definition of molecular abnormality in transthyretin (prealbumin). J Clin Invest 74:104–119.

Schneider HM, Storkel S, Will W (1980) [Amyloid of islets of Langerhans and its relation to diabetes mellitus (author's transl.).] Dtsch Med Wochenschr 105:1143–1147.

Schulman BA, Kim PS, Dobson CM, Redfield C (1997) A residue-specific NMR view of the non-cooperative unfolding of a molten globule. Nat Struct Biol 4:630–634.

Schultz CP (2000) Illuminating folding intermediates. Nat Struct Biol 7:7–10.

Schwarzinger S, Wright PE, Dyson HJ (2002) Molecular hinges in protein folding: the urea-denatured state of apomyoglobin. Biochemistry 41:12681–12686.

Schweers O, Schonbrunn-Hanebeck E, Marx A, Mandelkow E (1994) Structural studies of tau protein and Alzheimer paired helical filaments show no evidence for beta-structure. J Biol Chem 269:24290–24297.

Shimizu S, Chan HS (2002) Origins of protein denatured state compactness and hydrophobic clustering in aqueous urea: inferences from nonpolar potentials of mean force. Proteins 49:560–566.

Shoemaker BA, Portman JJ, Wolynes PG (2000) Speeding molecular recognition by using the folding funnel: the fly-casting mechanism. Proc Natl Acad Sci U S A 97:8868–8873.

Shortle D (1996a) The denatured state (the other half of the folding equation) and its role in protein stability. FASEB J 10:27–34.

Shortle D (1996b) Structural analysis of non-native states of proteins by NMR methods. Curr Opin Struct Biol 6:24–30.

Shortle D (2002) The expanded denatured state: an ensemble of conformations trapped in a locally encoded topological space. Adv Protein Chem 62:1–23.

Shortle D, Ackerman MS (2001) Persistence of native-like topology in a denatured protein in 8 M urea. Science 293:487–489.

Singleton AB, Farrer M, Johnson J, Singleton A, Hague S, Kachergus J, Hulihan M, Peuralinna T, Dutra A, Nussbaum R, Lincoln S, Crawley A, Hanson M, Maraganore D, Adler C, Cookson MR, Muenter M, Baptista M, Miller D, Blancato J, Hardy J, Gwinn-Hardy K (2003) Alpha-Synuclein locus triplication causes Parkinson's disease. Science 302:841.

Singleton A, Gwinn-Hardy K, Sharabi Y, Li ST, Holmes C, Dendi R, Hardy J, Crawley A, Goldstein DS (2004) Association between cardiac denervation and parkinsonism caused by alpha-synuclein gene triplication. Brain 127:768–772.

Smith LJ, Fiebig KM, Schwalbe H, Dobson CM (1996) The concept of a random coil: residual structure in peptides and denatured proteins. Fold Des 1:R95–R106.

Souillac PO, Uversky VN, Millett IS, Khurana R, Doniach S, Fink AL (2002a) Elucidation of the molecular mechanism during the early events in immunoglobulin light chain amyloid fibrillation: evidence for an off-pathway oligomer at acidic pH. J Biol Chem 277:12666–12679.

Souillac PO, Uversky VN, Millett IS, Khurana R, Doniach S, Fink AL (2002b) Effect of association state and conformational stability on the kinetics of immunoglobulin light chain amyloid fibril formation at physiological pH. J Biol Chem 277: 12657–12665.

Spillantini MG, Goedert M (2000) The alpha-synucleinopathies: Parkinson's disease, dementia with Lewy bodies, and multiple system atrophy. Ann N Y Acad Sci 920:16–27.

Spillantini MG, Schmidt ML, Lee VM, Trojanowski JQ, Jakes R, Goedert M (1997) Alpha-Synuclein in Lewy bodies. Nature 388:839–840.

Spillantini MG, Crowther RA, Jakes R, Hasegawa M, Goedert M (1998) Alpha-Synuclein in filamentous inclusions of Lewy bodies from Parkinson's disease and dementia with Lewy bodies. Proc Natl Acad Sci U S A 95:6469–6473.

Sunde M, Serpell LC, Bartlam M, Fraser PE, Pepys MB, Blake CC (1997) Common core structure of amyloid fibrils by synchrotron x-ray diffraction. J Mol Biol 273:729–739.

Sunde M, McGrath KC, Young L, Matthews JM, Chua EL, Mackay JP, Death AK (2004) TC-1 is a novel tumorigenic and natively disordered protein associated with thyroid cancer. Cancer Res 64:2766–2773.

Svensson M, Jonasson P, Freskgard PO, Jonsson BH, Lindgren M, Martensson LG, Gentile M, Boren K, Carlsson U (1995) Mapping the folding intermediate of human carbonic anhydrase: II. Probing substructure by chemical reactivity and spin and fluorescence labeling of engineered cysteine residues. Biochemistry 34: 8606–8620.

Tafer H, Hiller S, Hilty C, Fernandez C, Wuthrich K (2004) Nonrandom structure in the urea-unfolded *Escherichia coli* outer membrane protein X (OmpX). Biochemistry 43:860–869.

Tanford C (1961) Physical Chemistry of Macromolecules. New York: Wiley.

Tanford C (1968) Protein denaturation. Adv Protein Chem 23:121–282.

Tcherkasskaya O, Uversky VN (2001) Denatured collapsed states in protein folding: example of apomyoglobin. Proteins 44:244–254.

Tcherkasskaya O, Davidson EA, Uversky VN (2003) Biophysical constraints for protein structure prediction. J Proteome Res 2:37–42.

Teplow DB (1998) Structural and kinetic features of amyloid beta-protein fibrillogenesis. Amyloid 5:121–142.

Tompa P (2002) Intrinsically unstructured proteins. Trends Biochem Sci 27: 527–533.

Tompa P (2011) Unstructural biology coming of age. Curr Opin Struct Biol 21: 419–425.

Tompa P, Fuxreiter M (2008) Fuzzy complexes: polymorphism and structural disorder in protein–protein interactions. Trends Biochem Sci 33:2–8.

Trojanowski JQ, Lee VM (2003) Parkinson's disease and related alpha-synucleinopathies are brain amyloidoses. Ann N Y Acad Sci 991:107–110.

Turoverov KK, Kuznetsova IM, Uversky VN (2010) The protein kingdom extended: ordered and intrinsically disordered proteins, their folding, supramolecular complex formation, and aggregation. Prog Biophys Mol Biol 102:73–84.

Ueda K, Fukushima H, Masliah E, Xia Y, Iwai A, Yoshimoto M, Otero DA, Kondo J, Ihara Y, Saitoh T (1993) Molecular cloning of cDNA encoding an unrecognized component of amyloid in Alzheimer disease. Proc Natl Acad Sci U S A 90:11282–11286.

Ureta DB, Craig PO, Gomez GE, Delfino JM (2007) Assessing native and non-native conformational states of a protein by methylene carbene labeling: the case of *Bacillus licheniformis* beta-lactamase. Biochemistry 46:14567–14577.

Uversky VN (2002a) What does it mean to be natively unfolded? Eur J Biochem 269:2–12.

Uversky VN (2002b) Cracking the folding code: Why do some proteins adopt partially folded conformations, whereas other don't? FEBS Lett 514:181–183.

Uversky VN (2002c) Natively unfolded proteins: a point where biology waits for physics. Protein Sci 11:739–756.

Uversky VN (2003a) A protein-chameleon: conformational plasticity of alpha-synuclein, a disordered protein involved in neurodegenerative disorders. J Biomol Struct Dyn 21:211–234.

Uversky VN (2003b) Protein folding revisited. A polypeptide chain at the folding–misfolding–nonfolding cross-roads: Which way to go? Cell Mol Life Sci 60:1852–1871.

Uversky VN (2008) Amyloidogenesis of natively unfolded proteins. Curr Alzheimer Res 5:260–287.

Uversky VN (2009) Intrinsic disorder in proteins associated with neurodegenerative diseases. Front Biosci 14:5188–238.

Uversky VN (2011a) Intrinsically disordered proteins from A to Z. Int J Biochem Cell Biol 43:1090–1103.

Uversky VN (2011b) Multitude of binding modes attainable by intrinsically disordered proteins: a portrait gallery of disorder-based complexes. Chem Soc Rev 40:1623–1634.

Uversky VN (2013) Intrinsic disorder-based protein interactions and their modulators. Curr Pharm Des 19:4191–4213.

Uversky VN, Dunker AK (2010) Understanding protein non-folding. Biochim Biophys Acta 1804:1231–1264.

Uversky VN, Fink AL (2004) Conformational constraints for amyloid fibrillation: the importance of being unfolded. Biochim Biophys Acta 1698:131–153.

Uversky VN, Fink AL (2005) Pathways to amyloid fibril formation: partially folded intermediates in fibrillation of unfolded proteins. *In*: JD Sipe, ed. Amyloid Proteins: The Beta Pleated Sheet Conformation and Disease. Weinheim, Germany: Wiley-VCH, pp. 247–265.

Uversky VN, Ptitsyn OB (1994) "Partly folded" state, a new equilibrium state of protein molecules: four-state guanidinium chloride-induced unfolding of beta-lactamase at low temperature. Biochemistry 33:2782–2791.

Uversky VN, Ptitsyn OB (1996a) Further evidence on the equilibrium "pre-molten globule state": four-state guanidinium chloride-induced unfolding of carbonic anhydrase B at low temperature. J Mol Biol 255:215–228.

Uversky VN, Ptitsyn OB (1996b) All-or-none solvent-induced transitions between native, molten globule and unfolded states in globular proteins. Fold Des 1: 117–122.

Uversky VN, Semisotnov GV, Pain RH, Ptitsyn OB (1992) "All-or-none" mechanism of the molten globule unfolding. FEBS Lett 314:89–92.

Uversky VN, Karnoup AS, Segel DJ, Seshadri S, Doniach S, Fink AL (1998a) Anion-induced folding of staphylococcal nuclease: characterization of multiple equilibrium partially folded intermediates. J Mol Biol 278:879–894.

Uversky VN, Winter S, Galzitskaya OV, Kittler L, Lober G (1998b) Hyperphosphorylation induces structural modification of tau-protein. FEBS Lett 439:21–25.

Uversky VN, Gillespie JR, Millett IS, Khodyakova AV, Vasiliev AM, Chernovskaya TV, Vasilenko RN, Kozlovskaya GD, Dolgikh DA, Fink AL, Doniach S, Abramov VM

(1999a) Natively unfolded human prothymosin alpha adopts partially folded collapsed conformation at acidic pH. Biochemistry 38:15009–15016.

Uversky VN, Talapatra A, Gillespie JR, Fink AL (1999b) Protein deposits as the molecular basis of amyloidosis: II. Localized amyloidosis and neurodegenerative disorders. Med Sci Monitor 5:1238–1254.

Uversky VN, Talapatra A, Gillespie JR, Fink AL (1999c) Protein deposits as the molecular basis of amyloidosis: I. Systemic amyloidoses. Med Sci Monitor 5:1001–1012.

Uversky VN, Gillespie JR, Fink AL (2000) Why are "natively unfolded" proteins unstructured under physiologic conditions? Proteins 41:415–427.

Uversky VN, Lee HJ, Li J, Fink AL, Lee SJ (2001a) Stabilization of partially folded conformation during alpha-synuclein oligomerization in both purified and cytosolic preparations. J Biol Chem 276:43495–43498.

Uversky VN, Li J, Fink AL (2001b) Evidence for a partially folded intermediate in alpha-synuclein fibril formation. J Biol Chem 276:10737–10744.

Uversky VN, Li J, Fink AL (2001c) Trimethylamine-*N*-oxide–induced folding of alpha-synuclein. FEBS Lett 509:31–35.

Uversky VN, Li J, Fink AL (2001d) Pesticides directly accelerate the rate of alpha-synuclein fibril formation: a possible factor in Parkinson's disease. FEBS Lett 500:105–108.

Uversky VN, Li J, Fink AL (2001e) Metal-triggered structural transformations, aggregation, and fibrillation of human alpha-synuclein: a possible molecular NK between Parkinson's disease and heavy metal exposure. J Biol Chem 276: 44284–44296.

Uversky VN, Li J, Bower K, Fink AL (2002a) Synergistic effects of pesticides and metals on the fibrillation of alpha-synuclein: implications for Parkinson's disease. Neurotoxicology 23:527–536.

Uversky VN, Yamin G, Souillac PO, Goers J, Glaser CB, Fink AL (2002b) Methionine oxidation inhibits fibrillation of human alpha-synuclein in vitro. FEBS Lett 517: 239–244.

Uversky VN, Oldfield CJ, Dunker AK (2005) Showing your ID: intrinsic disorder as an ID for recognition, regulation and cell signaling. J Mol Recognit 18:343–384.

Uversky VN, Roman A, Oldfield CJ, Dunker AK (2006) Protein intrinsic disorder and human papillomaviruses: increased amount of disorder in E6 and E7 oncoproteins from high risk HPVs. J Proteome Res 5:1829–1842.

Uversky VN, Oldfield CJ, Dunker AK (2008) Intrinsically disordered proteins in human diseases: introducing the D2 concept. Annu Rev Biophys 37:215–246.

Uversky VN, Oldfield CJ, Midic U, Xie H, Xue B, Vucetic S, Iakoucheva LM, Obradovic Z, Dunker AK (2009) Unfoldomics of human diseases: linking protein intrinsic disorder with diseases. BMC Genomics 10 Suppl 1:S7.

Vacic V, Oldfield CJ, Mohan A, Radivojac P, Cortese MS, Uversky VN, Dunker AK (2007a) Characterization of molecular recognition features, MoRFs, and their binding partners. J Proteome Res 6:2351–2366.

Vacic V, Uversky VN, Dunker AK, Lonardi S (2007b) Composition Profiler: a tool for discovery and visualization of amino acid composition differences. BMC Bioinformatics 8:211.

von Bergen M, Barghorn S, Jeganathan S, Mandelkow EM, Mandelkow E (2006) Spectroscopic approaches to the conformation of tau protein in solution and in paired helical filaments. Neurodegener Dis 3:197–206.

Vucetic S, Xie H, Iakoucheva LM, Oldfield CJ, Dunker AK, Obradovic Z, Uversky VN (2007) Functional anthology of intrinsic disorder: 2. Cellular components, domains, technical terms, developmental processes, and coding sequence diversities correlated with long disordered regions. J Proteome Res 6:1899–1916.

Vulliet R, Halloran SM, Braun RK, Smith AJ, Lee G (1992) Proline-directed phosphorylation of human tau protein. J Biol Chem 267:22570–22574.

Wallgren M, Aden J, Pylypenko O, Mikaelsson T, Johansson LB, Rak A, Wolf-Watz M (2008) Extreme temperature tolerance of a hyperthermophilic protein coupled to residual structure in the unfolded state. J Mol Biol 379:845–858.

Wang HM, Yu C (2011) Investigating the refolding pathway of human acidic fibroblast growth factor (hFGF-1) from the residual structure(s) obtained by denatured-state hydrogen/deuterium exchange. Biophys J 100:154–164.

Wang Y, Shortle D (1995) The equilibrium folding pathway of staphylococcal nuclease: identification of the most stable chain–chain interactions by NMR and CD spectroscopy. Biochemistry 34:15895–15905.

Wang Y, Shortle D (1996) A dynamic bundle of four adjacent hydrophobic segments in the denatured state of staphylococcal nuclease. Protein Sci 5:1898–1906.

Wang Y, Shortle D (1997) Residual helical and turn structure in the denatured state of staphylococcal nuclease: analysis of peptide fragments. Fold Des 2:93–100.

Wang Z, Plaxco KW, Makarov DE (2007) Influence of local and residual structures on the scaling behavior and dimensions of unfolded proteins. Biopolymers 86:321–328.

Ward JJ, Sodhi JS, McGuffin LJ, Buxton BF, Jones DT (2004) Prediction and functional analysis of native disorder in proteins from the three kingdoms of life. J Mol Biol 337:635–645.

Watanabe A, Hasegawa M, Suzuki M, Takio K, Morishima-Kawashima M, Titani K, Arai T, Kosik KS, Ihara Y (1993) In vivo phosphorylation sites in fetal and adult rat tau. J Biol Chem 268:25712–25717.

Westermark P, Wernstedt C, Wilander E, Hayden DW, O'Brien TD, Johnson KH (1987) Amyloid fibrils in human insulinoma and islets of Langerhans of the diabetic cat are derived from a neuropeptide-like protein also present in normal islet cells. Proc Natl Acad Sci U S A 84:3881–3885.

Wetzel R (1997) Domain stability in immunoglobulin light chain deposition disorders. Adv Protein Chem 50:183–242.

Wirmer J, Berk H, Ugolini R, Redfield C, Schwalbe H (2006) Characterization of the unfolded state of bovine alpha-lactalbumin and comparison with unfolded states of homologous proteins. Protein Sci 15:1397–1407.

Wisniewski KE, Dalton AJ, McLachlan C, Wen GY, Wisniewski HM (1985) Alzheimer's disease in Down's syndrome: clinicopathologic studies. Neurology 35:957–961.

Wright PE, Dyson HJ (1999) Intrinsically unstructured proteins: re-assessing the protein structure–function paradigm. J Mol Biol 293:321–331.

Wright PE, Dyson HJ (2009) Linking folding and binding. Curr Opin Struct Biol 19:31–38.

Xie H, Vucetic S, Iakoucheva LM, Oldfield CJ, Dunker AK, Obradovic Z, Uversky VN (2007a) Functional anthology of intrinsic disorder: 3. Ligands, post-translational modifications, and diseases associated with intrinsically disordered proteins. J Proteome Res 6:1917–1932.

Xie H, Vucetic S, Iakoucheva LM, Oldfield CJ, Dunker AK, Uversky VN, Obradovic Z (2007b) Functional anthology of intrinsic disorder: 1. Biological processes and functions of proteins with long disordered regions. J Proteome Res 6:1882–1898.

Xue B, Dunker AK, Uversky VN (2012) Orderly order in protein intrinsic disorder distribution: disorder in 3500 proteomes from viruses and the three domains of life. J Biomol Struct Dyn 30:137–149.

Yoon S, Welsh WJ (2005) Rapid assessment of contact-dependent secondary structure propensity: relevance to amyloidogenic sequences. Proteins 60:110–117.

Zagorski MG, Yang J, Shao H, Ma K, Zeng H, Hong A (1999) Methodological and chemical factors affecting amyloid beta peptide amyloidogenicity. Methods Enzymol 309:189–204.

Zarranz JJ, Alegre J, Gomez-Esteban JC, Lezcano E, Ros R, Ampuero I, Vidal L, Hoenicka J, Rodriguez O, Atares B, Llorens V, Gomez Tortosa E, del Ser T, Munoz DG, de Yebenes JG (2004) The new mutation, E46K, of alpha-synuclein causes Parkinson and Lewy body dementia. Ann Neurol 55:164–173.

Zerovnik E (2002) Amyloid-fibril formation. Proposed mechanisms and relevance to conformational disease. Eur J Biochem 269:3362–3371.

Zhang O, Forman-Kay JD (1995) Structural characterization of folded and unfolded states of an SH3 domain in equilibrium in aqueous buffer. Biochemistry 34: 6784–6794.

Zhang O, Forman-Kay JD (1997) NMR studies of unfolded states of an SH3 domain in aqueous solution and denaturing conditions. Biochemistry 36:3959–3970.

Zhang YL, Zhou JM, Tsou CL (1996) Sequential unfolding of adenylate kinase during denaturation by guanidine hydrochloride. Biochim Biophys Acta 1295: 239–244.

Zoghbi HY, Orr HT (1999) Polyglutamine diseases: protein cleavage and aggregation. Curr Opin Neurobiol 9:566–570.

2

RECRUITING UNFOLDING CHAPERONES TO SOLUBILIZE MISFOLDED RECOMBINANT PROTEINS

RAYEES U.H. MATTOO AND PIERRE GOLOUBINOFF

2.1 INTRODUCTION

Anfinsen demonstrated that under optimal conditions, artificially unfolded proteins can spontaneously refold into their native conformation without requiring external assistance from other molecules. This seminal finding implied that the amino acid sequence of a polypeptide suffices to determine its native biologically active conformation (Anfinsen, 1973). Yet Anfinsen also observed that following artificial unfolding by urea, the yield of recovered native proteins decreased significantly as the temperature and protein concentration increased, suggesting that alternative nonproductive misfolding and aggregation pathways can compete with the physiological native refolding pathway of proteins. Destabilizing mutations, high temperatures, and prolonged stresses in general can cause more frequent transient unfolding events in proteins, which can be followed by misfolding and aggregation events. Moreover, early misfolded and aggregated species can be toxic to animal cells, neurons in particular. Thus, compact fibrils and amyloids are a hallmark of degenerative pathologies such as Parkinson's and Alzheimer's diseases (Hinault et al., 2006; Uversky, 2008).

Protein Aggregation in Bacteria: Functional and Structural Properties of Inclusion Bodies in Bacterial Cells, First Edition. Edited by Silvia Maria Doglia and Marina Lotti.

As observed while cooking an egg, the higher and longer the exposure to denaturing temperatures, the higher will be the oligomeric state, stability, compactness, SDS resistance, β-sheet content, and hydrophobic exposure of the resulting protein aggregates (Bagriantsev et al., 2006; Khare and Dokholyan, 2007; Uversky, 2008; Vendruscolo et al., 2011). With the advent of recombinant DNA technology, gene overexpression in bacteria became a central tool in biomedical and fundamental research as well as for industry (Ventura and Villaverde, 2006). Yet overexpression of recombinant polypeptides in bacteria produce inactive polypeptide species tightly compacted within insoluble inclusion bodies (IBs). Whereas IB formation can be a simple way to isolate large amounts of relatively pure recombinant polypeptides, subsequent steps to convert them into soluble functional proteins are often limiting. Thus, the effectiveness of iterative rounds of IB solubilization and unfolding by urea, guanidinium-HCl, and/or mild detergents, followed by refolding to the native state, typically by slow dialysis in the presence of osmolytes at low temperatures, determines whether initial IB formation is a favorable or a counterproductive prerequisite for the effective massive production of a given recombinant protein (Hagel et al., 1971; Vallejo and Rinas, 2004).

To obtain higher cellular amounts of recombinant proteins in their soluble, natively folded functional state, several parameters can be modified, which, however, generally correlate with a reduction in polypeptide production rates by bacteria. Thus, to reduce IB formation, the promoter strength can be decreased, the culture media impoverished, and the codon usage rarefied. The growth temperature can also be decreased and salts and osmolytes can be added to the medium to increase viscosity of the folding environment (Diamant et al., 2001; de Marco et al., 2007; Bandyopadhyay et al., 2012). Additionally, increasing the number of members of the chaperone network can improve the native folding of aggregation-prone recombinant proteins, either by plasmid-encoded chaperone genes, or by increasing the fluidity of the plasma membrane by chemical fluidizers, or naturally by heat shock, increasing the bacterial production of endogenous chaperones. Yet chaperone overexpression can reduce yields of the recombinant protein of interest, artificial fluidizers can be poisonous and impair protein synthesis, and heat shock can increase the propensity of a labile protein to misfold and aggregate rather than reach its native state (Worrall and Goss, 1989; Strandberg and Enfors, 1991; de Marco et al., 2005).

In the next section we discuss the various mechanisms by which chemical chaperones, foldases, and unfolding molecular chaperones can prevent IB formation in bacteria and convert *in vivo*, stable misfolded recombinant polypeptides into soluble active proteins.

2.2 CHEMICAL CHAPERONES

Osmolytes are naturally occurring small organic molecules, such as trehalose, glycerol, free proline, glycine betaine, trimethylamineN-oxide (TMAO) and trimethylglycine, which can be specifically absorbed by cells from the surroundings against a gradient of concentrations, or synthetized by cells in response to salt dehydration and/or osmotic stress (Diamant et al., 2001, 2003; Bandyopadhyay et al., 2012; De Los Rios and Goloubinoff, 2012). Osmolytes have been shown to inhibit *in vitro* protein aggregation during refolding of unfolded/misfolded proteins and can be considered as chemical chaperones standing first in line of the cellular defenses against stress-induced protein aggregation. Thus, the accumulation of osmolytes, such as free amino acids (proline in particular) and sugars in the cytoplasm, or their presence in molar amounts in the *in vitro* refolding buffer during dialysis, generally results in higher refolding yields of various recombinant proteins from urea-unfolded IBs (Yancey et al., 1982; de Marco et al., 2005). Arginine, glutamic acid, and proline are among the most commonly used folding aids to recover soluble proteins from IBs (Tsumoto et al., 2010; Wu et al., 2011). The present understanding of the mechanism of action of many osmolytes can be attributed to their presumed ability to destabilize unstructured segments in polypeptides, thereby shifting the free-energy balance in favor of protein conformers with a higher content of secondary structures (Cho et al., 2011). Moreover, depending on the amino acid composition of polypeptides, various osmolytes may be preferentially attracted to, or repulsed from, different unstructured segments of the polypeptide. Thus, TMAO may interact more strongly with α-helices than with β-structures (Cho et al., 2011). As misfolded proteins generally accumulate more misfolded β-structures at the expense of native α-helices, the presence of TMAO could favor the refolding of unfolded polypeptides into stable native structures and disfavor misfolded structures (De Los Rios and Goloubinoff, 2012). Osmolytes alone can act as buffers of protein evolution by stabilizing mutants with new functions that have not yet reached an optimal folding pathway to the native state (Bandyopadhyay et al., 2012). Osmolytes such as glycine betaine, glycerol, proline, alone or at best in combination with molecular chaperones (e.g., GroEL, DnaK, and ClpB), can assist the de novo folding of polypeptides and stabilization of the native end products of the folding reaction (Diamant et al., 2001; Tokuriki and Tawfik, 2009). Thus, whereas under nonstressed conditions, polypeptides emerging from the ribosome, especially short single-domain polypeptides, may not need particular assistance and readily follow the folding path dictated by their primary amino acid sequence and reach the native state. Yet when mutated or subject to external stress, they may, instead, misfold and aggregate (Figure 2.1, cycle I).

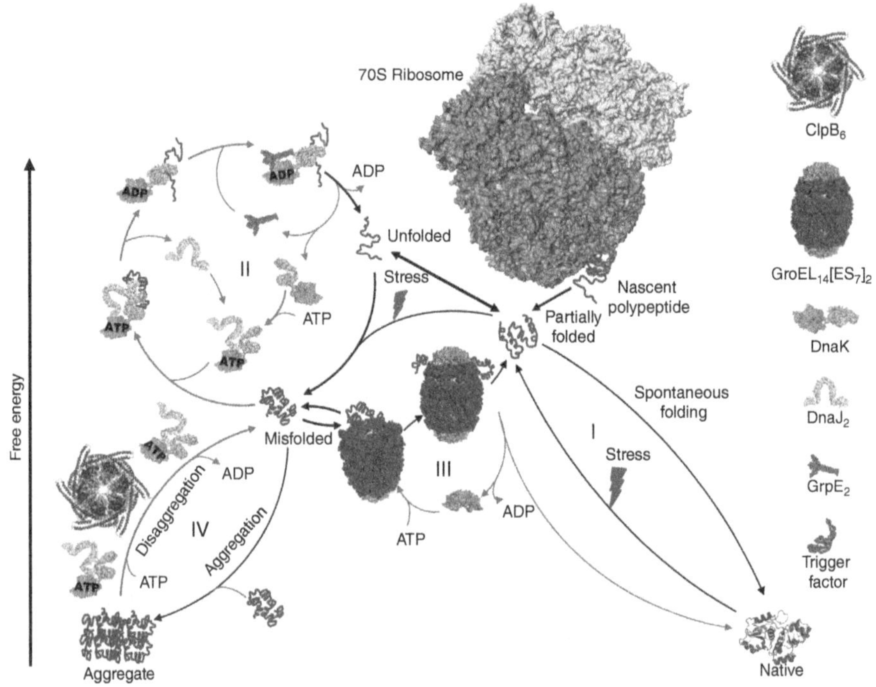

FIGURE 2.1 Schematic representation of the *E. coli* unfolding chaperone network promoting the native folding of aggregation-prone recombinant proteins. As nascent polypeptides (purple) emerge from the ribosome, the trigger factor, which is a PDI associated with the ribosome, assists in the sequential shaping of small native-like domains. Once released from the rhibosome and trigger factor, the partially folded proteins may reach the native state spontaneously, possibly assisted by chemical chaperones (cycle I). Yet, under stress or when mutated, they may instead misfold and become unfolded by the DnaK (KJE) system (cycle II), or by the GroEL/GroES chaperonin system (cycle III). In case the misfolded monomers escape unfolding by KJE or GroELS, they may form stable aggregates and IBs, but the KJE system can collaborate with ClpB to use the energy of ATP hydrolysis to unfold and solubilize individual polypeptides from the stable aggregates (cycle IV), thereby allowing them to reach their native state spontaneously. (*See insert for color representation of the figure.*)

2.3 PPIs AND PDIs ARE FOLDING ENZYMES

The native folding of some polypeptides may be limited by unfavorable distribution between *cis*- and *trans*-prolines and by the formation of the wrong disulfide bonds, requiring assistance of folding enzymes of peptidyl prolyl isomerases (PPIs) and protein disulfide isomerases (PDIs), respectively.

Native proteins have a strong bias in favor of *cis*-prolines, which may be lost in the unfolded state. Reacquisition of biased *cis*-prolines can be rate limiting for native folding and lead to aggregation (Brandts et al., 1977). Several classes of PPIases, such as cyclophilins, FK-506 binding proteins (FKBPs), and parvulins, can accelerate the *cis-trans* isomerization of prolines in refolding proteins and thus overcome rate-limiting steps in the folding pathway. Hence, the *Escherichia coli* trigger factor (TF), which is the first protein to interact with nascent chains exiting the ribosome (Deuerling and Bukau, 2004) (Figure 2.1), can catalyze prolyl-*cis-trans* isomerization of the nascent polypeptide while assisting in the sequential shaping of small protein domains (Stoller et al., 1995).

In the oxidative environment of the *endoplasmic reticulum* lumen, the cysteines of secreted proteins can wrongly associate with one another and lead to protein misfolding and aggregation. Thus, in his experiment with ribonuclease A, Anfinsen found that removing β-mercaptoethanol before removing the urea led to a dramatic decrease in the recovery of activity. This was attributed to the formation of random disulfide bridges between the eight cysteines present in the protein. Protein disulfide isomerases (PDIs) are folding enzymes, which, by virtue of their ability to disallow nonnative disulfide bonds can accelerate the formation of properly aligned disulfide bonds in native structures. However, because the bacterial cytoplasm is reduced and correspondingly less favorable to disulfide bonds than the *endoplasmic reticulum* lumen, the role of PDIs in the production of soluble recombinant proteins in the cytoplasm of bacteria is less relevant than for eukaryotes, where secreted proteins such as immunoglobins and pro-insulin need to be stabilized by properly aligned disulfide bonds.

2.4 MOLECULAR CHAPERONES

Behind the chemical chaperones and alongside the PPIs and PDIs foldases, the molecular chaperones form another line of active cellular defenses against stress- and mutation-induced protein misfolding and aggregation. Noticeably, although generally referred to misleadingly as being heat-shock proteins (Hsps), many molecular chaperones but not all are stress-inducible proteins. For example, of about 120 identified molecular chaperones, co-chaperones, and foldases that form the human genome, only a third are heat-induced (Finka et al., 2011). All molecular chaperones control cellular proteostasis by sharing the ability to screen the surfaces of potential misfolded or alternatively folded protein substrates, for nonnative structural elements, such as abnormal β-sheets associated with hydrophobic patches exposed to the aqueous phase (Hinault et al., 2006). They may act as passive "holding" chaperones upon mere binding to misfolding polypeptide intermediates, thus preventing further aggregations, and/or as active polypeptide unfoldases, which upon hydrolyzing ATP can pull apart stable aggregates and open stable misfolded structures in misfolded polypeptides, thereby promoting spontaneous folding to the native state (Nollen et al., 1999; Sharma et al., 2009). In addition to their role under stress, heat shock in particular, molecular chaperones may also carry housekeeping cellular functions, such as controlling the assembly and activity of native protein complexes (Morgan et al., 2001; Weiss et al., 2007); targeting and importing proteins into organelles; and blocking or activating various receptors, signaling proteins, and transcription factors. Importantly, chaperones are also involved in the selection of proteins to be targeted for degradation (Muchowski and Wacker, 2005).

Most molecular chaperones can be classified within five canonical families on the basis of sequence homologies and have been named according to their approximate molecular mass on SDS gels (in kilodaltons): (1) the small Hsps (IbpA/B), (2) the HSP90s (HtpG), (3) the HSP70s (DnaK), (4) the HSP104s (ClpB), and (5) the chaperonins HSP60 (GroEL) (*E. coli* orthologs shown in parentheses).

2.5 THE SMALL Hsps

The small Hsps (sHsps) are a family of ubiquitous chaperones found in bacteria and most cellular compartments of plants, fungi, and animal cells. They share a conserved 80- to 100-amino acid α-crystallin domain. The small Hsps do not hydrolyze ATP. They include mammalian Hsp27 and α-crystallins; organellar Hsp23, Hsp17, and Hsp16; and bacterial IbpA/B (*E. coli*) and Hsp17 (*Synechocystis*). IbpA and IbpB were originally isolated from inclusion bodies formed during the overproduction of recombinant proteins in *E. coli* cells and accordingly named *inclusion body binding protein A/B*. IbpA/B can also coaggregate with endogenous *E. coli* proteins and stress-misfolded proteins (Laskowska et al., 1996, 2003). Deletion of the *ibpA/B* operon resulted in a twofold increase in aggregated proteins and a 10-fold decrease in cell viability under stress conditions (Kuczynska-Wisnik et al., 2002). Indicating a possible overlap with other chaperone functions, overexpression of IbpB can alleviate thermo-sensitive point mutations of GroEL (Veinger et al., 1998). The oligomeric state and chaperone-like activity of IbpB is temperature responsive: Heat shock decreases IbpB's oligomeric size while increasing its chaperone-like activity (Jiao et al., 2005). By virtue of their association with misfolded proteins in IBs, the two *E. coli* sHsps (IbpA/B) are pivotal to the possible solubilization of IBs, as they can optimally present bound misfolded polypeptides to other unfolding chaperone systems, such as DnaK/DnaJ/GrpE and GroEL/GroES (Veinger et al., 1998) or to the disaggregating machinery ClpB-DnaK/DnaJ/GrpE (Mogk et al., 2003; Lethanh et al., 2005). In addition to preventing aggregation by passive binding, sHsps may also mediate the refolding of unfolded species to the native state (Lee and Vierling, 2000; Basha et al., 2004).

2.6 Hsp90

Hsp90s, named HtpG in bacteria and tumor necrosis factor receptor-associated protein (TRAP) in animal mitochondria, are among the most abundant proteins in human cells. They form dimers that can interact with a broad spectrum of misfolded, translocating, and native "client" proteins, particularly with polypeptides involved in signal transduction (Pearl and Prodromou, 2006). Structurally and mechanistically, it is not clear precisely where and how the various clients bind the Hsp90 dimers. The ATPase cycle is intimately coupled to the opening and closing of the dimer, which could act as a molecular clamp (Chadli et al., 2000; Prodromou et al., 2000), suggesting that Hsp90 may unfold its various misfolded or alternatively folded "clients," similar to other ATPase chaperones, (Walerych et al., 2010). Whereas eukaryotes require a functional cytoplasmic Hsp90 for viability (Borkovich et al., 1989; Versteeg et al., 1999), the *E. coli* HtpG knockout has no apparent phenotype. Thus, although HtpG is a highly conserved heat-inducible molecular chaperone, it is not a primary *E. coli* chaperone to improve the transitions between aggregated and native state in recombinant proteins.

2.7 Hsp70/Hsp40

The Hsp70 system, which in eubacteria is composed principally of DnaK and of its two co-chaperones DnaJ and GrpE, forms the central hub of the *E. coli* chaperone network (Hesterkamp and Bukau, 1998). Without ATP, DnaK and its co-chaperone DnaJ can prevent protein aggregation by mere passive "holding" of the labile proteins. Yet in the presence of DnaJ and ATP, prevention of aggregation by DnaK can be improved dramatically, suggesting an active process (Sharma et al., 2011). Moreover, the bacterial DnaK–DnaJ–GrpE (KJE) system can act as powerful polypeptide unfolding machinery that can catalytically unfold misfolded polypeptide monomers (Figure 2.1, cycle II) and disaggregate small soluble complexes of heat-denatured proteins (Diamant et al., 2000; Sharma et al., 2010). The mechanism by which the bacterial Hsp70 (KJE) system may facilitate substrate native folding comprises five main steps: (1) DnaJ and low-affinity ATP-DnaK bind a high-affinity misfolded polypeptide substrate. This triggers ATP hydrolysis in DnaK and the dissociation of the J-domain of DnaJ; (2) the ADP-DnaK "locks" onto the misfolded substrate and, by doing so, unfolds the substrate; (3) Unfolding of the polypeptide causes DnaJ dissociation from it (Hinault, 2010); (4) GrpE drives the release of ADP from DnaK and the release of the unfolded intermediate from DnaK; (5) the released intermediate spontaneously refolds in solution into a low-affinity native product, or misfolds again and re-enters the chaperone cycle as a high affinity substrate (Sharma et al., 2010) (Figure 2.1, cycle II). When, despite the unfolding activity of KJE, misfolded polypeptide became entangled and packed within IBs, the bacterial KJE system can recruit the disaggregating co-chaperone ClpB (Diamant et al., 2000) (Figure 2.1, cycle IV), which is a member of the AAA+superfamily (ATPases associated with a variety of cellular activities). The KJE system can activate and recruit the ClpB hexameric rings to the polypeptide aggregates (Oguchi et al., 2012). Individual polypeptides become disentangled upon being partially threaded in an ATP-dependent manner through the central channel of the ClpB hexameric ring (Haslberger et al., 2008) This may lead to dissociation of individual misfolded polypeptides from the aggregates, whose final refolding to the native state may, further, need partial unfolding by the KJE system (Goloubinoff et al., 1999) (Figure 2.1, cycle IV).

By virtue of its strong ATP-fueled unfolding and disaggregating activities, the KJE system is the most important chaperone machinery to maintain overexpressed recombinant proteins in a soluble form in *E. coli* cells and *in vitro*. It may also act in concert with other chaperones, such as GroEL, either upstream, by conditioning misfolded monomers to be fed to the GroEL cavity (Langer et al., 1992; Veinger et al., 1998), or downstream, by processing GroEL-released inactive species into natively refoldable species (Buchberger et al., 1996) (Figure 2.1, cycle III).

2.8 GroEL CHAPERONINS

GroEL is a complex of 14 identical 57-kDa subunits arranged as two back-to-back heptameric rings with two noncommunicating open cavities. The distal surfaces of the GroEL cavities exposed hydrophobic residues that nonnative polypeptide substrates can access from the external solution (Braig et al., 1994). GroES is a seven-membered ring of 10-kDa subunits that can cap one or both open ends of the GroEL cavities. The cage-like structure of GroEL is thought to mediate its chaperone activity (Horwich, 2011). Upon transiently sequestering stress-denatured or newly synthesized polypeptides in the cavity under GroES caps, GroEL is thought to prevent deleterious misfolding and aggregation events during the folding process (Figure 2.1, cycle III). Increased expression of both GroEL and GroES on a high-copy-number plasmid has been shown to increase the solubility and activity of recombinant cyanobacterial RubisCO (Goloubinoff et al., 1989). Moreover, co-expression on the same plasmid of up to four different complementing chaperone systems—IbpA/B, DnaK/DnaJ/GrpE, ClpB, and GroEL/GroES—can produce very high yields of soluble, natively folded recombinant proteins (de Marco et al., 2007).

2.9 CONCLUSIONS

A detailed understanding of the mechanisms by which chemical chaperones, folding enzymes, and various molecular chaperones can collaborate to prevent and actively revert misfolding and aggregation events in the cell is necessary for the successful massive production of soluble native recombinant proteins in bacteria. This can be achieved by reducing the synthesis rate, and thus the burden on the protein quality-control network, as well as by increasing the concentrations of chemical and protein chaperones in the cytoplasm. In addition to the advantages for biotechnology, finding new ways to increase the cellular levels of various molecular chaperones, particularly in animal cells, holds biomedical promises to find cures to protein-conformational diseases.

REFERENCES

Anfinsen, CB (1973) Principles that govern the folding of protein chains. Science 181:223–230.

Bagriantsev SN, Kushnirov VV, Liebman SW (2006) Analysis of amyloid aggregates using agarose gel electrophoresis. Methods Enzymol 412:33–48.

Bandyopadhyay A, Saxena K, Kasturia N, Dalal V, Bhatt N, Rajkumar A, Maity S, Sengupta S, Chakraborty, K (2012) Chemical chaperones assist intracellular folding to buffer mutational variations. Nat Chem Biol 8:238–245.

Basha E, Lee GJ, Demeler B, Vierling E (2004) Chaperone activity of cytosolic small heat shock proteins from wheat. Eur J Biochem 271:1426–1436.

Borkovich KA, Farrelly FW, Finkelstein DB, Taulien J, Lindquist S (1989) hsp82 is an essential protein that is required in higher concentrations for growth of cells at higher temperatures. Mol Cell Biol 9:3919–3930.

Braig K, Otwinowski Z, Hegde R, Boisvert DC, Joachimiak A, Horwich AL, Sigler PB (1994) The crystal structure of the bacterial chaperonin GroEL at 2.8 Å. Nature 371:578–586.

Brandts JF, Brennan M, Lung-Nan L (1977) Unfolding and refolding occur much faster for a proline-free proteins than for most proline-containing proteins. Proc Natl Acad Sci U S A 74:4178–4181.

Buchberger A, Schroder H, Hesterkamp T, Schonfeld HJ, Bukau B (1996) Substrate shuttling between the DnaK and GroEL systems indicates a chaperone network promoting protein folding. J Mol Biol 261:328–333.

Chadli A, Bouhouche I, Sullivan W, Stensgard B, McMahon N, Catelli MG, Toft DO (2000) Dimerization and N-terminal domain proximity underlie the function of the molecular chaperone heat shock protein 90. Proc Natl Acad Sci U S A 97:12524–12529.

Cho SS, Reddy G, Straub JE, Thirumalai D (2011) Entropic stabilization of proteins by TMAO. J Phys Chem B 115:13401–13407.

De Los Rios P, Goloubinoff P (2012). Protein folding: chaperoning protein evolution. Nat Chem Biol 8:226–228.

de Marco A, Vigh L, Diamant S, Goloubinoff P (2005) Native folding of aggregation-prone recombinant proteins in *Escherichia coli* by osmolytes, plasmid- or benzyl alcohol-overexpressed molecular chaperones. Cell Stress Chaperones 10:329–339.

de Marco A, Deuerling E, Mogk A, Tomoyasu T, Bukau B (2007) Chaperone-based procedure to increase yields of soluble recombinant proteins produced in *E. coli*. BMC Biotechnol 7:32.

Deuerling E, Bukau B (2004) Chaperone-assisted folding of newly synthesized proteins in the cytosol. Crit Rev Biochem Mol Biol 39:261–277.

Diamant S, Ben-Zvi AP, Bukau B, Goloubinoff P (2000) Size-dependent disaggregation of stable protein aggregates by the DnaK chaperone machinery. J Biol Chem 275:21107–21113.

Diamant S, Eliahu N, Rosenthal D, Goloubinoff P. (2001) Chemical chaperones regulate molecular chaperones in vitro and in cells under combined salt and heat stresses. J Biol Chem 276:39586–39591.

Diamant S, Rosenthal D, Azem A, Eliahu N, Ben-Zvi AP, Goloubinoff P (2003) Dicarboxylic amino acids and glycine-betaine regulate chaperone-mediated protein-disaggregation under stress. Mol Microbiol 49:401–410.

Finka A, Mattoo RU, Goloubinoff P (2011) Meta-analysis of heat- and chemically upregulated chaperone genes in plant and human cells. Cell Stress Chaperones 16:15–31.

Goloubinoff P, Gatenby AA, Lorimer GH (1989) GroE heat-shock proteins promote assembly of foreign prokaryotic ribulose bisphosphate carboxylase oligomers in *Escherichia coli*. Nature 337:44–47.

Goloubinoff P, Mogk A, Zvi AP, Tomoyasu T, Bukau B (1999) Sequential mechanism of solubilization and refolding of stable protein aggregates by a bichaperone network. Proc Natl Acad Sci U S A 96:13732–13737.

Hagel P, Gerding JJ, Fieggen W, Bloemendal H (1971) Cyanate formation in solutions of urea: I. Calculation of cyanate concentrations at different temperature and pH. Biochim Biophys Acta 243:366–373.

Haslberger T, Zdanowicz A, Brand I, Kirstein J, Turgay K, Mogk A, Bukau B (2008) Protein disaggregation by the AAA+chaperone ClpB involves partial threading of looped polypeptide segments. Nat Struct Mol Biol 15:641–650.

Hesterkamp T, Bukau B (1998) Role of the DnaK and HscA homologs of Hsp70 chaperones in protein folding in *E. coli*. EMBO J 17:4818–4828.

Hinault MP, Ben-Zvi A, Goloubinoff P (2006) Chaperones and proteases: cellular fold-controlling factors of proteins in neurodegenerative diseases and aging. J Mol Neurosci 30:249–265.

Hinault MP, Cuendet AF, Mattoo RU, Mensi M, Dietler G, Lashuel HA, Goloubinoff P (2010) Stable alpha-synuclein oligomers strongly inhibit chaperone activity of the Hsp70 system by weak interactions with J-domain co-chaperones. J Biol Chem 285: 38173–38182.

Horwich AL (2011). Protein folding in the cell: an inside story. Nat Med 17:1211–1216.

Jiao W, Qian M, Li P, Zhao L, Chang Z (2005) The essential role of the flexible termini in the temperature-responsiveness of the oligomeric state and chaperone-like activity for the polydisperse small heat shock protein IbpB from *Escherichia coli*. J Mol Biol 347:871–884.

Khare SD, Dokholyan NV (2007) Molecular mechanisms of polypeptide aggregation in human diseases. Curr Protein Pept Sci 8:573–579.

Kuczynska-Wisnik D, Kedzierska S, Matuszewska E, Lund P, Taylor A, Lipinska B, Laskowska E (2002) The *Escherichia coli* small heat-shock proteins IbpA and IbpB prevent the aggregation of endogenous proteins denatured in vivo during extreme heat shock. Microbiology 148:1757–1765.

Langer T, Lu C, Echols H, Flanagan J, Hayer MK, Hartl FU (1992) Successive action of DnaK, DnaJ and GroEL along the pathway of chaperone-mediated protein folding. Nature 356:683–689.

Laskowska E, Wawrzynow A, Taylor A (1996) IbpA and IbpB, the new heat-shock proteins, bind to endogenous *Escherichia coli* proteins aggregated intracellularly by heat shock. Biochimie 78:117–122.

Laskowska E, Kuczynska-Wisnik D, Bak M, Lipinska B (2003) Trimethoprim induces heat shock proteins and protein aggregation in *E. coli* cells. Curr Microbiol 47:286–289.

Lee GJ, Vierling E (2000) A small heat shock protein cooperates with heat shock protein 70 systems to reactivate a heat-denatured protein. Plant Physiol 122:189–198.

Lethanh H, Neubauer P, Hoffmann F (2005) The small heat-shock proteins IbpA and IbpB reduce the stress load of recombinant *Escherichia coli* and delay degradation of inclusion bodies. Microb Cell Fact 4:6.

Mogk A, Deuerling E, Vorderwulbecke S, Vierling E, Bukau B (2003) Small heat shock proteins, ClpB and the DnaK system form a functional triade in reversing protein aggregation. Mol Microbiol 50:585–595.

Morgan JR, Prasad K, Jin S, Augustine GJ, Lafer EM (2001) Uncoating of clathrin-coated vesicles in presynaptic terminals: roles for Hsc70 and auxilin. Neuron 32:289–300.

Muchowski PJ, Wacker JL (2005) Modulation of neurodegeneration by molecular chaperones. Nat Rev Neurosci 6:11–22.

Nollen EA, Brunsting JF, Roelofsen H, Weber LA, Kampinga HH (1999) In vivo chaperone activity of heat shock protein 70 and thermotolerance. Mol Cell Biol 19:2069–2079.

Oguchi Y, Kummer E, Seyffer F, Berynskyy M, Anstett B, Zahn R, Wade RC, Mogk A, Bukau B (2012) A tightly regulated molecular toggle controls AAA+disaggregase. Nat Struct Mol Biol.

Pearl LH, Prodromou C (2006) Structure and mechanism of the Hsp90 molecular chaperone machinery. Annu Rev Biochem 75:271–294.

Prodromou C, Panaretou B, Chohan S, Siligardi G, O'Brien R, Ladbury JE, Roe SM, Piper PW, Pearl LH (2000) The ATPase cycle of Hsp90 drives a molecular 'clamp' via transient dimerization of the N-terminal domains. EMBO J 19:4383–4392.

Sharma SK, Christen P, Goloubinoff P (2009) Disaggregating chaperones: an unfolding story. Curr Protein Pept Sci 10:432–446.

Sharma SK, De Los Rios P, Christen P, Lustig A, Goloubinoff P (2010) The kinetic parameters and energy cost of the Hsp70 chaperone as a polypeptide unfoldase. Nat Chem Biol 6:914–920.

Sharma SK, De Los Rios P, Goloubinoff P (2011) Probing the different chaperone activities of the bacterial HSP70–HSP40 system using a thermolabile luciferase substrate. Proteins 79:1991–1998.

Stoller G, Rucknagel KP, Nierhaus KH, Schmid FX, Fischer G, Rahfeld JU (1995) A ribosome-associated peptidyl-prolyl *cis/trans* isomerase identified as the trigger factor. EMBO J 14:4939–4948.

Strandberg L, Enfors SO (1991) Factors influencing inclusion body formation in the production of a fused protein in *Escherichia coli*. Appl Environ Microbiol 57:1669–1674.

Tokuriki N, Tawfik DS (2009) Chaperonin overexpression promotes genetic variation and enzyme evolution. Nature 459:668–673.

Tsumoto K, Abe R, Ejima D, Arakawa T (2010) Non-denaturing solubilization of inclusion bodies. Curr Pharm Biotechnol 11:309–312.

Uversky VN (2008) Amyloidogenesis of natively unfolded proteins. Curr Alzheimer Res 5:260–287.

Vallejo LF, Rinas U (2004) Strategies for the recovery of active proteins through refolding of bacterial inclusion body proteins. Microb Cell Fact 3:11.

Veinger L, Diamant S, Buchner J, Goloubinoff P (1998) The small heat-shock protein IbpB from *Escherichia coli* stabilizes stress-denatured proteins for subsequent refolding by a multichaperone network. J Biol Chem 273:11032–11037.

Vendruscolo M, Knowles TP, Dobson CM (2011) Protein solubility and protein homeostasis: a generic view of protein misfolding disorders. Cold Spring Harbor Perspect Biol 3.

Ventura S, Villaverde A (2006) Protein quality in bacterial inclusion bodies. Trends Biotechnol 24:179–185.

Versteeg S, Mogk A, Schumann W (1999) The *Bacillus subtilis* htpG gene is not involved in thermal stress management. Mol Gen Genet 261:582–588.

Walerych D, Gutkowska M, Klejman MP, Wawrzynow B, Tracz Z, Wiech M, Zylicz M, Zylicz A (2010) ATP binding to Hsp90 is sufficient for effective chaperoning of p53 protein. J Biol Chem 285:32020–32028.

Weiss YG, Bromberg Z, Raj N, Raphael J, Goloubinoff P, Ben-Neriah Y, Deutschman CS (2007) Enhanced heat shock protein 70 expression alters proteasomal degradation of IkappaB kinase in experimental acute respiratory distress syndrome. Crit Care Med 35:2128–2138.

Worrall DM, Goss NH (1989) The formation of biologically active beta-galactosidase inclusion bodies in *Escherichia coli*. Aust J Biotechnol 3:28–32.

Wu B, Nemeth JF, Janecki DJ, Jones B, Obmolova G, Malia TJ, Baker A, Bethea D, Elloso MM, Naso M, et al. (2011) Expression, refolding and purification of a human interleukin-17A variant. Cytokine 53:107–114.

Yancey PH, Clark ME, Hand SC, Bowlus RD, Somero GN (1982) Living with water stress: evolution of osmolyte systems. Science 217:1214–1222.

3

OSMOLYTES AS CHEMICAL CHAPERONES TO USE IN PROTEIN BIOTECHNOLOGY

Ario de Marco

3.1 INTRODUCTION

Naturally occurring intracellular osmolytes such as amino acids, sugars, and methylamines are compatible solutes that have a pivotal role in stabilizing protein structure and preserving their function under cellular stress conditions (Ou et al., 2001). Their protective effect apparently represents a major evolutionary advantage since osmolytes are produced and accumulated in almost any organism in response to stress conditions (Strøm and Kaasen, 1993; Singer and Lindquist, 1998a; Kandror et al., 2002; Burg and Ferraris, 2008). The spectrum of osmolytes comprises a relatively limited number of compound classes because there are fundamental constraints they must respect in terms of osmotic strength and the interaction with unfolded proteins for being compatible with macromolecular and cellular functions (Singer and Lindquist, 1998a; Roeßler and Müller, 2001). On the other hand, the stress conditions that different organisms have to cope with in different environmental conditions, their most critical heat-sensitive structures, and their physiologies are dissimilar, a reason that has led to the selection of molecules with different features, although all are characterized by "chemical chaperoning" activity (Singer and Lindquist, 1998b).

Protein Aggregation in Bacteria: Functional and Structural Properties of Inclusion Bodies in Bacterial Cells, First Edition. Edited by Silvia Maria Doglia and Marina Lotti.

Osmolytes such as glutamate, glycine betaine, dimethylsulfoniopropionate, trehalose, trimethylamine *N*-oxide (TMAO), and hydroxybutyrate protect bacteria from low temperatures, desiccation, urea accumulation, and high salinity (Gauthier and Le Rudulier, 1990; Gauthier et al., 1991; Cosquer et al., 1999; Kandror et al., 2002; McIntyre et al., 2007; Dreux et al., 2008; Bandyopadhyay et al., 2012; Soto et al., 2012). In the case of archaea that live in very challenging conditions, the selective pressure resulted in the development of synthetic pathways that enable the accumulation of a large array of unusual chemical chaperones capable of protecting cells and macromolecules from damage by external stresses (Lentzen and Schwarz, 2006). Extremophilic microorganisms use "extremolytes" to minimize biopolymer denaturation under conditions of low pH, high salinity, and elevated temperatures. In most cases, archaea produce osmolytes carrying a negative charge resulting from the addition of a carboxylate, phosphate, or sulfate group to molecules commonly present in bacteria (Martin et al., 1999), probably because anionic solutes must balance the abundance of inorganic cations in extremophiles (Roeßler and Müller, 2001). Consequently, derivatives of bacterial carbohydrates accumulate in archaea, such as 2-sulfotrehalose accumulates in *Natronococcus occultus*, diglycerol phosphate in *Archaeoglobus fulgidus*, glucosylglycerate in *Methanohalophilus portucalensis*, and dimyoinositol phosphate (DIP) in *Methanococcus igneus* (Robertson et al., 1992; Ciulla et al., 1994; Desmarais et al., 1997; Martins et al., 1997).

Amino acids and chemically related molecules seem to be present more frequently as stabilizing additives in plants and mammalians than in microorganisms (Khan et al., 2010). Nevertheless, plants can also accumulate trehalose in response to stress conditions, and transgenic plants overexpressing microbial genes for improving trehalose biosynthesis efficiency better tolerate drought, salt, and cold conditions (Iordanescu and Imai, 2008). The positive effects of glycine betaine have been exploited as well for generating transgenic plants with improved tolerance to salt and low temperatures (Holmström et al., 2000), whereas the role of proline in protecting plants from membrane leakage and oxidative stress has recently been reviewed (Hayat et al., 2012). There is also an increasing interest in the clinical applications of chemical chaperones (Rajan et al., 2011). A set of osmolytes able to mediate the relocation of E-cadherin, a condition potentially useful for inhibiting metastatic development, has been suggested for cancer therapy (Figueiredo et al., 2011). Trehalose has shown a remarkable capacity to inhibit fibrillar aggregation of insulin and β-amyloid peptides *in vitro* (Arora et al., 2004; Liu et al., 2005), and its therapeutic administration succeeded in mitigating neurodegenerative progression in a murine system (Rodriguez-Navarro et al., 2010). Since polysaccharides such as trehalose are not toxic even at high concentrations, their neuroprotective effect is now under evaluation for clinical applications (Nelson et al., 2012a). At present, the limiting factor for the *in vivo* pharmacological assessment of osmolytes as a means intended for the alleviation of diseases related to pathological protein folding is obtaining a reliable delivery system to assure a sufficient concentration in the desired organs (Rajan et al., 2011). The accumulation of more homogeneous data will enable us to evaluate the health consequences of the overall osmolyte accumulation in mammalians since the results available concerning, for example, the physiological effects of osmolytes introduced into an organism as a consequence of dietary habits, are contradictory (Ueland, 2011; Lenky et al., 2012).

3.2 PROTEIN-DESTABILIZING CONDITIONS AND COUNTERACTING MECHANISMS: SHARED OR INDEPENDENT ROUTES?

Some authors (Kaushik and Bhat, 2003; Mukaiyama et al., 2008) inferred from their experimental data that any protein, regardless of its robustness, would be stabilized by osmolytes, and this belief contributed to the interest in osmolytes as biotechnologically relevant reagents to use for improving protein folding and stabilization during recombinant expression (Figure 3.1). Indeed, these conclusions were drawn from *in vitro* experiments and did not consider some theoretical questions. The first is whether different stress conditions leading to protein aggregation trigger common mechanisms of structure destabilization. This is clearly not the case, and an amazing diversity of aggregates formed upon induction by various chemical and physical types of stress has been documented (Natalello et al., 2009). The picture is complicated by the fact that it is also true that the same osmolyte can protect from a variety of abiotic stresses (Singer and Lindquist, 1998a; Garg et al., 2002; Jang et al., 2003) and increase the thermostability of the proteins *in vitro* (Schultz et al., 2007), an observation that might suggest a common stabilizing mechanism and lead to the hypothesis of reversing any destabilizing condition using a single chemical chaperone. More systematic surveys showed that, actually, the accumulation of soluble protein seems to be case dependent. Not every osmolyte affects the solubility of a specific protein in the same manner, so a trial-and-error approach is necessary to identify the combinations that result in superior yields with respect to other methods, such as the induction of heat-shock proteins or the fusion of the target protein to solubilizing carriers (de Marco et al., 2005; Schultz et al., 2007; Zhou et al., 2012).

Even though some generic biophysical traits are common to different osmolyte classes (Mukaiyama et al., 2008), as described below, osmolytes exert their action by specific, probably multiple, concentration-dependent mechanisms (Natalello et al., 2008), are host- and stress-adapted [as illustrated by the osmolyte specificities in archaea, there are convincing lines of evidence that they possess substrate specialization (Bandyopadhyay et al., 2012)], and their effectiveness depends on such physical features as the length of the polypeptidic chain (Kaushik and Bhat, 2003; Hu et al., 2010). Finally, *in vivo* chemical chaperones interact with molecular chaperones, a condition that needs fine regulation to prevent activity inhibition (Singer and Lindquist, 1998a, b) and that should be optimized when using the recombinant protein expression systems. Taken together, it is necessary to have an overview of the models describing osmolyte mechanisms of action and of their interaction with substrates.

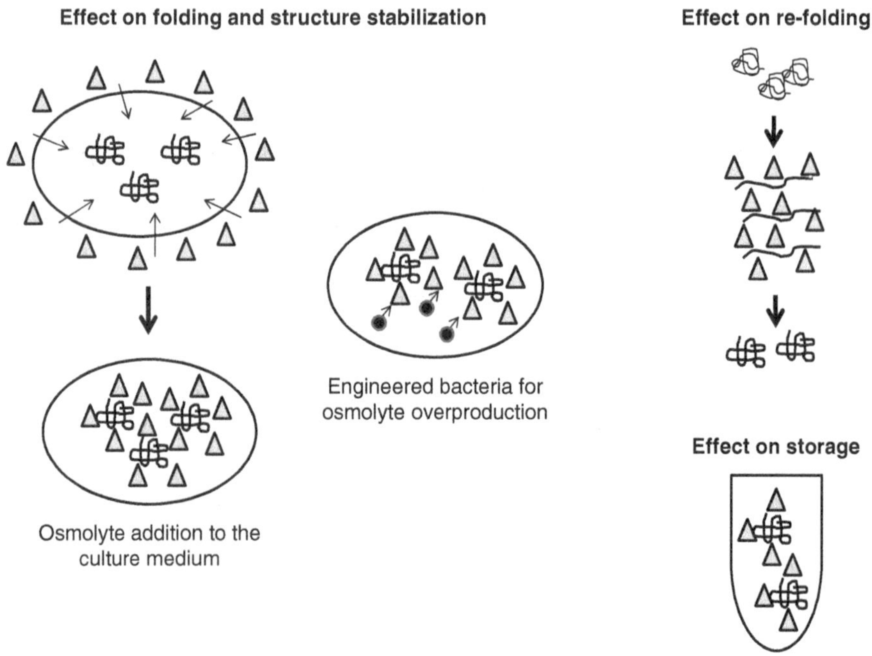

FIGURE 3.1 Protein production steps positively influenced by osmolytes. The presence of osmolytes can be profitable at any phase during protein production. Such molecules can favor productive folding and impair the aggregation of newly synthetized polypeptides, increase the re-folding efficiency of unfolded proteins, and stabilize reagents during storage.

3.3 PROPOSED MOLECULAR MECHANISMS FOR OSMOLYTE ACTIVITIES

The experimentally identified preferential specificity of the various osmolytes for some stress conditions and particular residues within the substrates does not exclude common mechanisms of action. The favored binding of a native protein rather than its unfolded form has been considered sufficient to explain the stabilizing effect of trehalose on folded proteins (Table 3.1) (Xie and Timasheff, 1997). More generally, the stabilizing osmolytes have been thought to exert their protective activity because they are strongly excluded by the protein surface; that is, they possess favorable interactions with solvent water and unfavorable interactions with a protein surface. This condition of preferential solvation should also protect (partially) unfolded polypeptides, which have greater exposed backbones than those of folded proteins, from further denaturation (Kaushik and Bhat, 2003; Canchi et al., 2012). The same mechanism should also improve protein folding by water sequestration and favor the creation of a hydrophobic environment around each single polypeptide, forcing it to reach the most stable conformation (Khan et al., 2010; Sarma and Paul, 2013). Although this model can explain some experimental data, recent results have shown that the classical solvent exchange concept according to which the osmolyte-dependent aggregation prevention of polypeptides is obtained by water exclusion is rather typical of denaturing molecules such as urea (Auton et al., 2008a, b). Protective osmolytes may display a more complex and differential pattern of interactions with protein backbone and side chains leading to simultaneous osmolyte-dependent preferential exclusion at some specific groups and to hydration at others.

TABLE 3.1 Biotechnologically Relevant Chaperoning and Stabilization Activities Promoted by the Most Commonly Used Osmolytes

Osmolytes	Stabilizing Effects on Polypeptides
Proline	Protein stabilization; storage
Glycine betaine	Protein stabilization; inhibition fibrillar aggregation
Arginine	Refolding adjuvant; storage
K^+-Glutamate	Protein stabilization
DIP	Protein stabilization; storage
Trehalose	Inhibition fibrillar aggregation; protein stabilization; storage; cryoprotective effect
TMAO	Protein stabilization; cryoprotective effect; inhibition fibrillar aggregation; inhibition pressure-induced denaturation

The demonstration that osmolytes possess the chemical features necessary and sufficient to stabilize proteins could suggest that they are always beneficial for cell homeostasis. Nevertheless, experiments elucidating the osmolyte mode of actions were performed mostly *in vitro*, and the results appeared strongly influenced by experimental conditions (Ferreon et al., 2012; Seelinger et al., 2013). The *in vivo* environment is more complex because of the simultaneous interactions among many molecules of different species, and there are indications that osmolyte concentration is a factor that can influence differently already folded proteins and polypeptides that must refold (Fan et al., 2012). Therefore, the observation that the constitutive osmolyte production is very low, but stress inducible, might be explained not only by energetic considerations but assuming that for physiological reasons the optimal osmolyte concentrations present inside a cell must be adapted to stress periods as well as to post-stress recovery. It is a matter of fact that yeast mutants missing the capacity to degrade trehalose when stress conditions cease cannot survive (Singer and Lindquist, 1998a) and that the trehalose-degrading enzyme-neutral trehalase Nth1 is coexpressed as an inactive form, together with the enzymes Tps1 and Tps2 necessary for trehalose synthesis in response to heat stress (Zähringer et al., 2000), a fact that suggests that the osmolyte accumulation must remain transient and that trehalose excess must be removed rapidly and effectively.

The molecular mechanism that can explain why a molecule that is protective under stress conditions becomes detrimental during recovery after stress has been elucidated by Singer and Lundquist (1998b). Trehalose initially stabilizes the native conformation of proteins challenged by environmental conditions causing protein unfolding. When stress conditions are too extreme for the protective capacity of the osmolyte and proteins are denatured progressively, trehalose stabilizes the polypeptides present in the nonnative state, impairing their aggregation. These intermediates represent a convenient source of protein for post-stress cell recovery and are the preferential substrate for the molecular chaperone refolding machinery. At this point, trehalose bound to the unfolded and inactive polypeptides can interfere with the refolding because it competes with the molecular chaperones and needs to be rapidly hydrolyzed.

There is also another level of possible interaction between chemical and molecular chaperones. It has been demonstrated for Hsp90 in an *in vitro* system that the naturally occurring osmolytes K^+-glutamate and glycine betaine could strongly and specifically activate the molecular chaperone and counteract destabilization of its oligomeric form in the presence of high salt and dilution conditions (Diamant et al., 2003). Osmolytes apparently interfere with the structure of Hsp90 and influence the kinetics of its transition between closed and open conformations (Street et al., 2010). It was speculated that betaine at low concentrations could also assist molecular chaperones by supporting local folding of bound target proteins (Diamant et al., 2001). Chemical and molecular chaperone-protective activities can probably overlap at different levels for stabilizing macromolecules during stress conditions. In the case of the glutamine synthetase, it was demonstrated that glutamate stabilized the *Bacillus subtilis* enzyme at elevated temperatures, but was ineffective in increasing the T_m value of the *Methanococcus jannaschi* isoform from its intrinsic value of less than 60°C to the growth temperature of the organism (85°C), whereas interaction with the molecular chaperone GroEL or the ribosomal protein L2 resulted in a T_m gain of more than 25°C and could explain the structural stability of the enzyme *in vivo* (Neelon et al., 2005).

Beside these direct cross activities between chemical and molecular chaperones, osmolytes could interfere with cell physiology by stabilizing regulative proteins characterized by the presence within their sequence of regions with elevated intrinsic disorder that must undergo tight regulation (Liu et al., 2006). This effect is probably concentration dependent, and its consequences have not yet been investigated sufficiently.

Indeed, artificially added chemical chaperones accumulate at elevated concentrations inside cells and are considered compatible for the organism, despite evidence suggesting the inhibitory effects of trehalose and betaine during protein refolding (Singer and Lindquist, 1998b; Diamant et al., 2001). *In vitro* experiments showed the positive effect of proline on macromolecule stabilization at low osmolyte concentrations, whereas the same additive had a negative effect at high concentrations (Hayat et al., 2012). In artificially recreated crowding systems resembling those of cell cytoplasm, osmolytes exert more effective protein stability protection than in diluted environments. In contrast, diluted solutions are more convenient for refolding activities (Fan et al., 2012). In the case of betaine, the addition of slightly different concentrations induced completely divergent physical consequences when a GFP–GST construct was used in an *in vitro* model (Natalello et al., 2008). Large aggregates of partially unfolded protein could first be induced in the presence of 5 mM betaine and dissolved successively into unfolded monodispersed monomers by increasing the osmolyte concentration to 20 mM. However, there is no report evaluating *in vivo* the relationship between osmolyte concentration and the effect on the substrate proteins.

An intriguing hypothesis concerning the buffering capacity of chemical chaperones suggests that one of the osmolyte crucial roles *in vivo* would be of assisting protein evolution (Bandyopadhyay et al., 2012). According to this idea, osmolytes could stabilize genetic variations of proteins otherwise prone to aggregate by a mechanism that prevents entropic traps for flexible intermediates and in the long term allows the insertion of further mutations that should serve to recover stability and assure new functions. The osmolyte-dependent buffering effect assures the target protein solubility of metastable constructs without influencing their expression rate. The experimental data show that osmolytes can buffer *in vivo* metastable mutants of proteins that—as wild types—do not depend on osmolyte presence to fold into their native state. The protective effect of osmolytes is not simply quantitative, but qualitative in the sense that each osmolyte seems to buffer specifically a subset of mutations, a feature probably related to the different mechanisms of action displayed by these molecules. Specifically, TMAO and proline seem to stabilize buried and superficial residues, respectively. Therefore, collaborative effects of osmolyte combinations could be necessary to assist protein evolution, and interactions with molecular chaperones cannot be excluded, as indicated by the case of GroEL-dependent thermal stabilization of glutamine synthetase (Neelon et al., 2005).

3.4 OSMOLYTES AND EXPRESSION OF RECOMBINANT PROTEINS

There are examples in which the engineering of bacteria and plants with genes that favor the expression and accumulation of chemical chaperones reduced the sensitivity to stress conditions and protein aggregation (Holmström et al., 2000; Sakamoto and Murats, 2000; Kandror et al., 2002; Jang et al., 2003; Schultz et al., 2007). *In vitro* experiments showed that osmolyte addition increases the midpoint of the thermal unfolding transition of the macromolecules in solution, a parameter that correlates empirically with their enhanced structural stability (Huang et al., 2006; King et al., 2011). Therefore, rendering the cytoplasmic environment of bacteria used for recombinant expression more protective for synthetized polypetides can conceivably be achieved by inducing the accumulation of protective osmolytes. Toxic effects such as the one observed in yeast, where trehalose is degraded after the termination of stress conditions (Singer and Lindquist, 1998b), should be irrelevant in cells used for recombinant expression that are harvested immediately. Different osmolyte-dependent approaches have been developed for boosting the recombinant expression of proteins: (1) the organism used for recombinant expression is engineered to assure the accumulation of naturally occurring osmolytes that will stabilize the cellular folding machinery in a nonspecific manner (de Marco et al., 2005; Ignatova and Gierasch, 2006; Schultz et al., 2006); (2) the accumulation of osmolytes present in the same organism of the target protein is favored to improve their specific interaction (Diamant et al., 2003; Schultz et al., 2007); and (3) osmolytes recovered from extremophiles are added to the recombinant expression systems because they have evolved under extreme stress conditions and therefore should be more protective than additives derived from organisms growing under standard physiological conditions (Schultz et al., 2006, 2007).

The results show that chemical chaperones can induce *in vivo* protein refolding after stress conditions (Diamant et al., 2003) and increase the yields of recombinant soluble proteins when their accumulation in bacteria is induced by osmotic stress (Schultz et al., 2006, 2007). Bacteria have also been engineered for overexpressing trehalose. Although the experiments were performed with a small number of different proteins, the results indicated that the mutated strain capable of accumulating the osmolyte in its cytoplasm was superior to the wild-type bacteria grown in media containing trehalose in terms of soluble recombinant protein production (Schultz et al., 2006). Finally, combinations of techniques aimed at increasing the accumulation of both chemical and molecular chaperones have been used to boost the recombinant expression of soluble proteins in *E. coli* (de Marco et al., 2005; Ascione et al., 2012). No apparent advantage have been detected, but these negative results could simply result from the difficulty in obtaining productive combinations given the highly complex *in vivo* interactions between molecules belonging to the folding machinery (Singer and Lundquist, 1998a).

3.5 BIOTECHNOLOGICAL RELEVANCE OF OSMOLYTES FOR PRESERVING PURIFIED PROTEINS

Osmolytes belonging to different classes have been used successfully to stabilize purified proteins at both low and high temperatures. TMAO was particularly effective in preserving the native fold of bacterial azurin, even in frozen water, due to its capacity to bind to ice, thereby inhibiting protein adsorption to the solid phase (Strambini and Gonnelli, 2008). On the other hand, the midpoint of the thermal unfolding transition was increased, the function was preserved, and the protein aggregation process was diminished and reversed by supplying additives such as trehalose and DIP (Lamosa et al., 2000; Faria et al., 2004; Attanasio et al., 2007; Schultz et al., 2007; Corradini et al., 2013). When several additives within different chemical classes were compared for their stabilizing efficiency in lyophilized formulations, optimal recovery of native protein in the initial dried solid was obtained in the presence of trehalose and sucrose, which formed an amorphous phase and which had the capacity to hydrogen-bond to the dried protein (Kreilgaard et al., 1999). Antibodies and antibody fragments represent a constantly growing portion of the molecules used in diagnostics and therapy. Single-chain fragments have been refolded effectively in the presence of arginine (Kudou et al., 2011), whereas proline, arginine, histidine, and sorbitol succeeded in inhibiting and reversing the aggregation process of both full-length IgGs and Fab fragments (Chen et al., 2003; Kashanian et al., 2008; Nelson et al., 2012b). Amino acids are often used as additives for protection from heat stress and from freeze–thaw cycles in liquid formulation (Arakawa et al., 2007) and arginine for extracting folded proteins from insoluble pellets (Tsumoto et al., 2004). Osmolytes such as TMAO, betaine, and sarcosine have also been used successfully as precipitants for the preparation of protein crystals and showed an elevated protective effect as cryoprotective agents (Marshall et al., 2012).

3.6 CONCLUSIONS

Osmolytes possess the capacity to stabilize proteins by favoring their folding, and can even rescue and disaggregate aggregates originating from partially misfolded protein intermediates. Consequently, their usefulness for biotechnological and therapeutic applications *in vivo* was assessed. Despite some clear success, the results have often been considered deceptive. This judgment is, however, the logical result of poor initial comprehension of the molecular mechanisms by which osmolytes exert their activity. Chemical chaperones have long been considered as universal stabilizer molecules. However, now we know that they possess substrate specificity, change properties according to their concentration and environment crowding, and overlap with the action of molecular chaperones in a way that can be both synergistic and negative. Therefore, we must renounce the naive belief that the simple addition of one osmolyte to the culture medium can increase the yields of any recombinant protein produced in bacteria. At the same time, the theoretical knowledge has allowed the appreciation of how osmolyte combinations and the coupled action of chemical and molecular chaperones could be exploited positively. This overall information indicates that osmolytes can be precious reagents for stabilizing macromolecules, but an ad hoc strategy should be developed for each individual need. Automated screening systems for identifying optimal culture conditions have already been validated (Vincentelli et al., 2005) and may represent the optimal option for comparing different stabilizer combinations and choosing the most suitable condition for up-scaling trials.

REFERENCES

Arakawa T, Tsumoto K, Kita Y, Chang B, Ejiama D (2007) Biotechnology applications of amino acids in protein purification and formulation. Amino Acids 33:587–605.

Arora A, Ha C, Park CB (2004) Inhibition of insulin amyloid formation by small stress molecules. FEBS Lett 564:121–125.

Ascione G, de Pascale D, De Santi C, Pedone C, Dathan NA, Monti SM (2012) Native expression and purification of hormone-sensitive lipase from *Psychrobacter* sp. TA144 enhances protein stability and activity. Biochem Biophys Res Commun 420:542–546.

Attanasio F, Cascio C, Fisichella S, Nicoletti VG, Pignataro B, Savarino A, Rizzarelli E (2007) Trehalose effects on alpha-crystallin aggregates. Biochem Biophys Res Commun 354:899–905.

Auton M, Bolen DW, Rösgen J (2008a) Structural thermodynamics of protein preferential solvation: osmolyte solvation of proteins, aminoacids, and peptides. Proteins 73:802–813.

Auton M, Rösgen J, Sinev M, Holthauzen LMF, Bolen DW (2008b) Osmolyte effects on protein stability and solubility: a balancing act between backbone and side-chains. Biophys Chem 159:90–99.

Bandyopadhyay A, Saxena K, Kasturia N, Dalal V, Bhatt N, Rajkumar A, Maity S, Sengupta S, Chakraborty K (2012) Chemical chaperones assist intracellular folding to buffer mutational variations. Nat Chem Biol 8:238–245.

Burg MB, Ferraris JD (2008) Intracellular organic osmolytes: function and regulation. J Biol Chem 283:7309–7313.

Canchi DR, Jayasimha P, Rau DC, Makhatadze GI, Garcia AE (2012) Molecular mechanism for the preferential exclusion of TMAO from protein surfaces. J Phys Chem B116:12095–12104.

Chen B, Bautista R, Yu K, Zapata GA, Mulkerrin MG, Chamow SM (2003) Influence of histidine on the stability and physical properties of a fully human antibody in aqueous and solid forms. Pharmacol Res 20:1952–1960.

Ciulla RA, Burggraf S, Stetter KO, Roberts MF (1994) Occurrence of di-myo-inositol1,1′-phosphate in *Methanococcus igneus*. Appl Environ Microbiol 176:3177–3187.

Corradini D, Strekalova EG, Stanley HE, Gallo P (2013) Microscopic mechanism of protein cryopreservation in an aqueous solution with trehalose. Sci Rep 3:1218.

Cosquer A, Pichereau V, Pocard JA, Minet J, Cormier M, Bernard T (1999) Nanomolar levels of dimethylsulfoniopropionate, dimethylsulfonioacetate, and glycine betaine are sufficient to confer osmoprotection to *Escherichia coli*. Appl Environ Microbiol 65:3304–3311.

de Marco A, Vigh L, Diamant S, Goloubinoff P (2005) Native folding of aggregation-prone recombinant proteins in *Escherichia coli* by osmolytes, plasmid- or benzyl alcohol–overexpressed molecular chaperones. Cell Stress Chaperones 10:329–339.

Desmarais D, Jablonski PE, Fedarko NS, Roberts MF (1997) 2-Sulfotrehalose, a novel osmolyte in haloalkaliphilic archaea. J Bacteriol 179:3146–3153.

Diamant S, Eliahu N, Rosenthal D, Goloubinoff P (2001) Chemical chaperones regulate molecular chaperones in vitro and in cells under combined salt and heat stresses. J Biol Chem 276:39586–39591.

Diamant S, Rosenthal D, Azem A, Eliahu N, Ben-Zvi AP, Goloubinoff P (2003) Dicarboxylic amino acids and glycine-betaine regulate chaperone-mediated protein disaggregation under stress. Mol Microbiol 49:401–410.

Dreux N, Albagnac C, Sleator RD, Hill C, Carlin F, Morris CE, Nguyen C (2008) Glycine betaine improves *Listeria monocytogenes* tolerance to desiccation on parsley leaves independent of the osmolyte transporters BetL, Bbu and OpuC. J Appl Microbiol 104:1221–1227.

Fan YQ, Lee J, Oh S, Liu HJ, Li C, Luan YS, Yang JM, Zhou HM, Lü ZR, Wang YL (2012) Effects of osmolytes on human brain-type creatine kinase folding in dilute solutions and crowding systems. Int J Biol Macromol 51:845–858.

Faria TO, Lima JC, Bastos M, Maçanita AL, Santos H (2004) Protein stabilization by osmolytes from hyperthermophiles: effect of mannosylglycerate on the thermal unfolding of recombinant nuclease a from *Staphylococcus aureus* studied by picosecond time-resolved fluorescence and calorimetry. J Biol Chem 279:48680–48691.

Ferreon ACM, Moosa MM, Gambin Y, Deniz AA (2012) Counteracting chemical chaperone effects on the single-molecule alpha–synuclein structural landscape. Proc Natl Acad Sci U S A 109:17826–17831.

Figueiredo J, Simoes-Correia J, Sderberg O, Suriano G, Seruca R (2011) ADP-ribosylation factor 6 mediates E-cadherin recovery by chemical chaperones. PloSOne 6:e23188.

Garg AK, Kim JK, Owens TG, Ranwala AP, Choi YD, Kochian LV, Wu RJ (2002) Trehalose accumulation in rice plants confers high tolerance levels to different abiotic stresses. Proc Natl Acad Sci U S A 99:15898–15903.

Gauthier MJ, Le Rudulier D (1990) Survival in seawater of *Escherichia coli* cells grown in marine sediments containing glycine betaine. Appl Environ Microbiol 56:2915–2918.

Gauthier MJ, Flatau GN, Le Rudulier D, Clement RL, Combarro MP (1991) Intracellular accumulation of potassium and glutamate specifically enhances survival of *Escherichia coli* in marine seawater. Appl Environ Microbiol 56:2915–2918.

Hayat S, Hayat Q, Alyemeni MN, Wani AS, Pichtel J, Ahmad A (2012) Role of proline under changing environment: a review. Plant Signal Behav 7:11.

Holmström KO, Somersalo S, Mandal A, Palva TE, Welin B (2000) Improved tolerance to salinity and low temperature in transgenic tobacco producing glycine betaine. J Exp Bot 51:177–185.

Hu CY, Kokubo H, Lynch GC, Bolen DW, Pettitt BM (2010) Backbone additivity in the transfer model of protein solvation. Protein Sci 19:1011–1022.

Huang H, Liu J, de Marco A (2006) Induced fit of passenger proteins fused to archaea maltose binding proteins. Biochem Biophys Res Commun 344:25–29.

Kandror O, DeLeon A, Goldberg AL (2002) Trehalose synthesis is induced upon exposure of *Escherichia coli* to cold and is essential for viability at low temperatures. Proc Natl Acad Sci U S A 99:9727–9732.

Kashanian S, Paknejad M, Ghobadi S, Omidfar K, Ravan H (2008) Effect of osmolytes on the conformational stability of mouse monoclonal antidigoxin antibody in long-term storage. Hybridoma 27:99–106.

Kaushik JK, Bhat R (2003) Why is trehalose an exceptional protein stabilizer? J Biol Chem 278:26458–25465.

Khan SH, Ahmad N, Ahmad F, Kumar R (2010) Naturally occurring organic osmolytes: from cell physiology to disease prevention. IUBMB Life 62:891–895.

King AC, Woods M, Liu W, Lu Z, Gill D, Krebs MRH (2011) High-throughput measurement, correlation analysis, and machine-learning predictions for pH and thermal stabilities of Pfizer-generated antibodies. Prot Sci 20:1546–1557.

Kreilgaard L, Frokjaer S, Flink JM, Randolph TW, Carpenter JF (1999) Effect of additives on the stability of *Humicula lanuginosa* lipase during freeze-drying and storage in the dried solid. J Pharm Sci 88:281–290.

Kudou M, Ejima D, Sato H, Yumioka R, Arakawa T, Tsumoto K (2011) Refolding single-chain antibody (scFv) using lauroyl-L-glutamate as a solubilization detergent and arginine as a refolding additive. Protein Expr Purif 77:68–74.

Ignatova Z, Gierasch LM (2006) Inhibition of protein aggregation in vitro and in vivo by a natural osmoprotectant. Proc Natl Acad Sci U S A 103:13357–13361.

Iordanescu M, Imai R (2008) Trehalose biosynthesis in response to abiotic stresses. J Integr Plant Biol 50:1223–1229.

Jang IC, Oh SJ, Seo JS, Choi WB, Song SI, Kim CH, Kim YS, Seo HS, Choi HD, Nahm BH, Kim JK (2003) Expression of a bifunctional fusion of the *Escherichia coli* genes for trehalose-6-phosphate synthase and trehalose-6-phosphate phosphatase in transgenic rice plants increases trehalose accumulation and abiotic stress tolerance without stunting growth. Plant Physiol 131:516–524.

Lamosa P, Burke A, Peist R, Huber R, Liu MY, Silva G, Rodrigues-Pousada C, LeGall J, Maycock C, Santos H (2000) Thermostabilization of proteins by diglycerol phosphate, a new compatible solute from the hyperthermophile *Archaeoglobus fulgidus*. Appl Environ Microbiol 66:1974–1979.

Lenky CC, McEntyre CJ, Lever M (2012) Measurement of marine osmolytes in mammalian serum by liquid chromatography-tandem mass spectrometry. Anal Biochem 420:7–12.

Lentzen G, Schwarz T (2006) Extremolytes: natural compounds from extremophiles for versatile applications. Appl Microbiol Biotechnol 72:623–634.

Liu R, Barkhordarian H, Emadi S, Park CB, Sierks MR (2005) Trehalose differentially inhibits aggregation and neurotoxicity of beta-amyloid 40 and 42. Neurobiol Dis 20:74–81.

Liu J, Peruma NB, Oldfield CJ, Sue EW, Uversky VN, Dunker AK (2006) Intrinsic disorder in transcription factors. Biochemistry 45:6873–6888.

Marshall H, Venkat M, Seng NS, Cahn J, Juers DH (2012) The use of trimethylamine *N*-oxide as a primary precipitating agent and related methylamine osmolytes as cryoprotective agents for macromolecular crystallography. Acta Crystallogr D 68:69–81.

Martin DD, Ciulla RA, Roberts MF (1999) Osmoadaptation in archaea. Appl Environ Microbiol 65:1815–1825.

Martins LO, Huber R, Huber H, Stetter KO, da Costa MS, Santos H (1997) Organic solutes in hyperthermophilic archaea. Appl Environ Microbiol 63:896–902.

McIntyre HJ, Davies H, Hore TA, Miller SH, Dufour JP, Ronson CW (2007) Trehalose biosynthesis in *Rhizobium legumonosarum* bv. *trifolii* and its role in desiccation tolerance. Appl Environ Microbiol 73:3984–3992.

Mukaiyama A, Koga Y, Takano K, Kanaya S (2008) Osmolyte effect on the stability and folding of a hyperthermophilic protein. Proteins 71:110–118.

Natalello A, Santarella R, Doglia SM, de Marco A (2008) Physical and chemical perturbations induce the formation of protein aggregates with different structural features. Protein Expr Purif 58:356–366.

Natalello A, Liu J, Ami D, Doglia SM, de Marco A (2009) The osmolyte betaine promotes protein misfolding and disruption of protein aggregates. Proteins 75:509–517.

Neelon K, Schreier HJ, Meekins H, Robinson PM, Roberts MF (2005) Compatible soluble effects on thermostability of glutamine synthetase and aspartate transcarbamoylase from *Methanococcus jannaschii*. Biochim Biophys Acta 1753:164–173.

Nelson ED, Ramberg JE, Best T, Sinnot RA (2012a) Neurologic effects of exogenous saccharides: a review of controlled human, animal, and in vitro studies. Nutr Neurosci 15:149–162.

Nelson AD, Hoffmann MM, Parks CA, Dasari S, Schrum AG, Gil D (2012b) IgG Fab fragments forming bivalent complexes by a conformational mechanism that is reversible by osmolytes. J Biol Chem PMID: 23109335.

Ou WB, Park YD, Zhou HM (2001) Molecular mechanisms for osmolyte protection of creatine kinase against guanidine denaturation. Eur J Biochem 268:5901–5911.

Rajan RS, Tsumoto K, Tokunaga M, Tokunaga H, Kita Y, Arakawa T (2011) Chemical and pharmacological chaperones: application for recombinant protein production and protein folding diseases. Curr Med Chem 18:1–15.

Robertson DE, Lai MC, Gunsalus RP, Roberts MF (1992) Composition, variation, and dynamics of major osmotic solutes *Methanohalophilus* Strain FDF1. Appl Environ Microbiol 58:2438–2443.

Rodriguez-Navarro JA, Rodriguez L, Casarejos MJ, Solano RM, Gomez A, Perucho J, Cuervo AM, Garcia de Yebenes J, Mena MA (2010) Trehalose ameliorates dopaminergic and tau pathology in parkin deleted/tau overexpressing mice through autophagy activation. Neurobiol Dis 39:423–438.

Roeßler M, Müller V (2001) Osmoadaptation in bacteria and archaea: common principles and differences. Environ Microbiol 3:743–754.

Sakamoto A, Murata N (2000) Genetic engineering of glycine betaine synthesis in plants: current status and implications for enhancement of stress tolerance. J Exp Bot 51:81–88.

Sarma R, Paul S (2013) Crucial importance of water structure modification on trimethylamine *N*-oxide counteracting effect at high pressure. J Phys Chem B117:677–689.

Schultz T, Martinez L, de Marco A (2006) The evaluation of factors that cause aggregation during recombinant protein expression in *E. coli* is simplified by the employment of an aggregation-sensitive reporter. Microb Cell Fact 5:28.

Schultz T, Liu J, Capasso P, de Marco A (2007) The solubility of recombinant proteins expressed in *Escherichia coli* is increased by *otsA* and *otsB* cotransformation. Biochem Biophys Res Commun 355:234–239.

Seelinger J, Estel K, Erwin N, Winter R (2013) Cosolvent effects on the fibrillation reaction of human IAPP. Phys Chem Chem Phys 15:8902–8907.

Singer MA, Lindquist S (1998a) Thermotolerance in *Saccharomyces cerevisiae*: the yin and yang of trehalose. Trends Biotechnol 16:460–468.

Singer MA, Lindquist S (1998b) Multiple effects of trehalose on protein folding in vitro and in vivo. Mol Cell 1:639–648.

Soto G, Setten L, Lisi C, Maurelis C, Mozzicafreddo M, Cuccioloni M, Angeletti M, Ayub ND (2012) Hydroxybutyrate prevents protein aggregation in the halotolerant bacterium *Pseudomonas* sp. CT13 under abiotic stress. Extremophiles 16:455–462.

Strambini GB, Gonnelli M (2008) Singular efficacy of trimethylamine N-oxide to counter protein destabilization in ice. Biochemistry 47:3322–3331.

Street TO, Krukenberg KA, Rösgen J, Bolen DW, Agard DA (2010) Osmolyte-induced conformational changes in the Hsp90 molecular chaperone. Protein Sci 19:57–65.

Strøm AR, Kaasen I (1993) Trehalose metabolism in *Escherichia coli*: stress protection and stress regulation of gene expression. Mol Microbiol 8:205–210.

Tsumoto K, Umetsu M, Kumagai I, Ejima D, Philo JS, Arakawa T (2004) Role of arginine in protein refolding, solubilization, and purification. Biotechnol Prog 20:1301–1308.

Ueland PM (2011) Choline and betaine in health and disease. J Inherit Metab Dis 34:3–15.

Vincentelli R, Canaan S, Offant J, Cambillau C, Bignon C (2005) Automated and expression solubility screening of His-tagged proteins in 96-well format. Anal Biochem 346:77–84.

Xie G, Timasheff SN (1997) The thermodynamic mechanism of protein stabilization by trehalose. J Biol Chem 64:25–43.

Zähringer H, Thevelein JM, Nwaka S (2000) Induction of neutral trehalase Nth1 by heat and osmotic stress is controlled by STRE elements and Msn2/Msn4 transcription factors: variations of PKA effect during stress and growth. Mol Microbiol 35:397–406.

Zhou K, Zou R, Stephanopoulos G, Too HP (2012) Enhancing solubility of deoxyxylulose phosphate pathway enzymes for microbial isoprenoid production. Microb Cell Fact 11:148.

4

INCLUSION BODIES IN THE STUDY OF AMYLOID AGGREGATION

ANNA VILLAR-PIQUÉ AND SALVADOR VENTURA

4.1 INTRODUCTION

After their synthesis at the ribosomes, most proteins need to attain a globular, compact, and specific three-dimensional conformation in a biologically relevant time frame in order to perform their physiological functions, since only properly folded protein structures can interact selectively with their molecular targets (Daggett and Fersht, 2009). Misfolded species or folding intermediates often display an intrinsic tendency to interact aberrantly, which results in their self-assembly into insoluble aggregates that in many cases become toxic and interfere with important cellular pathways. Accordingly, protein misfolding and deposition into human tissues is linked to the pathology of an increasing number of disorders, from neurodegenerative diseases such as Alzheimer's, Parkinson's, or the transmissible spongiform encephalopathies to nonneurodegenerative systemic and localized amyloidosis, as senile systemic amyloidosis, type 2 diabetes, or even some types of cancer (Chiti and Dobson, 2006). Irrespective of the nature of the proteins involved in these disorders, the aggregates usually adopt the form of amyloid fibrils, all sharing a common cross-β sheet structure in which the β-strands are oriented perpendicularly to the axis of the fibril (Fernandez-Busquets et al., 2008). Importantly, the formation of amyloid-like structures is not restricted to a subset of pathological proteins but, rather, appears to be a

Protein Aggregation in Bacteria: Functional and Structural Properties of Inclusion Bodies in Bacterial Cells, First Edition. Edited by Silvia Maria Doglia and Marina Lotti.

generic property of polypeptides across all kingdoms of life, including bacteria (Dobson, 2003; Sabate et al., 2010).

Protein aggregation is also a major bottleneck during the biotechnological production and purification of proteins, hindering the commercialization of protein-based drugs (Ventura and Villaverde, 2006). Despite it having been observed recurrently that recombinant protein expression in bacterial hosts often leads to aggregation of the target protein into compact aggregated structures known as inclusion bodies (IBs), only recently this phenomenon has been shown to share striking similarities with the process of amyloid fibril formation by pathogenic proteins. As we discuss in this chapter, this evidence is paving the way to the use of bacteria as fascinating model systems to study protein interactions leading to the formation of toxic and infective amyloids as well as their use to screen for molecules able to interfere with these contacts.

4.2 STRUCTURE OF IBs

The IBs formed during the recombinant expression of proteins unrelated in sequence, secondary structure content, three-dimensional conformation, size, or origin all appear, upon isolation, as refractile, porous, hydrated, and very dense particles. They normally have a spherical-like shape (in some cases appearing as short cylinders and paracrystalline forms) with a diameter range between 0.2 and 1.2 μm, probably determined by the shape and size of the bacterial cell, the protein intrinsic characteristics, and the microenvironmental conditions during the expression process (Wang, 2009), but overall they display poor morphological variety. It has been shown that certain genes, such as the protease clpP, play an active role in the control of IB shape. Accordingly, clpP deletion results in unusual tear-shaped particles displaying enhanced surface/volume ratios (García-Fruitós et al., 2010). IBs are located either in the periplasmatic space, when a signal peptide is fused to the target protein (Arie et al., 2006), or in the cytoplasm of the expression hosts. In this last case, the aggregates are usually localized at the cell poles (Rokney et al., 2009). DAPI staining of bacterial DNA has revealed that IBs are in fact interspersed with the bacterial nucleoid (Villar-Piqué et al., 2012a). Under electronic microscopy, IBs seem to have an amorphous electrodense structure without any visible element of regularity (Carrió and Villaverde, 2005). This visual information has sustained the traditional idea that IBs are protein deposits lacking any functionality other than acting as "dust balls" to harvest misfolded polypeptides and avoid their interaction with other cellular components. Thus, for a long time IBs have been considered to be unstructured aggregates and have remained poorly characterized. Nevertheless, recently, this classical vision has changed, and accumulating evidence of an organized molecular structure allows us to describe the IBs formed by many different polypeptides as amyloid-like aggregates (de Groot et al., 2009; García-Fruitós et al., 2011).

4.2.1 Amyloid-like Nature of IBs

In recent years, the existence of amyloid-like features in bacterial aggregates has been reported by several independent groups (Carrió et al., 2005; Ono et al., 2006; Wang et al., 2008). Our group, using an aggregation-prone version of β-galactosidase, whose overexpression in *Escherichia coli* resulted in both soluble and aggregated species, was the first to suggest this possibility. We showed that the aggregated material was able to seed specifically the deposition of monomeric and soluble species, which become increasingly incorporated in the aggregated seeds, showing that protein deposition in bacteria is a selective process and that specific interactions rather than the previously proposed unspecific contacts should be at the basis of IB formation. To explore the degree of molecular organization of these IBs, we used infrared spectroscopy with a resulting spectrum indicative of extended intermolecular β-sheet structures. Furthermore, specific amyloid dyes were used (Thioflavin-T (Th-T) and Congo Red (CR)) and the typical increase in the emission signal for amyloid structures was detected in those IBs as well. Importantly, the IBs formed by different and unrelated proteins shared similar conformational properties, suggesting for the first time that the presence of amyloid-like features could be a common characteristic of bacterial protein deposits (Carrió et al., 2005). Afterward, the Riek group provided a conclusive demonstration of the generic presence of amyloid-like conformations in IBs by characterizing the structural properties of the bacterial aggregates formed by three proteins representative of the different native folds: all α-helix, α-helix/β-sheet, and all β-sheet. Using alternative techniques such as x-ray diffraction, circular dichroism, or NMR, they could demonstrate the presence of specific well-protected β-sheets in the aggregates, surrounded by less structured protein regions (Wang et al., 2008).

Recognition of the amyloid-like nature of IBs by the scientific community was hindered by their apparent amorphous appearance and the lack of any fibrillar structure. It was assumed that this was due to the inner localization of the amyloid cross-β sheets in the core of the aggregate, while the rest of the protein remained unstructured or even displayed a native-like conformation at the surface of the IB. Fortunately, proteolytic treatment with proteinase K has permitted us to visualize the existence of fibrillar regions of IBs with morphology and dimensions compatible with those of amyloid fibrils formed by the same protein. This enzyme displays a high proteolytic activity with respect to globular or disordered conformations, but is poorly active against densely packed cross-β regions (Wilson et al., 2007). Hence, the addition of proteinase K to purified IBs has become a useful technique to visualize fibrillar structure inside bacterial aggregates, by either electronic (Wang et al., 2008) or atomic force microscopy (Morell et al., 2008), representing the ultimate and unequivocally proof of amyloid-like content in IBs. Therefore, it now seems clear that protein overexpression in bacteria results in the formation of nonnative intermolecular contacts that glue the polypeptide chains to form IBs. The amyloid-like nature of IBs relies essentially on the fact that, as happens during the formation of fibrillar assemblies, the construction of these bacterial aggregates depends on the establishment of interactions occurring through homologous protein patches in a nucleation-dependent manner (García-Fruitós et al., 2011). Hence, IBs are composed primarily of the recombinant protein, although other proteins can also be embedded in them, such as molecular chaperones (Carrió and Villaverde, 2005). The conformational quality of IBs varies depending on the protein expressed, as its intrinsic aggregation propensity determines both the assembly rate and the conformational and stability features of the aggregate (de Groot and Ventura, 2006a). In this way, despite the fact that the main secondary structure elements in many IBs are nonnative β-sheets, which form the core of the aggregate and probably act to sequester misfolded species or slow-folding intermediates into the IBs, it has been reported that these regions might coexist with disordered fragments and native-like conformations. This last observation has a place of prominent importance in the field of biotechnology, as the range of possible utilities of IBs containing active proteins expands from their use as biocatalysts to the *in vivo* engineering of inert nanostructured materials (García-Fruitós et al., 2012).

4.2.2 Detection and Characterization of Amyloid Conformations Inside IBs

The list of techniques available to characterize the amyloid-like nature of IBs is in a state of continuous expansion. Only the most widely used are addressed briefly in this section and summarized in Figure 4.1. Moreover, in Table 4.1 we list the various approaches and protein models for which positive results have been reported. On the one hand, the typical amyloid dyes constitute a fast method to report amyloid-like properties in bacterial aggregates. Among them, Th-T, Thioflavin-S (Th-S), and CR are the most commonly employed. Upon binding to amyloid-like structures, they undergo spectroscopic changes induced by the presence of the highly repetitive and regular cross-β-sheet motif. These dyes can be used in purified IBs (Carrió et al., 2005) or cell extracts (Ono et al., 2006). On the other hand, low-resolution biophysical approaches have been employed in numerous studies to confirm the amyloid content of IBs. One of the first structural techniques applied to IBs was infrared spectroscopy. The resulting spectra in the amide I region are usually dominated by a main band around 1621 to 1628 cm^{-1}, indicative of a prominent intermolecular β-sheet content (Doglia et al., 2008; Wang, 2009). Also, the use of circular dichroism spectroscopy in the far-ultraviolet region is very helpful to confirm the presence of this secondary structure element, detected as a minimum around 217 nm, coexisting with signals corresponding to those of unstructured conformations (Wang et al., 2008). Perhaps, x-ray diffraction constitutes the most robust technique to demonstrate unequivocally the presence of β-sheet arranged in an amyloid-like fashion in IBs. The diffraction patterns of various IBs have been shown to consist of two rings, typical for the cross-β-sheet structure in amyloid fibrils, with a major reflection at 4.7 Å, corresponding to the spacing between the strands in the β-sheet and a diffused reflection at about 10 Å, interpreted as the distance between adjacent β-sheets. The two reflections are circular and not orthogonal as in the cross-β structure in highly ordered amyloids, indicating a lower alignment of the cross-β components in IBs, which is consistent with the presence of alternative conformations in these aggregates.

More recent studies have also used high-resolution techniques for the accurate mapping of the structured core of bacterial aggregates. Some examples are ^{13}C–^{13}C solid-state NMR, used to define the structure of HET-s prion IBs (Wasmer et al., 2009), and hydrogen/deuterium exchange–based solution-state NMR, which permits us to identify the solvent-protected backbone amide protons involved in intermolecular hydrogen bonds (Wang et al., 2008; Dasari et al., 2011). Finally, the kinetic and thermodynamic stabilities of IBs can be analyzed by adapting available physicochemical methods aimed at studying protein folding and stability. Among them, IBs denaturation experiments with urea (Upadhyay et al., 2012) or guanidine-HCl (Espargaro et al., 2008b), whose solubilization curves reflect their resistance to chemical unfolding, provide insights on the cooperativity of the interactions sustaining the aggregate structure. Proteinase K digestion curves are also useful to measure the aggregates compactness and stability (Upadhyay et al., 2012); the kinetic solubilization constants derived from proteolytic experiments usually correlate with the stabilities calculated from chemical denaturation assays (de Groot and Ventura, 2006a). Finally, atomic force microscopy tip indentation is a methodology that evaluates the mechanical properties of IBs and allows us to identify the presence and quantify the stability of the amyloid core, as the fibrillar material inside them is the primary responsible of the resistance exerted by the IBs in front of this rupture force (Villar-Piqué et al., 2012a).

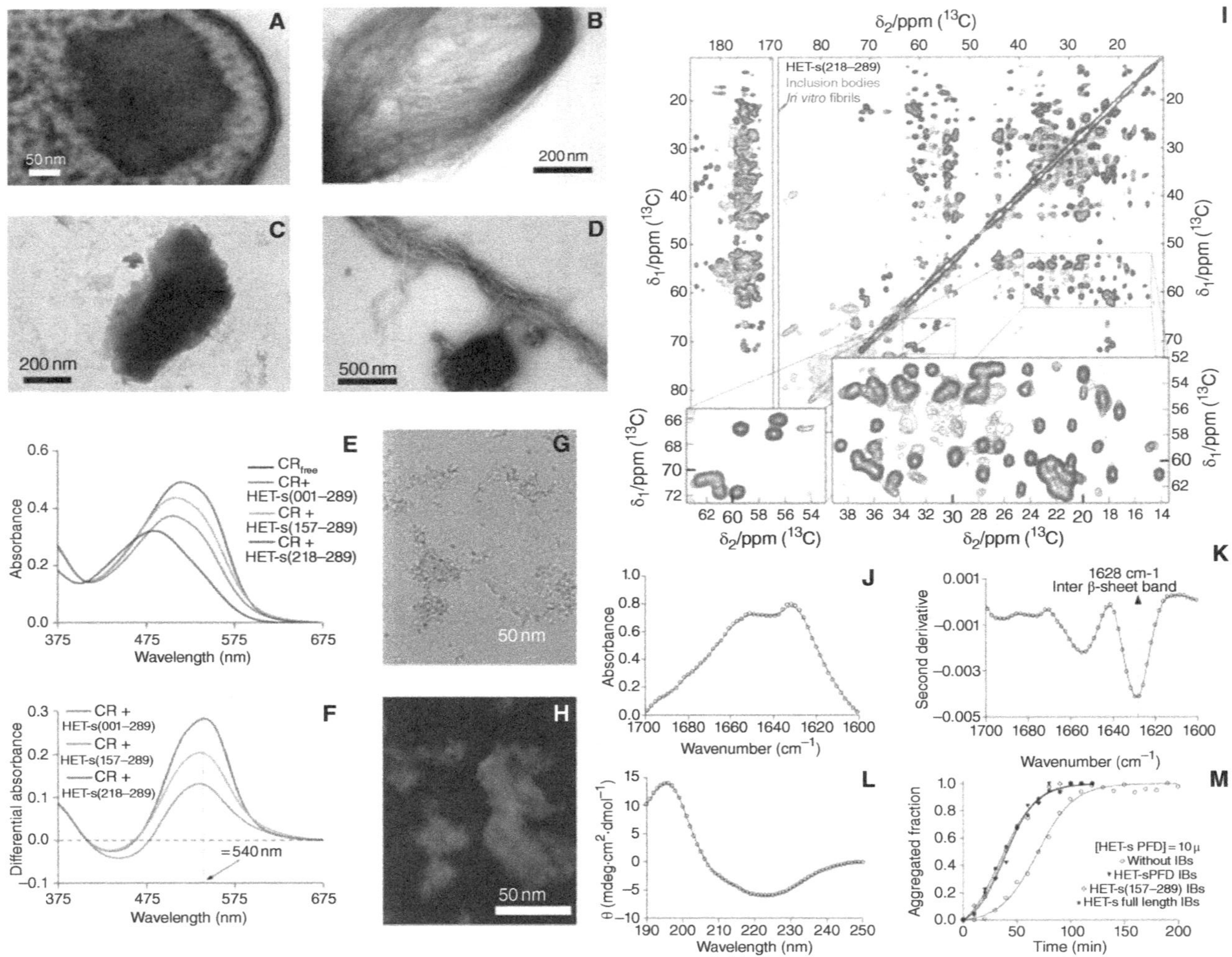
A
50 nm
B
200 nm
C
200 nm
D
500 nm
E
Absorbance
Wavelength (nm)
CR free
CR+ HET-s(001–289)
CR + HET-s(157–289)
CR + HET-s(218–289)
F
Differential absorbance
Wavelength (nm)
CR + HET-s(001–289)
CR + HET-s(157–289)
CR + HET-s(218–289)
=540 nm
G
50 nm
H
50 nm
I
δ2/ppm (13C)
δ1/ppm (13C)
HET-s(218–289)
Inclusion bodies
In vitro fibrils
J
Absorbance
Wavenumber (cm−1)
K
1628 cm-1
Inter β-sheet band
Second derivative
Wavenumber (cm−1)
L
θ (mdeg·cm2·dmol−1)
Wavelength (nm)
M
Aggregated fraction
Time (min)
[HET-s PFD] = 10 μ
Without IBs
HET-sPFD IBs
HET-s(157–289) IBs
HET-s full length IBs

4.3 FORMATION OF IBs

Protein expression in bacteria is a fine-tuned process whereby subtle imbalances in protein homeostasis can trigger massive protein misfolding and aggregation. Although living cells are endowed with powerful protein quality machinery composed primarily of molecular chaperones and proteases, protein deposition is a common event during heterologous overexpression, especially when eukaryotic proteins are produced in microbial cell factories. This effect can be attributed to two main factors. The first factor is the absence of adequate protein post-translational modifications, such as glycosilations, in the bacterial cytosol. Importantly, the discovery that bacteria possess, in fact, both N- and O-linked glycosylation pathways displaying a certain similarity to their eukaryotic counterparts is opening new opportunities to manipulate these pathways in order to engineer proteins with improved solubility features (Nothaft and Szymanski, 2010). The second factor is the fast and high transcription levels attained during recombinant protein expression. On the one hand, the accumulation of folding intermediates saturates the protein quality machinery. On the other hand, it is now clear that a correlation exists between the abundance of a protein in its natural environment and its intrinsic aggregation propensity, in such a way that because protein aggregation is a second- or higher-order reaction and therefore strongly dependent on protein concentrations, abundant proteins are endorsed with higher intrinsic solubility than poorly expressed ones, which can support a higher intrinsic aggregation tendency (Tartaglia and Vendruscolo, 2009; de Groot and Ventura, 2010; Castillo et al., 2011). Protein production tends to maximize the expression levels, bypassing this evolutive constraint, resulting in most cases in proteins being expressed at much higher concentrations than the physiological concentrations. This imbalance leads indefectibly to protein aggregation.

FIGURE 4.1 Physicochemical techniques to detect and characterize the amyloid-like structure in bacterial IBs formed by *Podospora anserina* prion HET-s. A to D: Visualization of IBs under transmission electronic microscopy. Proteolytic digestion with proteinase K enables the apparition of fibrillar structure (D). E to H: CR binding to IBs. Spectroscopic changes promoted by the IBs amyloid structure and analyzed by absorbance (E and F) and CR staining of IBs visualized under wide-field (G) and fluorescence microscopy (H). I: ^{13}C–^{13}C solid-state NMR, showing that HET-s IBs display a spectrum similar to that of HET-s *in vitro* amyloid fibrils. J to L: Secondary structure analysis of IBs, confirming the presence of intermolecular β-sheets by means of infrared spectroscopy (J and K) and circular dichroism (L). M: Seeding capacity of HET-s IBs in the *in vitro* amyloid fibrils formation. [Reproduced with permission from García-Fruitós et al. (2011).] (*See insert for color representation of the figure.*)

TABLE 4.1 Proteins Forming Amyloid-like IBs in Bacteria[a]

	Disordered		α-Fold		α- and β-Fold		β-Fold:	Prion Proteins		
	Aβ Peptide	VP1	ESAT-6	Human Growth Hormone	BMP2-(13–74)	L-Asparaginase II	MOG (ECD)	HET-s	Ure2p	Sup35
Amyloid dyes		✓	✓	✓	✓	✓	✓	✓	✓	✓
Infrared spectroscopy	✓	✓		✓		✓		✓	✓	✓
Circular dichroism	✓		✓					✓		
Seeding ability	✓	✓						✓	✓	✓
Solid-state NMR	✓							✓		
H/D exchange solution-state NMR	✓		✓		✓		✓			
AFM tip identation	✓									
Denaturation curves	✓	✓		✓		✓		✓	✓	✓
Proteolytic curves	✓	✓		✓		✓				
X-ray diffraction			✓		✓		✓			

Atomic force microscopy	✓									
Electronic microscopy	✓	✓	✓	✓	✓	✓		✓		✓
Reference[b]	1–5	4, 6, 7	8	9	8	9	8	10, 11	12	12, 13

[a]Regardless of their native conformations, all *in vivo*–formed aggregates display amyloid properties, as demonstrated by different techniques.

[b]1. M. Dasari, A. Espargaro, R. Sabate, et al., Chembiochem 12(3):407 (2011).
2. N.S. de Groot and S. Ventura, FEBS Lett 580(27):6471 (2006).
3. A. Espargaro, R. Sabate, and S. Ventura, FEBS Lett 582(25–26):3669 (2008b).
4. M. Morell, R. Bravo, A. Espargaro, et al., Biochim Biophys Acta 1783(10):1815 (2008).
5. A. Villar-Piqué, N.S. de Groot, R. Sabate et al., J Mol Biol 421(2–3):270 (2012a).
6. M. Carrió, N. Gonzalez-Montalban, A. Vera, et al., J Mol Biol 347(5):1025 (2005).
7. O. Cano-Garrido, E. Rodriguez-Carmona, C. Diez-Gil, et al., Acta Biomater 9(4):6134-42 (2013).
8. L. Wang, S.K. Maji, M.R. Sawaya, et al., PLoS Biol 6(8):e195 (2008).
9. A.K. Upadhyay, A. Murmu, A. Singh, et al., PLoS One 7(3):e33951 (2012).
10. C. Wasmer, L. Benkemoun, R. Sabate, et al., Angew Chem Int Ed Engl 48(26):4858 (2009).
11. R. Sabate, A. Espargaro, S.J. Saupe, et al., Microb Cell Fact 8:56 (2009).
12. A. Espargaro, A. Villar-Piqué, R. Sabate, et al., Microb Cell Fact 11(1):89 (2012b).
13. B. Ono, M. Kubota, H. Kimiduka, et al., Biosci Biotechnol Biochem 70(12):2813 (2006).

4.3.1 *In Vivo* Formation Kinetics

During recombinant protein expression, the formation of IBs starts when a threshold in the levels of misfolded or unfolded species in the cytoplasm of the bacterial cell is exceeded (Ignatova and Gierasch, 2004). This process has been suggested to occur in two steps. First, misfolded or folding intermediate species aggregate in multiple small foci randomly localized in the cytoplasm, and afterward, they are directed selectively to the poles of the cell, where they are fused to form a large aggregate tethered to the cell pole. It has been suggested that both processes occur in an energy-dependent manner (Rokney et al., 2009), the last step requiring the proton motive force and DnaK and DnaJ chaperones. Nonetheless, this last point is controversial, as another work indicates that the polar localization of intracellular IBs is circumstantial, in such a way that aggregation occurs in DNA-free space and therefore that nucleoid occlusion is necessary and sufficient for the polar localization of IBs, thus suggesting that the deposition of protein aggregates at the poles of *E. coli* occurs through a passive mechanism (Winkler et al., 2010). Despite these discrepancies, what seems clear is that a bacillar bacterial cell contains at most, one or two IBs, which are distributed between the two daughter cells after division. Actually, the presence of more than two IBs in a cell is indicative of a reproductive defect and the impairment of binary fission (Lindner et al., 2008).

During *in vivo* IB assembly, a kinetic competition takes place between protein folding and aggregation, which reflects a balance between the formation of intramolecular contacts, leading to the protein native state and the establishment of intermolecular contacts that lead to protein deposition. A recent wide human proteome analysis has revealed that regions relevant to folding are also important to aggregation and, more interestingly, that aggregation propensity is anticorrelated with folding propensity (Tartaglia and Vendruscolo, 2010). This competition between folding and aggregation has been demonstrated by fusioning a fluorescent reporter to aggregation-prone peptides in such a way that the final aggregate fluorescence can be correlated directly with the *in vivo* aggregation rate (de Groot and Ventura, 2006b). In a recent work, we have been able to visualize, in real time, IB formation *in vivo* under confocal microscopy for proteins displaying differential intrinsic aggregation propensity, and thus confirm this theory (Villar-Piqué et al., 2012a).

4.3.2 Molecular Determinants of IB Aggregation

It is widely accepted at present that the ability to self-assemble into amyloid-like structures is an inherent property of proteins (Chiti and Dobson, 2006). However, the tendency to aggregate varies dramatically depending on the intrinsic characteristics of the polypeptide chains and the environmental conditions, such as pH, ionic strength, or protein concentration (DuBay et al., 2004). The amyloid-like nature of IBs permits us to study the molecular determinants in their formation in a manner that is analogous to that used for the study of the determinants involved in the *in vitro* assembly of amyloid fibrils (Winkelmann et al., 2010). In this sense, several mutagenesis studies have revealed the essential role played by specific side-chain properties, such as hydrophobicity, secondary structure propensity, or net charge in amyloid formation (Polverino de Laureto et al., 2003). A deeply studied molecule is the amyloid-β peptide, whose central hydrophobic cluster, consisting of five residues (LVFFA), plays a crucial role in its nonnative deposition (Esler et al., 1996). Substitutions in this stretch with polar or charged amino acids decrease the *in vitro* aggregation propensity of the complete 42-residue sequence. The same applies for IB formation, where it exists an absolute correlation between the intrinsic physicochemical properties of the side chains in the central position of this sequence stretch (Phe19 in the wild-type sequence) and the intracellular aggregation propensity of the mutated peptide in bacteria (de Groot et al., 2006).

Usually, the cross-β-sheet core of the amyloid fibrils or amyloid-like IBs is composed of short hydrophobic stretches called "hot spots" (HSs) with a high aggregation propensity (Conchillo-Sole et al., 2007), confirming the accepted idea that not all of the polypeptide sequence is equally relevant for protein aggregation (Lopez de la Paz and Serrano, 2004; Ventura et al., 2004). These amyloidogenic regions modulate the aggregation propensity of the entire polypeptide chain, thus regulating the *in vivo* IB formation rate (de Groot and Ventura, 2006b). However, the effect of HSs driving to aggregation can be counteracted by the presence of "gatekeepers" (Figure 4.2). These short regions are enriched in residues that disfavor the formation of β-structure, such as proline or charged residues (Reumers et al., 2009). They usually flank HSs and are thought to promote the sequence-specific binding of several molecular chaperones (Rousseau et al., 2006). These sequential determinants of aggregation are absolutely crucial in the case of intrinsically disordered proteins, whose lack of a defined three-dimensional structure provokes the complete exposure of amyloidogenic regions, if they are present. Hence, their aggregation rate is basically modulated by the intrinsic aggregation propensity of the primary sequence. This effect underlies aggregation in yeast prions where unstructured regions in soluble form and the prion-forming domain embedded in the cross-β core of the amyloids overlap (Chiti and Dobson, 2006).

In globular proteins, nonstructured regions are also relevant to aggregation, even if, in some cases, disordered and flexible loops can prevent it by means of the entropy penalty associated with the required conformational change (De Simone et al., 2012). Moreover, despite the fact that protein deposition is highly influenced by the aggregation propensity of the primary

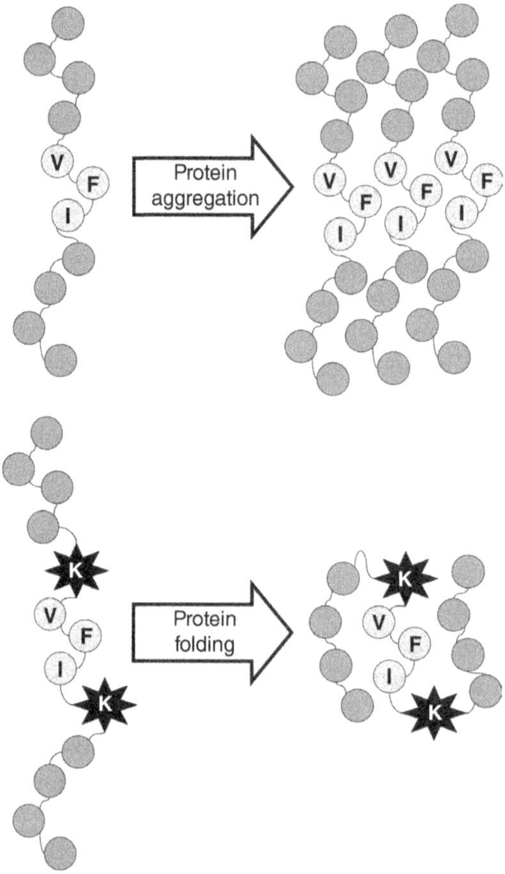

FIGURE 4.2 Representation of hot spots and gatekeepers. Above: Conceptual schema of protein aggregation driven by solvent-exposed HS (represented as dotted residues). Below: The presence of gatekeepers (represented as black star-shaped residues) favors protein folding against aggregation reaction. Acquisition of the native state prevents detrimental self-assembly since HSs remain buried in the inner protein core.

sequence, in the case of globular proteins is also regulated by their conformational stability (Chiti and Dobson, 2006). In natively folded proteins, HSs or amyloidogenic stretches are usually buried in the hydrophobic core, due to their functional role in the hydrophobic collapse during folding, which also acts as an evolutionary mechanism to prevent protein self-assembly (Lin and Zewail, 2012). Therefore, a population of at least partial unfolded conformers is required to uncover HSs and trigger aggregation. A favorable scenario for that reaction occurs during recombinant expression (synthesis, transport, and folding) when the nascent polypeptide is still partially folded and the bulk of protein input saturates the protein quality machinery of the cell (Hendrick and Hartl, 1993), leaving aggregation-prone regions free to establish nonnative intermolecular contacts. Nevertheless, even when the protein has attained its native state, the strong correlation observed between the aggregation propensity of globular proteins and their conformational stability suggests that local fluctuations of the native state might also leave aggregation-prone regions transiently exposed and prone to self-assembly (Calloni et al., 2005; Espargaro et al., 2008a; Chiti and Dobson, 2009). In this sense, stabilization of the native state has been proposed as a therapeutic approach in conformational diseases (Liu and Bitan, 2012) and has been shown to be a successful strategy to increase solubility in the recombinant production of globular proteins (Castillo et al., 2010).

4.3.3 Sequence Specificity in IB Formation

The sequence-specific nature of the contacts leading to *in vivo* IB formation has long been discussed (Hart et al., 1990; Speed et al., 1996). Among the first evidences to come along was simultaneous coexpression of two different proteins in *E. coli*. This resulted in the appearance of two types of IBs in a single cell, differing in morphology and composition and separable by differential centrifugation (Hart et al., 1990). More recently, our group reported on the spatial distribution inside IBs when two different polypeptides are expressed in the same cell. Using two amyloidogenic proteins fused to fluorescent reporters, we made use of FRET measurements to determine the lack of significant colocalization of both proteins in IBs. Interestingly, we demonstrated that although the aggregates were made up of both proteins, their relative aggregation propensities determined their relative positions in the IB, being the most aggregation-prone one, the one embedded in the inner core (Morell et al., 2008). This kinetic segregation of protein aggregation confirmed that interactions leading to the formation of the aggregate occur preferentially between homologous sequences. This property has also been demonstrated by means of seeding experiments. In independent works, purified bacterial IBs have been used to seed amyloid fibril assemblies (Carrio and Vill, 2005; Fernandez-Tresguerres et al., 2010; Dasari et al., 2011), demonstrating the specific amyloid-like nature of IBs (Figure 4.1 M). This specificity was best visualized using cross-seeding reactions. As an example, the IBs formed by amyloid peptides Aβ40 and Aβ42 were able to seed any of the two soluble monomers. However, the IBs formed by the fungal prion HET-s did not affect either the nucleation or elongation rate of the amyloid-formation processes of Aβ peptides (Dasari et al., 2011). Again, the cross-seeding between both Aβ peptides suggested that their aggregation reactions are nucleated through common regions, in agreement with many data demonstrating that this is the case for the corresponding amyloid fibrils (Kuperstein et al., 2010).

4.4 IBs AS THE SIMPLEST MODEL FOR *IN VIVO* AMYLOID TOXICITY

The similarity between bacterial IBs and amyloid fibrils suggests the possibility of exploiting these cheap and easily obtainable protein aggregates as a model in the study of amyloid aggregation, especially for that related to human disorders where protein deposits are pathogenic. Thus, it is likely that bacteria, despite being a prokaryotic and simple organism, might provide us with a valuable tool to dissect the mechanisms associated with aggregation-associated toxicity (García-Fruitós et al., 2011). The intrinsic toxicity of IBs is still an open question, with evidence pointing to opposite theories. A detoxification role of IBs has been proposed based on the idea that oligomers and prefibrillar aggregates are the toxic agents in the cellular environment (Kirkitadze et al., 2002). Hence, a mature protein deposit may act as an aggregation nucleus sequestering the toxic, still-soluble species. However, when the IBs formed by amyloid proteins are added to neurons, they turn out to be highly toxic (de Groot et al., 2009). Analysis of the fitness cost of protein aggregation for living bacterial cells has allowed us to explain this apparent contradiction.

4.4.1 The Fitness Cost of Amyloid Aggregation

Albeit several examples of functional amyloid structures have been described (Fowler et al., 2007), to a large extent, protein deposits are assumed to be deleterious for a cell. Accordingly, protein aggregation is evolutionarily constrained. Organisms have developed different strategies to cope with this deleterious reaction, and at the same time, polypeptide chains have been shaped to diminish the risk of nonnative self-assembly (Monsellier and Chiti, 2007; Pontarotti et al., 2012). Still, since aggregation-prone regions are also essential for folding, the stability of the native state, or the formation of quaternary structures, they cannot be avoided completely in protein sequences. As a consequence, most proteins should cope with a basal intrinsic aggregation risk.

Although the effect exerted by protein aggregates in bacterial physiology is still unclear, there exists evidence that points to a toxic gain of function of proteins upon aggregation in the bacterial cytosol. Among the first data supporting this view were those provided by a study addressing the assembly of the yeast Sup35 prion protein in the cytoplasm of bacteria. They described the formation of amyloid-like aggregates whose presence results in a decreased colony-forming ability once the bacterial population reaches the stationary phase (Ono et al., 2006). It has been demonstrated that during bacterial division, protein aggregates are asymmetrically segregated, giving rise to two cells with differential intracellular content (Stewart et al., 2005). Asymmetric division results in an old pole daughter cell inheriting the majority of deposits and displaying an aged phenotype, which is reflected in a much lower division rate (Lindner et al., 2008; Fernandez-Tresguerres et al., 2010; Winkler et al., 2010). Whether aggregation burden is the cause or the consequence of this aged phenotype is still a question to be answered. If protein aggregation impairs cell fitness in bacteria, division would act asymmetrically as an evolutionary mechanism to keep the damage out of the new cells, promoting rejuvenation of the bacterial population.

Supporting this view, using a large set of amyloidogenic peptides with different intrinsic aggregation tendencies, we have recently reported a highly significant correlation between the aging effect in bacteria and the aggregation propensity of the peptide. We have observed that the peptides with a higher intrinsic aggregation tendency form more compact and resistant IBs as a consequence of their faster assembly rate, promoting division defects to a larger extent (Villar-Piqué et al., 2012a). These results indicate that the toxic effect promoted by protein aggregates in bacteria depends on their formation kinetics and/or conformational quality, despite the fact that the mechanism underlying this damage is yet to be discovered. A recent work has converged to demonstrate the existence of a fitness cost for aggregation in bacteria by means of mutagenesis experiments in gatekeeper residues. As explained above, these stretches counteract the aggregation propensity of the HSs (Figure 4.2). Using a recombinant fusion of a short aggregation-prone peptide with GFP, the authors demonstrate that amino acid substitutions in the gatekeeper residues flanking the aggregation moiety have a severe effect on the aggregation burden of the cell, promoting differences in cell fitness and affecting the proteostatic response (Beerten et al., 2012).

4.4.2 Citotoxicity of Amyloid IBs

The addition of amyloid material to mammalian cells results in cell metabolism impairment and death, demonstrating the toxic nature of the protein structures embedded in these aggregates. It is now clear that the most toxic conformations correspond to the prefibrillar amyloid oligomers rather than the mature fibrils, independently if they are or are not assemblies of disease-related proteins (Bucciantini et al., 2002; Baglioni et al., 2006; Sakono and Zako, 2010). Therefore, the amyloid-like nature of IBs has raised the question of their possible citotoxicity for mammalian cells and tissues. We addressed this question using Alzheimer's disease–related peptides Aβ40 and Aβ42. These peptides were expressed in bacteria, resulting in the formation of insoluble aggregates displaying physicochemical features similar to those of the corresponding amyloid fibrils. Their addition in mammalian cell cultures induced metabolic damage, as measured by means of the cell proliferation MTT assay. Interestingly enough, this cytotoxicity was higher for IBs formed by Aβ42, which display higher amyloidogenicity and contain more oligomeric structures than Aβ40, demonstrating that the amyloid propensity correlates with *in vivo* toxicity (Dasari et al., 2011). Furthermore, another work carried out with β-galactosidase fused to an amyloidogenic protein of foot-and-mouth disease virus (VP1) reported that *in vitro* thermal aggregates are not toxic, in contrast to purified IBs produced in bacteria (González-Montalbán et al., 2007). As the former are more compact and structured than IBs, these results converge with the idea that prefibrillar aggregates are more toxic than mature fibrils. Recent work using different ataxin variants links the cell fitness cost caused by various recombinant proteins to the number of oligomeric species, thus suggesting that the mechanisms of toxicity in bacteria and eukaryotic cells are somehow conserved evolutionarily (Invernizzi et al., 2012).

4.4.3 Infectious Properties of IBs

Amyloid fibrils of prion proteins are characterized by their self-perpetuating and infectivity nature. The expression of these proteins in bacterial cells results in the formation of IBs with amyloid-like properties similar to those of their amyloid fibrils in their natural environment (Sabate et al., 2009; Espargaro et al., 2012b). Regardless of the relevance of this structural information, the newest, most interesting property of these IBs is their infective nature. This is the case for HET-s, a *Podospora anserina* prion whose IBs were used to transfect prion-free *P. anserina* strains, resulting in acquisition of the prion phenotype at levels comparable to that induced by HET-s amyloid fibrils (Wasmer et al., 2009). The infectious properties of bacterial IBs have also been demonstrated for the yeast prion domain of Sup35 (Sup35-NM) in two independent studies. In the first study, the protein expressed was a variant of Sup35-NM, displaying an increased propensity to convert into the prion conformation (Garrity et al., 2010), whereas in the second study, wild-type Sup35-NM was used instead (Espargaro et al., 2012b) (Figure 4.3). In both cases, the aggregates were used to infect yeast prion-free strains and induced the prion phenotype. Interestingly enough, the infectivity of these bacterial aggregates can be tuned easily by controlling conditions during protein production. These experiments unequivocally demonstrate that the bacterial cytoplasm allows the formation of prion aggregates with infectious and transmissible capabilities and, because the infectious properties of amyloids depend on highly specific structural characteristics, provide perhaps the most conclusive data in favor of the amyloid nature of IBs.

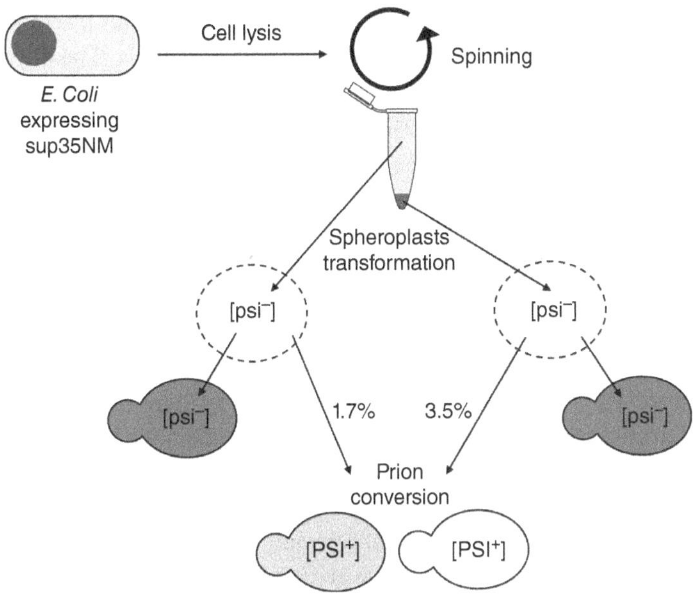

FIGURE 4.3 The prion domain of the yeast prion Sup35 (Sup35-NM) is expressed in *E. coli* cells forming amyloid-like intracellular IBs. After cell lysis, soluble and insoluble fractions are separated by centrifugation, and both are used to transform spheroplasts of prion-free yeast stains (phenotype [psi-]). Upon growing on ¼ YPD medium plates, cells converted to prion conformation (phenotype [PSI+]) appear as white or pink, whereas [psi-] remaining cells appear as red. Transformation with the insoluble fraction results in a higher prion conversion efficiency, demonstrating that at least part of the protein embedded in these bacterial IBs acquires a transmissible prion conformation. (*See insert for color representation of the figure.*)

4.5 USING IBs TO SCREEN FOR AMYLOID INHIBITORS

The increasing impact in our aging society of amyloid-linked neurodegenerative diseases such as Alzheimer's or Parkinson's disorders is pushing the discovery of new approaches to identify molecules able to interfere with aggregation pathways in order to halt the progress of these devastating pathologies or even to prevent their appearance. Screening methods for amyloid inhibitors are performed traditionally employing synthetic peptides. However, their use presents several disadvantages: The peptides are expensive; they tend to be highly insoluble, causing reproducibility problems in the assays; and most important, the inhibitors are selected under physiologically nonrelevant conditions. Also, the early detection of soluble prefibrillar aggregates is difficult, as these assays usually measure aggregation by means of turbidity or by using amyloid dyes, and early aggregates are invisible to these tests.

A considerable advance in this area has come along with the implementation of *in vivo* systems based on the use of bacteria models. The amyloid-like structure of bacterial aggregates has allowed their exploitation as a cheap and powerful tool for screening assays. In this way, the demonstration that the assembly rate of amyloidogenic proteins into IBs is modulated by external factors provides a means to test the effect of small compounds on the aggregation of these polypeptides.

Waldo and co-workers pioneered the use of GFP as a folding reporter, demonstrating that the activity of a fluorescent tag fused to a protein depends on the solubility of the latter. Therefore, the fluorescence of cells expressing such protein chimera correlates with the aggregation propensity of the protein (Waldo et al., 1999). This approach has been exploited for the high-throughput screen of amyloid peptide aggregation inhibitors using 96-well plates. On the one hand, it has been employed using living bacterial cells (Kim et al., 2006). The main advantage of this system is the possibility of studying the effect of the compounds tested in a biological relevant environment. Nonetheless, this method is limited by the ability of these compounds to enter the cells, introducing bias on the properties of the screened library. Recently, we have overcome this problem using purified IBs instead of complete cells. As it happens *in vivo*, we have shown that the *in vitro* refolding of these aggregates is sensible to external conditions, such as the presence of aggregation promoters or inhibitors, which can be detected easily reading the final recovered GFP fluorescence after refolding (Villar-Piqué et al., 2012b) (Figure 4.4). Despite the simplicity of this assay, it has been shown to distinguish between aggregation pathways and to provide a qualitative assessment of amyloid assembly modulators, such as metallic ions. Beyond fluorescent tag-based methods, intracellular aggregation can also be measured using other engineered proteins. One interesting example is the system based on the Tat protein export pathway, which uses the β-lactamase enzyme as a reporter. A ternary fusion between this enzyme, the Tat signal peptide, and the target protein is able to cross the inner membrane only if the latter is folded properly, thus conferring ampicillin resistance to the cell (Fisher et al., 2006). On the contrary, the aggregation of the target protein prevents the periplasmatic localization of the β-lactamase, sensitizing cells to the antibiotic. This genetic selection method can be used as well as a bacterial cell–based screening assay for aggregation inhibitors (Lee et al., 2009). Finally, our group has recently developed a Th-S-based tool to detect amyloid-like conformations *in vivo*. Bacterial cells containing IBs are stained with this dye for experiments involving spectroscopic changes upon binding to the amyloid structures inside bacterial aggregates. The coupling of this procedure with the high-speed, multiparametric data acquisition and analysis properties of flow cytometry permits its use in the wide-scale analysis of large bacterial populations (Espargaro et al., 2012a). This fast, quantitative,

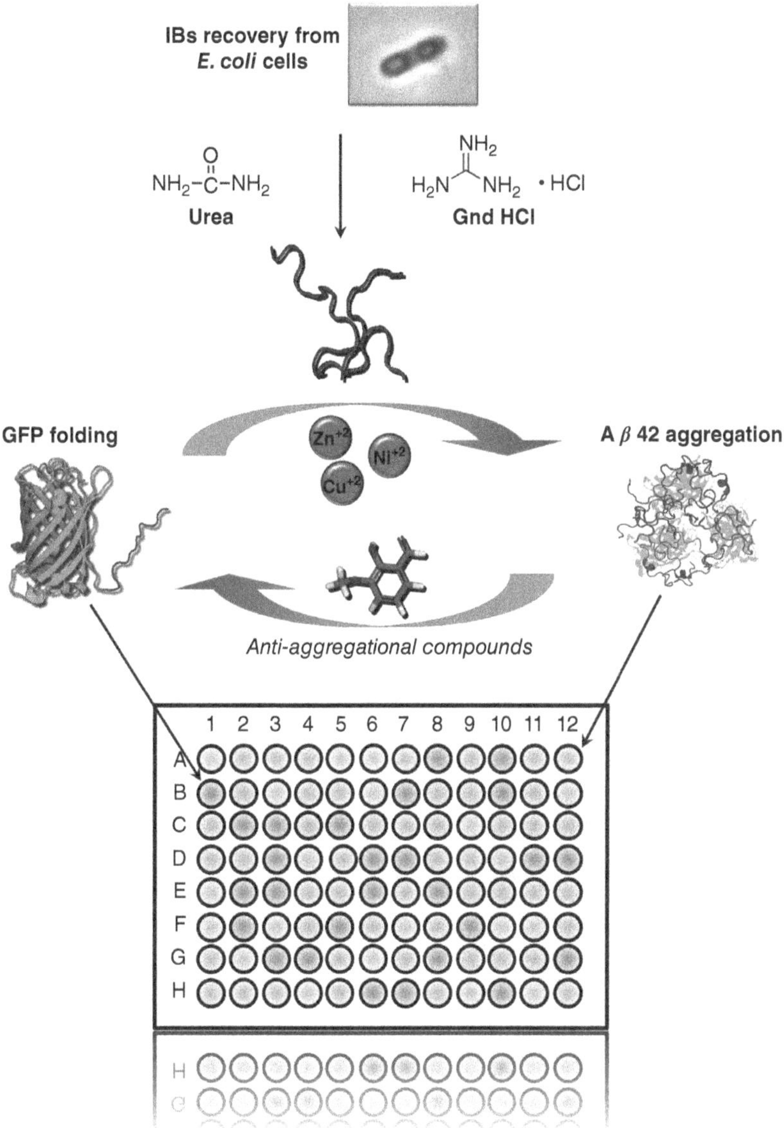

FIGURE 4.4 Schematic representation of a screening assay for amyloid modulators based on Aβ42-GFP IBs. IBs recovered and purified from bacterial cells are chemically denatured and, subsequently, *in vitro* refolded. The method is based on the assumption that the kinetic competition between Aβ42 aggregation and GFP folding occurring during IB formation in the cell can be reproduced in an *in vitro* environment. Thus, the presence of molecular modulators of amyloid aggregation in the refolding buffer is detected straightforwardly by means of the GFP fluorescence recovered after the reaction: Aggregation enhancers decrease the fluorescence recovered, while aggregation inhibitors promote high fluorescence recovery. [Reproduced with permission from Villar-Piqué et al. (2012b).] (*See insert for color representation of the figure.*)

high-throughput method has the potential to be used in screening assays for amyloid inhibitors, being an *in vivo* noninvasive tool that overcomes the use of fusions to protein reporters, thus allowing us to study natural, unmodified amyloidogenic proteins.

4.6 CONCLUSIONS

In this chapter we illustrate how an increasing number of studies are taking advantage of the amyloid-like nature of the intracellular aggregates formed in bacteria for a number of different purposes, ranging from the mechanistic analysis of the process of protein aggregation in an *in vivo*-like environment, to the identification of novel antiamyloidogenic compounds. In the near future it is likely that we will witness the use of bacterial systems to understand and model the evolutionary constraint imposed by aggregation on polypeptidic sequences or the use of proteomic tools to understand the cellular response to specific amyloid conformers. Overall, bacteria provide us with a golden lab tool for the study of amyloids.

REFERENCES

Arie JP, Miot M, et al. (2006) Formation of active inclusion bodies in the periplasm of *Escherichia coli*. Mol Microbiol 62(2):427–437.

Baglioni S, Casamenti F, et al. (2006) Prefibrillar amyloid aggregates could be generic toxins in higher organisms. J Neurosci 26(31):8160–8167.

Beerten J, Jonckheere W, et al. (2012) Aggregation gatekeepers modulate protein homeostasis of aggregating sequences and affect bacterial fitness. Protein Eng Des Sel 25(7):357–366.

Bucciantini M, Giannoni E, et al. (2002) Inherent toxicity of aggregates implies a common mechanism for protein misfolding diseases. Nature 416(6880):507–511.

Calloni G, Zoffoli S, et al. (2005) Investigating the effects of mutations on protein aggregation in the cell. J Biol Chem 280(11):10607–10613.

Cano-Garrido O, Rodríguez-Carmona E, et al. (2013) Supramolecular organization of protein-releasing functional amyloids solved in bacterial inclusion bodies. Acta Biomater 9(4):6134–6142.

Carrió MM, Villaverde A (2005) Localization of chaperones DnaK and GroEL in bacterial inclusion bodies. J Bacteriol 187(10):3599–3601.

Carrió M, González-Montalbán N, et al. (2005) Amyloid-like properties of bacterial inclusion bodies. J Mol Biol 347(5):1025–1037.

Castillo V, Espargaro A, et al. (2010) Deciphering the role of the thermodynamic and kinetic stabilities of SH3 domains on their aggregation inside bacteria. Proteomics 10(23):4172–4185.

Castillo V, Grana-Montes R, et al. (2011) The aggregation properties of *Escherichia coli* proteins associated with their cellular abundance. Biotechnol J 6(6): 752–760.

Chiti F, Dobson CM (2006) Protein misfolding, functional amyloid, and human disease. Annu Rev Biochem 75:333–366.

Chiti F, Dobson CM (2009) Amyloid formation by globular proteins under native conditions. Nat Chem Biol 5(1):15–22.

Conchillo-Sole O, de Groot NS, et al. (2007) AGGRESCAN: a server for the prediction and evaluation of "hot spots" of aggregation in polypeptides. BMC Bioinformatics 8:65.

Daggett V, Fersht AR (2009) Protein folding and binding: moving into unchartered territory. Curr Opin Struct Biol 19(1):1–2.

Dasari M, Espargaro A, et al. (2011) Bacterial inclusion bodies of Alzheimer's disease beta-amyloid peptides can be employed to study native-like aggregation intermediate states. Chembiochem 12(3):407–423.

de Groot NS, Ventura S (2006a) Effect of temperature on protein quality in bacterial inclusion bodies. FEBS Lett 580(27):6471–6476.

de Groot NS, Ventura S (2006b) Protein activity in bacterial inclusion bodies correlates with predicted aggregation rates. J Biotechnol 125(1):110–113.

de Groot NS, Ventura S (2010) Protein aggregation profile of the bacterial cytosol. PLoS One 5(2):e9383.

de Groot NS, Aviles FX, et al. (2006) Mutagenesis of the central hydrophobic cluster in Abeta42 Alzheimer's peptide: side-chain properties correlate with aggregation propensities. FEBS J 273(3):658–668.

de Groot NS, Sabate R, et al. (2009) Amyloids in bacterial inclusion bodies. Trends Biochem Sci 34(8):408–416.

De Simone A, Kitchen C, et al. (2012) Intrinsic disorder modulates protein self-assembly and aggregation. Proc Natl Acad Sci U S A 109(18):6951–6956.

Dobson CM (2003) Protein folding and misfolding. Nature 426(6968):884–890.

Doglia SM, Ami D, et al. (2008) Fourier transform infrared spectroscopy analysis of the conformational quality of recombinant proteins within inclusion bodies. Biotechnol J 3(2):193–201.

DuBay KF, Pawar AP, et al. (2004) Prediction of the absolute aggregation rates of amyloidogenic polypeptide chains. J Mol Biol 341(5):1317–1326.

Esler WP, Stimson ER, et al. (1996) Point substitution in the central hydrophobic cluster of a human beta-amyloid congener disrupts peptide folding and abolishes plaque competence. Biochemistry 35(44):13914–13921.

Espargaro A, Castillo V, et al. (2008a) The in vivo and in vitro aggregation properties of globular proteins correlate with their conformational stability: the SH3 case. J Mol Biol 378(5):1116–1131.

Espargaro A, Sabate R, et al. (2008b) Kinetic and thermodynamic stability of bacterial intracellular aggregates. FEBS Lett 582(25–26):3669–3673.

Espargaro A, Sabate R, et al. (2012a) Thioflavin-S staining coupled to flow cytometry: a screening tool to detect in vivo protein aggregation. Mol Biosyst 8(11):2839–2844.

Espargaro A, Villar-Piqué A, et al. (2012b) Yeast prions form infectious amyloid inclusion bodies in bacteria. Microb Cell Fact 11(1):89.

Fernandez-Busquets X, de Groot NS, et al. (2008) Recent structural and computational insights into conformational diseases. Curr Med Chem 15(13):1336–1349.

Fernandez-Tresguerres ME, de la Espina SM, et al. (2010) A DNA-promoted amyloid proteinopathy in *Escherichia coli*. Mol Microbiol 77(6):1456–1469.

Fisher AC, Kim W, et al. (2006) Genetic selection for protein solubility enabled by the folding quality control feature of the twin-arginine translocation pathway. Protein Sci 15(3):449–458.

Fowler DM, Koulov AV, et al. (2007) Functional amyloid—from bacteria to humans. Trends Biochem Sci 32(5):217–224.

García-Fruitós E, Seras-Franzoso J, et al. (2010) Tunable geometry of bacterial inclusion bodies as substrate materials for tissue engineering. Nanotechnology 21(20):205101.

García-Fruitós E, Sabate R, et al. (2011) Biological role of bacterial inclusion bodies: a model for amyloid aggregation. FEBS J 278(14):2419–2427.

García-Fruitós E, Vazquez E, et al. (2012) Bacterial inclusion bodies: making gold from waste. Trends Biotechnol 30(2):65–70.

Garrity SJ, Sivanathan V, et al. (2010) Conversion of a yeast prion protein to an infectious form in bacteria. Proc Natl Acad Sci U S A 107(23):10596–10601.

González-Montalbán N, Villaverde A, et al. (2007) Amyloid-linked cellular toxicity triggered by bacterial inclusion bodies. Biochem Biophys Res Commun 355(3):637–642.

Hart RA, Rinas U, et al. (1990) Protein composition of *Vitreoscilla* hemoglobin inclusion bodies produced in *Escherichia coli*. J Biol Chem 265(21):12728–12733.

Hendrick JP Hartl FU (1993) Molecular chaperone functions of heat-shock proteins. Annu Rev Biochem 62:349–384.

Ignatova Z, Gierasch LM (2004) Monitoring protein stability and aggregation in vivo by real-time fluorescent labeling. Proc Natl Acad Sci U S A 101(2):523–528.

Invernizzi G, Aprile FA, et al. (2012) The relationship between aggregation and toxicity of polyglutamine-containing ataxin-3 in the intracellular environment of *Escherichia coli*. PLoS One 7(12):e51890.

Kim W, Kim Y, et al. (2006) A high-throughput screen for compounds that inhibit aggregation of the Alzheimer's peptide. ACS Chem Biol 1(7):461–469.

Kirkitadze MD, Bitan G, et al. (2002) Paradigm shifts in Alzheimer's disease and other neurodegenerative disorders: the emerging role of oligomeric assemblies. J Neurosci Res 69(5):567–577.

Kuperstein I, Broersen K, et al. (2010) Neurotoxicity of Alzheimer's disease Abeta peptides is induced by small changes in the Abeta42 to Abeta40 ratio. EMBO J 29(19):3408–3420.

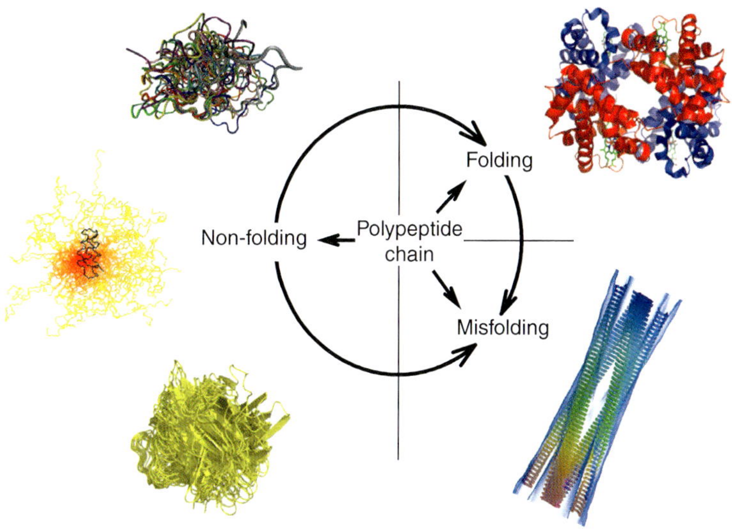

FIGURE 1.1 Fate of a newly synthesized polypeptide chain in a cell.

Protein Aggregation in Bacteria: Functional and Structural Properties of Inclusion Bodies in Bacterial Cells, First Edition. Edited by Silvia Maria Doglia and Marina Lotti.

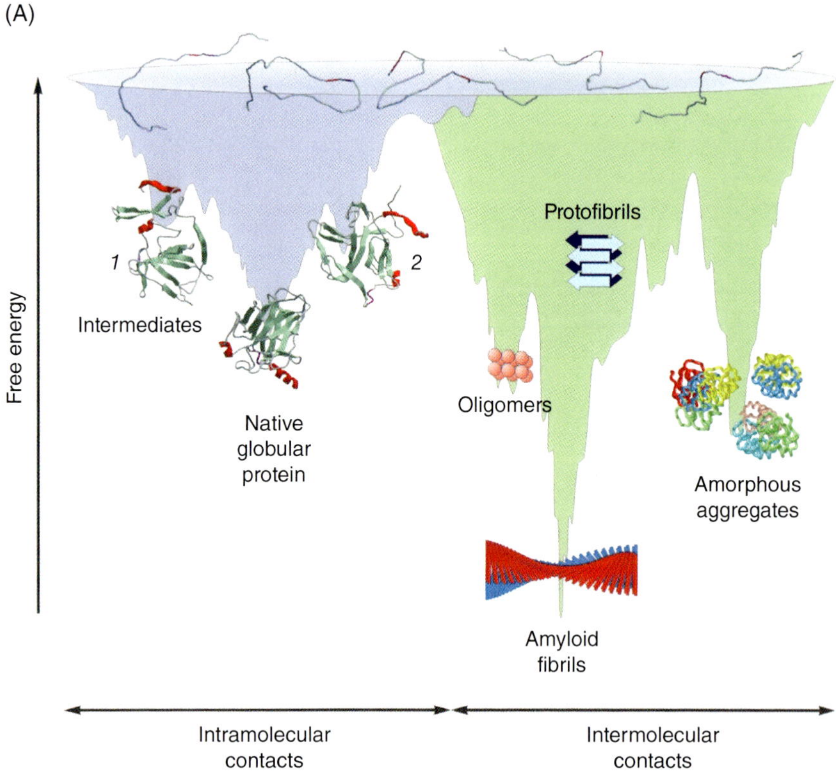

FIGURE 1.2 The energy landscape model, illustrating the formation of native globular and intrinsically disordered proteins, supramolecular complexes, amorphous aggregates, and amyloid fibrils. (A) Globular proteins. In the globular protein folding, the increase in the free energy associated with the folding-induced entropy decrease is compensated by the formation of specific intramolecular contacts. Local free-energy minima at the energy landscape correspond to the formation of partially folded intermediates. Intermolecular contacts of partially folded protein molecules can result in the formation of oligomers, amorphous aggregates, or amyloid fibrils

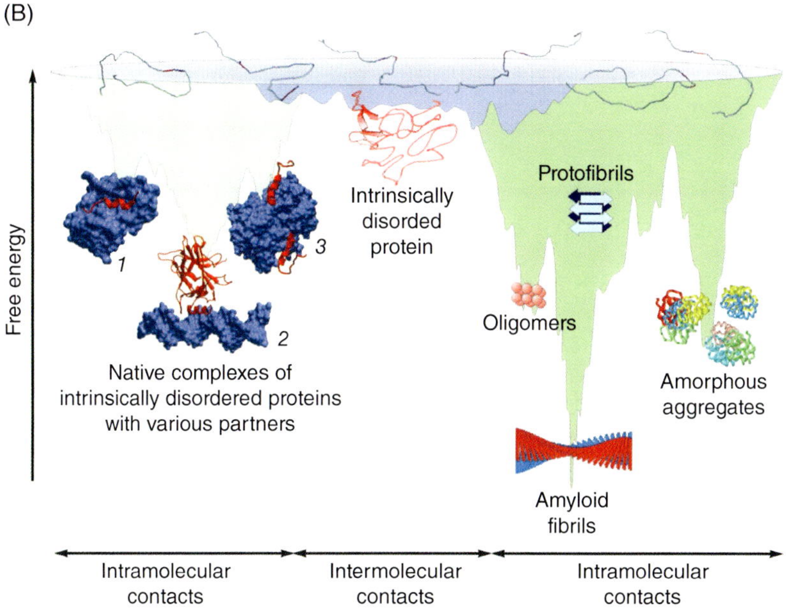

FIGURE 1.2 (*Continued*) (B) Intrinsically disordered proteins. Many native proteins with distinctive biological functions lack compact globular structure in aqueous solutions. Disordered segments of these proteins can gain ordered structure at the interaction with specific binding partners in a case if the free energy of such complexes is lower than the free energies of the intrinsically disordered protein and its partner. The propensity of native completely or partially disordered proteins to interact with various partners determines their biological functions in recognition of various binding partners (ligands, nucleic acids, and other proteins), in regulation of almost all cellular processes, and in signal transduction. In contrast to the folded globular proteins, which have to unfold to become amyloidogenic, disordered proteins seem always to be ready for such intermolecular interactions. 1, 2, and 3 represent native complexes of intrinsically disordered proteins with various partners. This figure is based on the energy funnel model developed for globular proteins (Schultz, 2000; Jahn and Radford, 2005). [Adapted from Turoverov et al. (2010).]

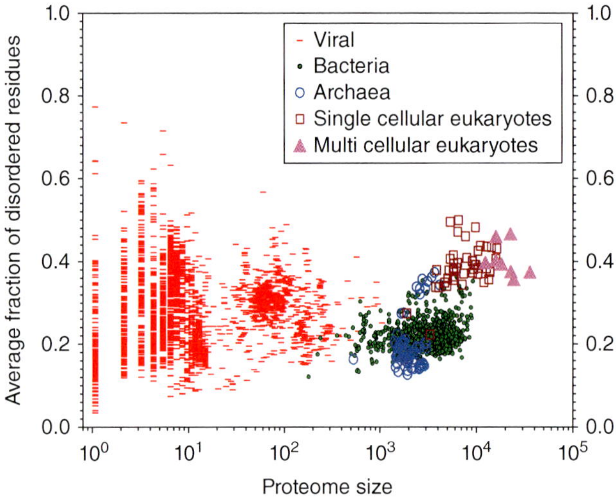

FIGURE 1.5 Correlation between the intrinsic disorder content and proteome size for 3484 species from viruses, archaea, bacteria, and eukaryotes. Each symbol indicates a species. There are a total of five groups of species: viruses (red horizontal marks), bacteria (small green circles), archaea (open blue circles), unicellular eukaryotes (open brown squares), and multicellar eukaryotes (pink triangles). The proteome size is the number of proteins in the proteome of that species and is shown on a logarithmic scale. The average fraction of disordered residues is calculated by averaging the fraction of disordered residues of each sequence over all the sequences of that species. Disorder prediction is evaluated by PONDR-VSL2B. [Adapted from Xue et al. (2012).]

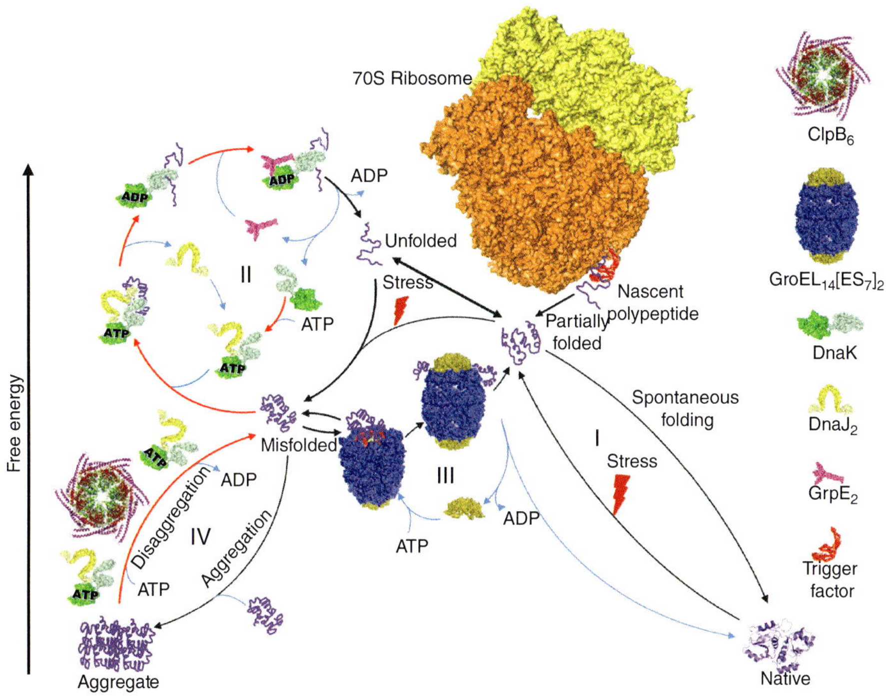

FIGURE 2.1 Schematic representation of the *E. coli* unfolding chaperone network promoting the native folding of aggregation-prone recombinant proteins. As nascent polypeptides (purple) emerge from the ribosome, the trigger factor, which is a PDI associated with the ribosome, assists in the sequential shaping of small native-like domains. Once released from the rhibosome and trigger factor, the partially folded proteins may reach the native state spontaneously, possibly assisted by chemical chaperones (cycle I). Yet, under stress or when mutated, they may instead misfold and become unfolded by the DnaK (KJE) system (cycle II), or by the GroEL/GroES chaperonin system (cycle III). In case the misfolded monomers escape unfolding by KJE or GroELS, they may form stable aggregates and IBs, but the KJE system can collaborate with ClpB to use the energy of ATP hydrolysis to unfold and solubilize individual polypeptides from the stable aggregates (cycle IV), thereby allowing them to reach their native state spontaneously.

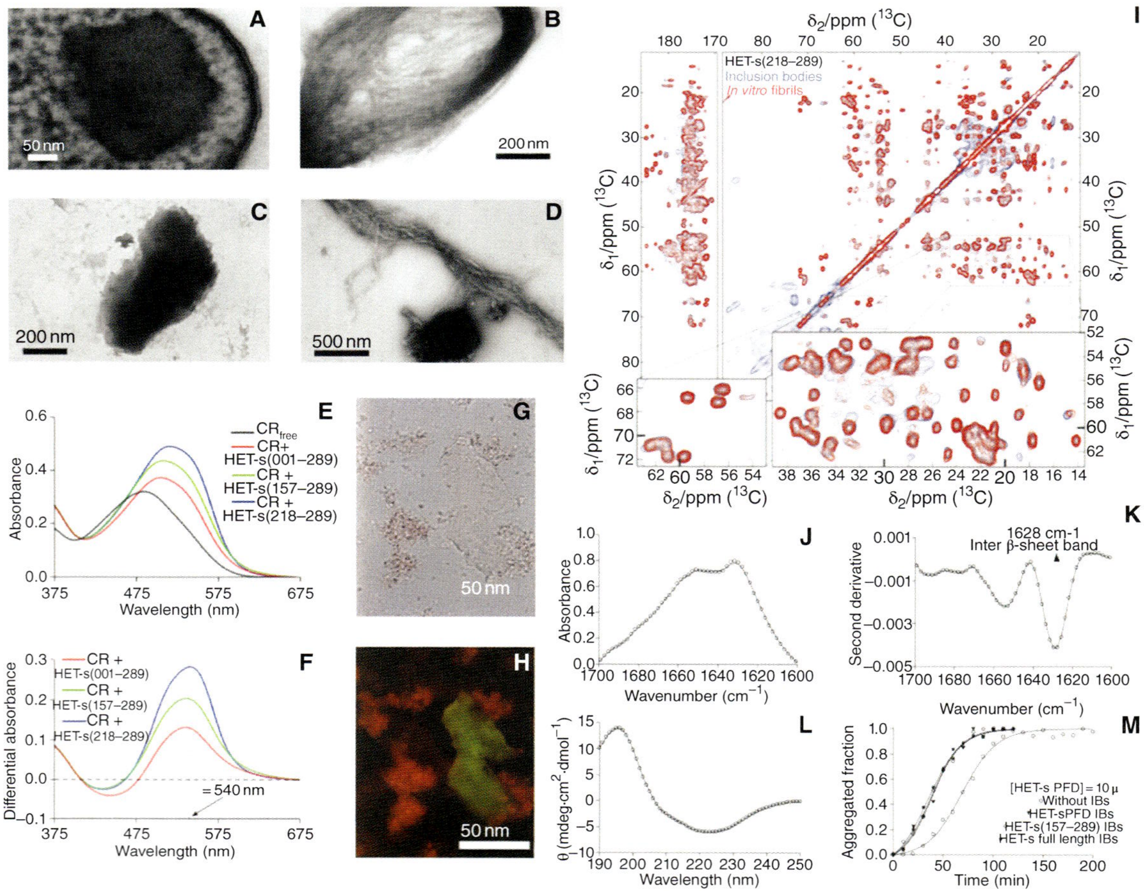

FIGURE 4.1 See text for caption.

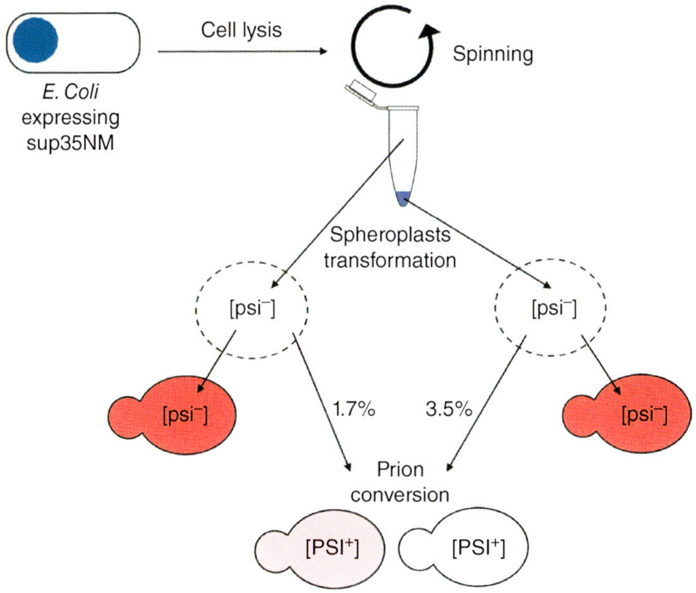

FIGURE 4.3 The prion domain of the yeast prion Sup35 (Sup35-NM) is expressed in *E. coli* cells forming amyloid-like intracellular IBs. After cell lysis, soluble and insoluble fractions are separated by centrifugation, and both are used to transform spheroplasts of prion-free yeast stains (phenotype [psi-]). Upon growing on ¼ YPD medium plates, cells converted to prion conformation (phenotype [PSI+]) appear as white or pink, whereas [psi-] remaining cells appear as red. Transformation with the insoluble fraction results in a higher prion conversion efficiency, demonstrating that at least part of the protein embedded in these bacterial IBs acquires a transmissible prion conformation.

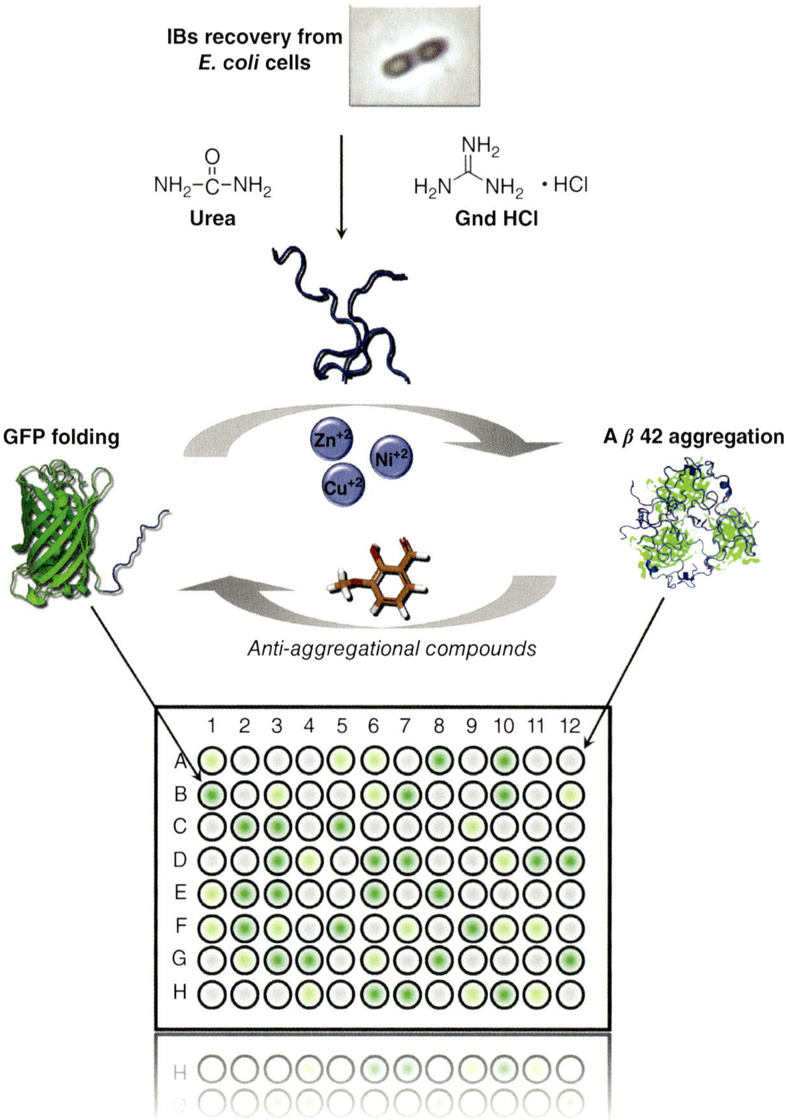

FIGURE 4.4 Schematic representation of a screening assay for amyloid modulators based on Aβ42-GFP IBs. IBs recovered and purified from bacterial cells are chemically denatured and, subsequently, *in vitro* refolded. The method is based on the assumption that the kinetic competition between Aβ42 aggregation and GFP folding occurring during IB formation in the cell can be reproduced in an *in vitro* environment. Thus, the presence of molecular modulators of amyloid aggregation in the refolding buffer is detected straightforwardly by means of the GFP fluorescence recovered after the reaction: Aggregation enhancers decrease the fluorescence recovered, while aggregation inhibitors promote high fluorescence recovery. [Reproduced with permission from Villar-Piqué et al. (2012b).]

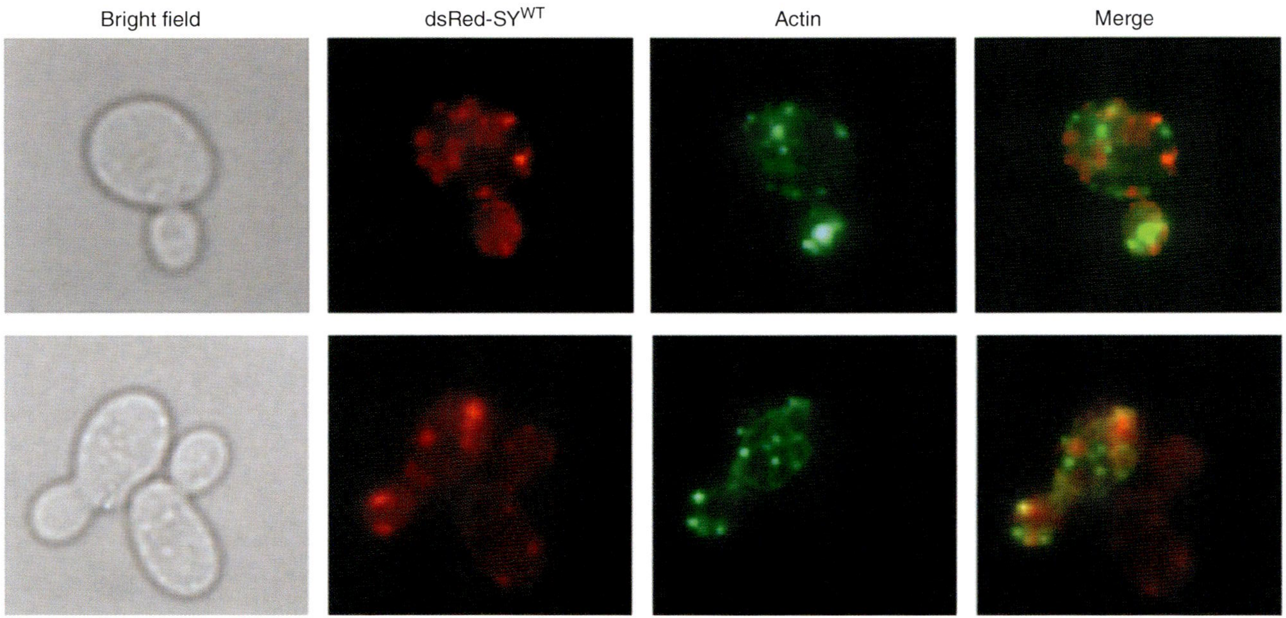

FIGURE 5.2 Transport of synphilin-1 inclusions along actin cables. Fluorescence microscopy images of late exponential yeast cells expressing fluorescently tagged synphilin-1 (dsRed-SY^{WT}) stained with Alexa Fluor 488-phalloidin to visualize actin patches and actin fibers.

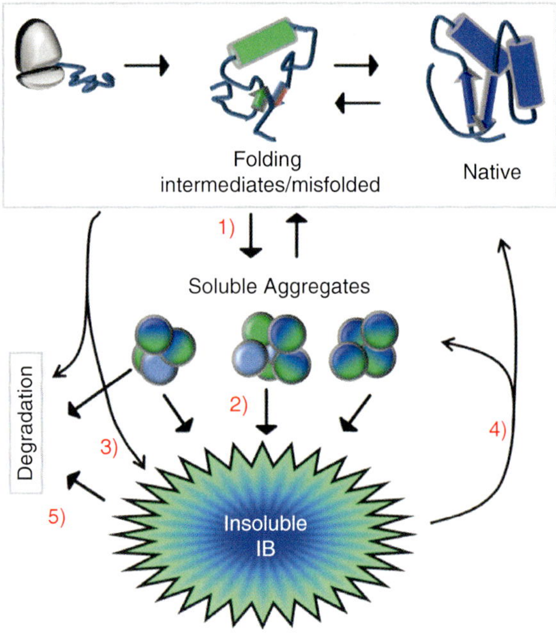

FIGURE 6.1 Scheme of inclusion body formation. See text for details.

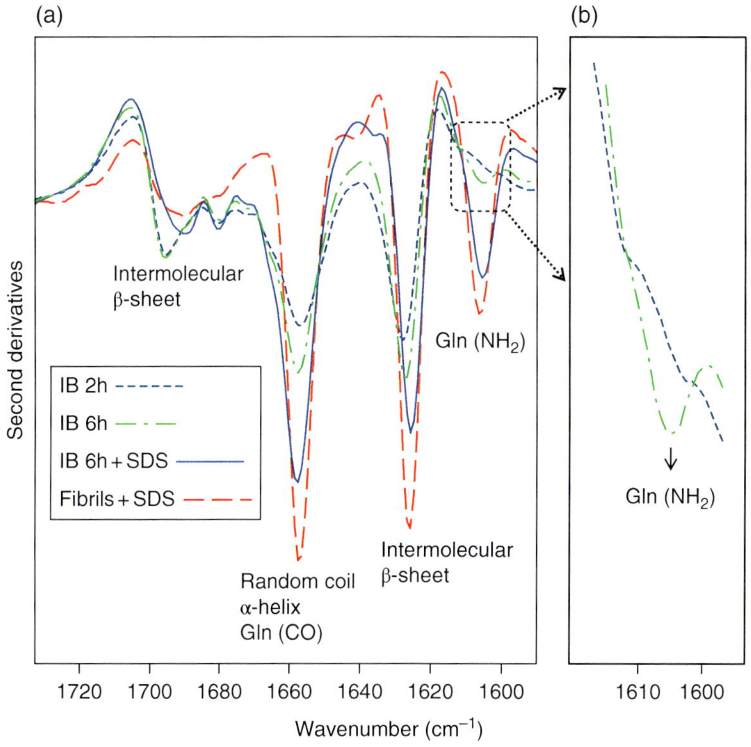

FIGURE 6.3 Ataxin-3 aggregation in form of IBs. (a) Second derivative infrared spectra in the amide I region of extracted IBs of AT3-Q55, collected after 2 h of induction in *E. coli* and after 6 h. The second derivatives spectra of AT3-Q55 fibrils, obtained *in vitro*, and of IBs at 6 h of induction are also reported after SDS treatments. (b) Second derivative spectra of IBs, extracted 2 h and 6 h after induction, are reported on an enlarged wavenumber scale in the spectral region of the glutamine NH_2 deformation modes. The band assignment to the secondary structure components is indicated. The second derivative spectra are normalized at the tyrosine peak at about 1515 cm^{-1}, to take into account possible differences in protein content.

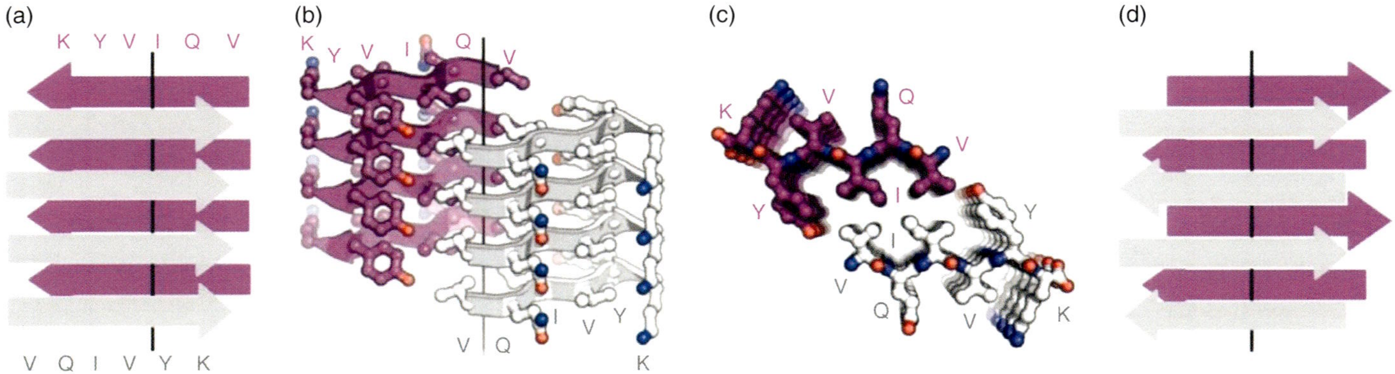

FIGURE 7.2 (a) Parallel cross-β-sheet structure of amyloid fibrils of the VQIVYK peptide from tau protein; (b) side view of VQIVYK amyloid fibrils with side chains; (c) top view of VQIVYK amyloid fibrils with sidechains; (d) antiparallel cross-β-sheet structure. [Parts (b) and (c) are reproduced with permission from the Nature Publishing Group.]

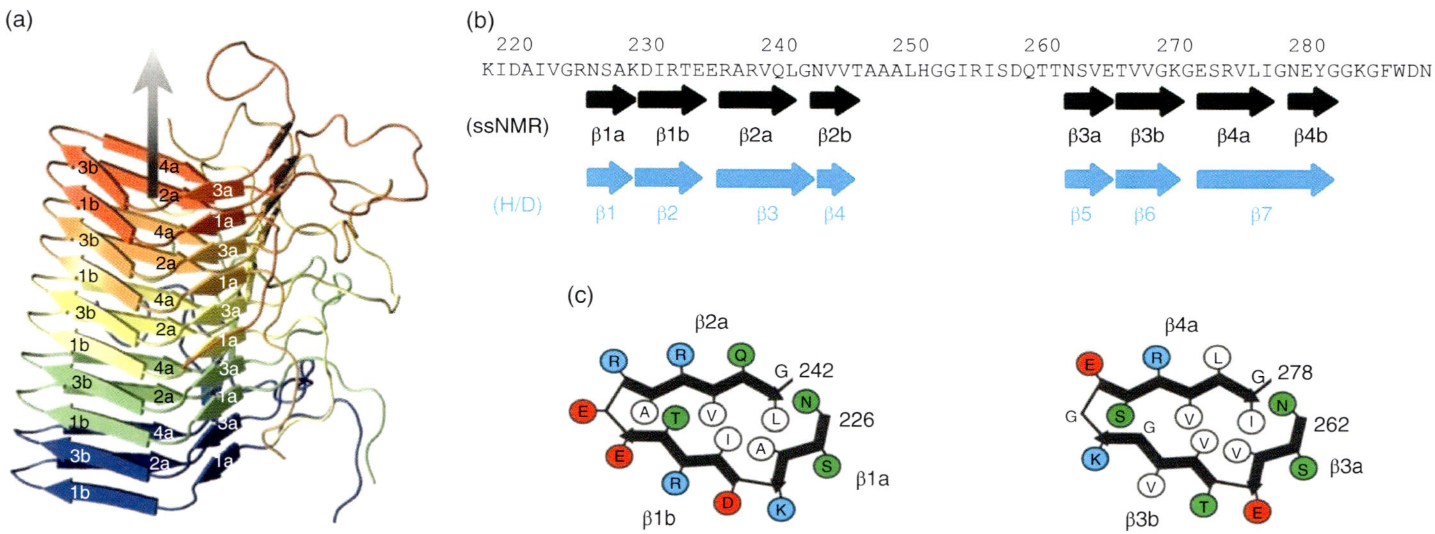

FIGURE 7.3 (a) Cross-β-sheet structure of HET-s(218–289) amyloid fibrils. Each color represents one HET-s(218–289) molecule, and its β-strands are labeled; (b) amino acid sequence of HET-s(218–289). Black arrow bars represent β-strands formed in the amyloid fibrils, as determined by solid-state NMR (ssNMR). Blue arrow bars represent β-strands determined with solution-state NMR and the H/D-exchange method (H/D); (c) residue arrangement around the cross-β-sheet structure: hydrophobic residues are colored white, acidic residues red, basic residues blue, and others green. [Parts (a) and (c) reproduced with permission from *Science*.]

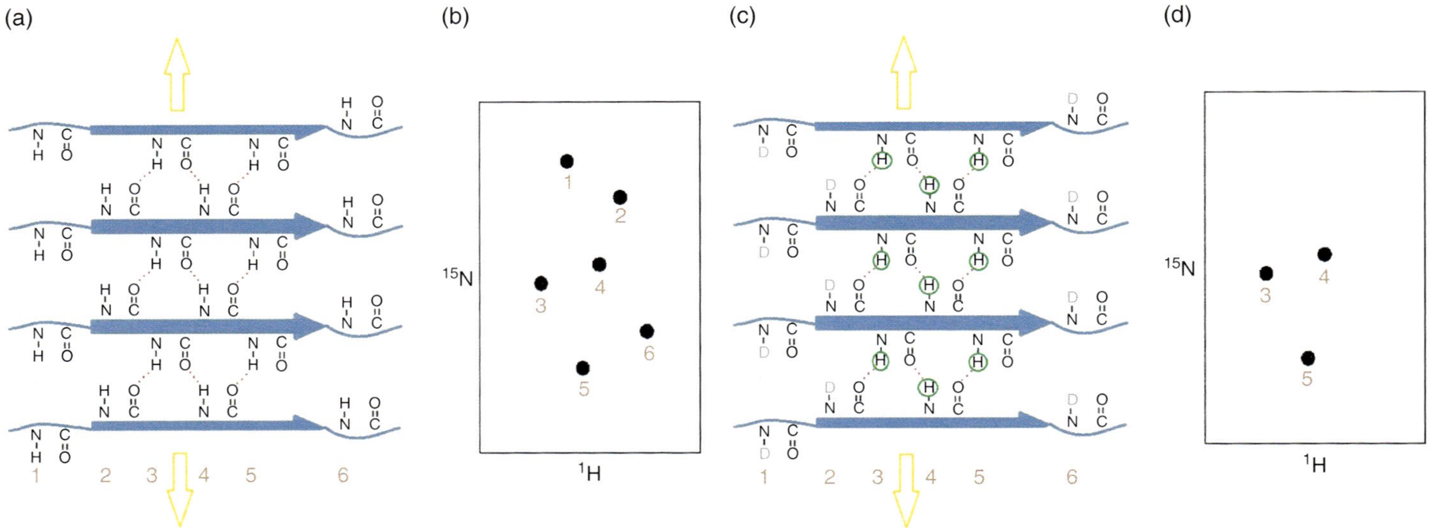

FIGURE 7.4 (a) Protein aggregates that contain cross-β-sheet structure. Each blue arrow bar and two end curves represent one protein molecule; the arrow bar represents β-strand in the cross-β-sheet, and curves represent the rest of the molecule. Backbone N–H and C=O pairs are labeled, and N–H pairs are numbered 1 through 6. Red dots represent hydrogen bonds, and yellow arrows indicate the growing direction of β-sheet. (b) The reference spectrum (NMR–HSQC spectrum). Each peak in the spectrum is assigned to a backbone N–H pair of the protein aggregates. (c) The protein aggregates after H/D-exchange. Hydrogens forming hydrogen bonds in the cross-β-sheet are protected against deuterium replacement (positions 3, 4, and 5). (d) The exchange spectrum (NMR–HSQC spectrum). Remaining peaks in the spectrum correspond to backbone N–H pairs in the cross-β-sheet (positions 3, 4, and 5).

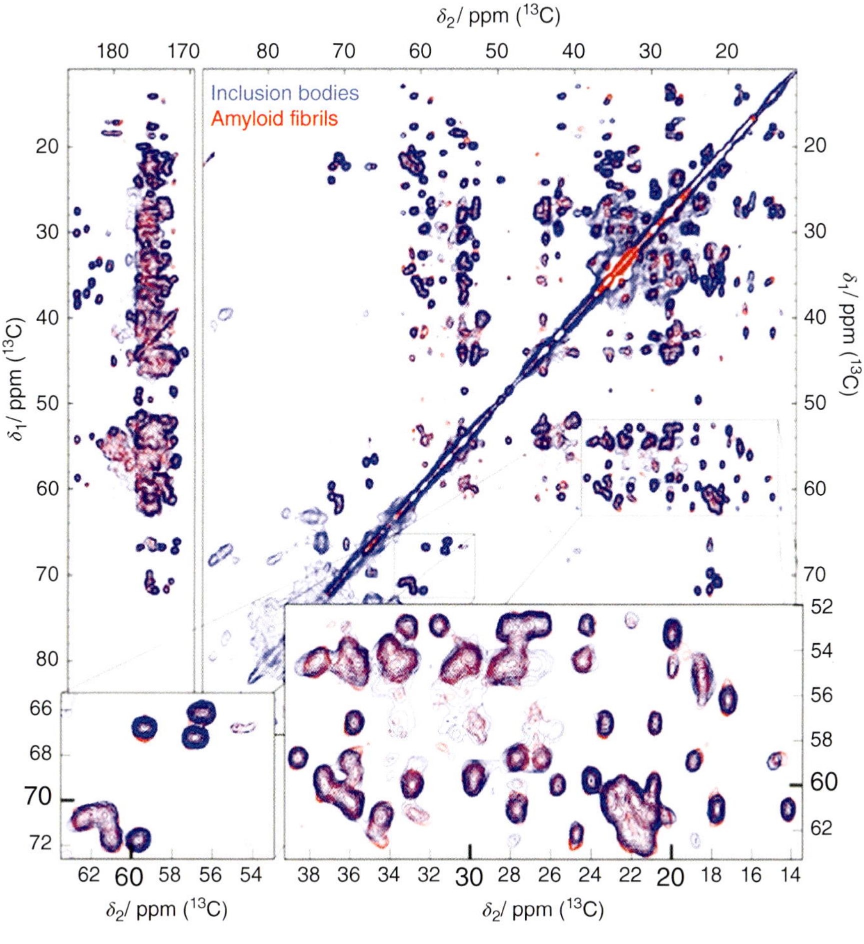

FIGURE 7.6 ^{13}C–^{13}C solid-state-NMR correlation spectrum (proton-driven spin diffusion with a mixing time of 50 ms) of purified HET-s(218–289) inclusion bodies (blue contours) compared with the same type of spectrum of HET-s(218–289) amyloid fibrils. All resonance peaks identified for the amyloid fibrils were also observed in the spectrum of purified inclusion bodies. The insets demonstrate that the linewidths of the two spectra are virtually identical, with no significant chemical-shift changes. (Reproduced with permission from Wiley-VCH Verlag GmbH.)

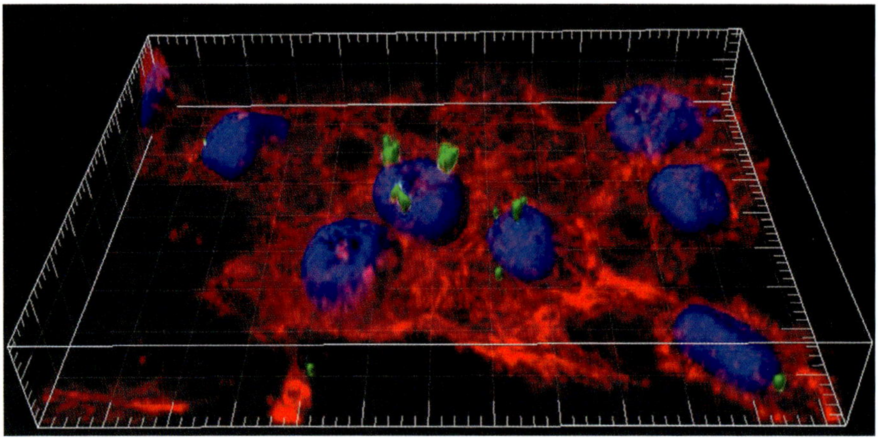

FIGURE 8.4 Confocal *xyz* stack of several sections generated with the Imaris 3D software (using the Isosurface module), showing penetration of GFP-based IBs into cultured HeLa cells. IBs are shown in green, nuclear DNA in blue, and the cell membranes in red. Experimental details can be found elsewhere (Vázquez et al., 2012).

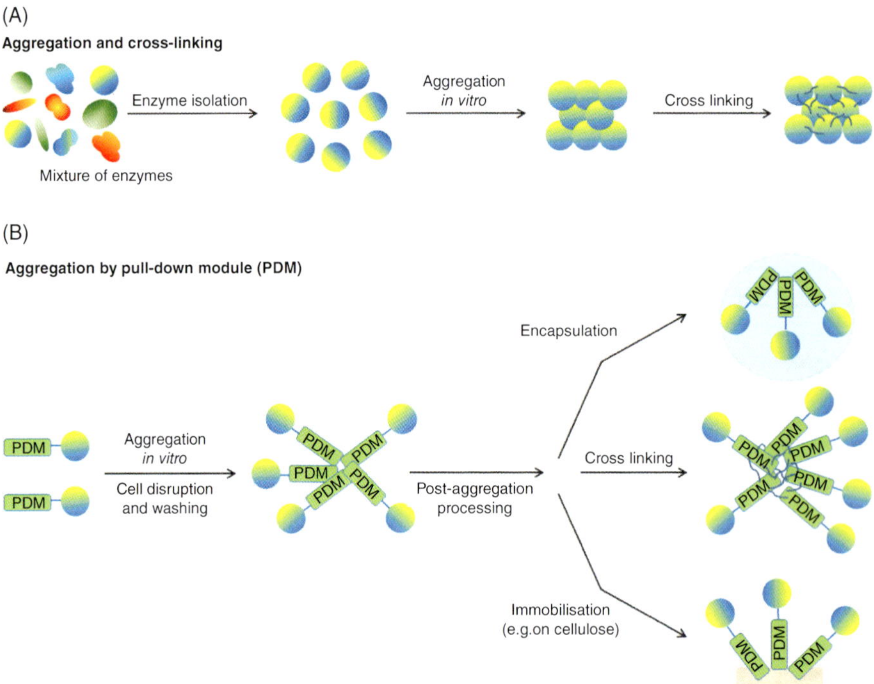

FIGURE 10.1 Preparation of insoluble enzyme aggregates using (A) CLEA technology or (B) aggregation by pull-down module.

Lee LL, Ha H, et al. (2009) Discovery of amyloid-beta aggregation inhibitors using an engineered assay for intracellular protein folding and solubility. Protein Sci 18(2):277–286.

Lin MM Zewail AH (2012) Hydrophobic forces and the length limit of foldable protein domains. Proc Natl Acad Sci U S A 109(25):9851–9856.

Lindner AB, Madden R, et al. (2008) Asymmetric segregation of protein aggregates is associated with cellular aging and rejuvenation. Proc Natl Acad Sci U S A 105(8):3076–3081.

Liu T, Bitan G (2012) Modulating self-assembly of amyloidogenic proteins as a therapeutic approach for neurodegenerative diseases: strategies and mechanisms. ChemMedChem 7(3):359–374.

Lopez de la Paz M, Serrano L (2004) Sequence determinants of amyloid fibril formation. Proc Natl Acad Sci U S A 101(1):87–92.

Monsellier E, Chiti F (2007) Prevention of amyloid-like aggregation as a driving force of protein evolution. EMBO Rep 8(8):737–742.

Morell M, Bravo R, et al. (2008) Inclusion bodies: specificity in their aggregation process and amyloid-like structure. Biochim Biophys Acta 1783(10): 1815–1825.

Nothaft H, Szymanski CM (2010) Protein glycosylation in bacteria: sweeter than ever. Nat Rev Microbiol 8(11):765–778.

Ono B, Kubota M, et al. (2006) Production of a polymer-forming fusion protein in *Escerichia coli* strain BL21. Biosci Biotechnol Biochem 70(12):2813–2823.

Polverino de Laureto P, Taddei N, et al. (2003) Protein aggregation and amyloid fibril formation by an SH3 domain probed by limited proteolysis. J Mol Biol 334(1):129–141.

Reumers J, Maurer-Stroh S, et al. (2009) Protein sequences encode safeguards against aggregation. Hum Mutat 30(3):431–437.

Rokney A, Shagan M, et al. (2009) *E. coli* transports aggregated proteins to the poles by a specific and energy-dependent process. J Mol Biol 392(3):589–601.

Rousseau F, Serrano L, et al. (2006) How evolutionary pressure against protein aggregation shaped chaperone specificity. J Mol Biol 355(5):1037–1047.

Sabate R, Espargaro A, et al. (2009) Characterization of the amyloid bacterial inclusion bodies of the HET-s fungal prion. Microb Cell Fact 8:56.

Sabate R, de Groot NS, et al. (2010) Protein folding and aggregation in bacteria. Cell Molec Life Sci 67(16):2695–2715.

Sakono M, Zako T (2010) Amyloid oligomers: formation and toxicity of Abeta oligomers. FEBS J 277(6):1348–1358.

Speed MA, Wang DI, et al. (1996) Specific aggregation of partially folded polypeptide chains: the molecular basis of inclusion body composition. Nat Biotechnol 14(10):1283–1287.

Stewart EJ, Madden R, et al. (2005) Aging and death in an organism that reproduces by morphologically symmetric division. PLoS Biol 3(2):e45.

Tartaglia GG Vendruscolo M (2009) Correlation between mRNA expression levels and protein aggregation propensities in subcellular localisations. Mol Biosyst 5(12):1873–1876.

Tartaglia GG, Vendruscolo M (2010) Proteome-level interplay between folding and aggregation propensities of proteins. J Mol Biol 402(5):919–928.

Upadhyay AK, Murmu A, et al. (2012) Kinetics of inclusion body formation and its correlation with the characteristics of protein aggregates in *Escherichia coli*. PLoS One 7(3):e33951.

Ventura S, Villaverde A (2006) Protein quality in bacterial inclusion bodies. Trends Biotechnol 24(4):179–185.

Ventura S, Zurdo J, et al. (2004) Short amino acid stretches can mediate amyloid formation in globular proteins: the Src homology 3 (SH3) case. Proc Natl Acad Sci U S A 101(19):7258–7263.

Villar-Piqué A, de Groot NS, et al. (2012a) The effect of amyloidogenic peptides on bacterial aging correlates with their intrinsic aggregation propensity. J Mol Biol 421(2–3):270–281.

Villar-Piqué A, Espargaro A, et al. (2012b) Using bacterial inclusion bodies to screen for amyloid aggregation inhibitors. Microb Cell Fact 11:55.

Villar-Piqué A, Ventura S (2012) Protein Aggregation Acts as Strong Constraint During Evolution. Evolutionary Biology: Mechanisms and Trends. Berlin: Springer, pp. 103–120.

Waldo GS, Standish BM, et al. (1999) Rapid protein-folding assay using green fluorescent protein. Nat Biotechnol 17(7):691–695.

Wang L (2009) Towards revealing the structure of bacterial inclusion bodies. Prion 3(3):139–145.

Wang, L, Maji SK, et al. (2008) Bacterial inclusion bodies contain amyloid-like structure. PLoS Biol 6(8):e195.

Wasmer C, Benkemoun L, et al. (2009) Solid-state NMR spectroscopy reveals that *E. coli* inclusion bodies of HET-s(218–289) are amyloids. Angew Chem Int Ed Engl 48(26):4858–4860.

Wilson LM, Mok YF, et al. (2007) A structural core within apolipoprotein C-II amyloid fibrils identified using hydrogen exchange and proteolysis. J Mol Biol 366(5):1639–1651.

Winkelmann J, Calloni G, et al. (2010) Low-level expression of a folding-incompetent protein in *Escherichia coli*: search for the molecular determinants of protein aggregation in vivo. J Mol Biol 398(4):600–613.

Winkler J, Seybert A, et al. (2010) Quantitative and spatio-temporal features of protein aggregation in *Escherichia coli* and consequences on protein quality control and cellular ageing. EMBO J 29(5):910–923.

5

PROTEIN AGGREGATION IN UNICELLULAR EUKARYOTES

MARINA CALDARA, JORIS WINDERICKX, AND VANESSA FRANSSENS

5.1 INTRODUCTION

Protein homeostasis, the proper balance among the synthesis, maturation, and degradation of proteins, is essential for a cell to survive. Disturbance of this balance as a consequence of a variety of factors, such as stress-induced denaturation, destabilizing missense mutations, or a lack of oligomeric assembly partners, results in misfolding of native proteins, which can lead eventually to toxicity and cell death. For that reason, cells have evolved complicated protein quality control (QC) systems. These systems include molecular chaperones, which detect misfolded proteins and facilitate their refolding and repair (Verghese et al., 2012). However, when proteins are beyond repair, they will be degraded by the ubiquitin–proteasome system or the autophagy–lysosome pathway (Tyedmers et al., 2010). Overwhelming or malfunctioning of the QC systems will lead to the formation of misfolded proteins and aggregates, a phenomenon closely linked to several human protein folding diseases. Accumulation of protein oligomers and aggregates, which are difficult to degrade, are leading to a gain in toxicity, causing a variety of diseases, such as diabetes and neurodegenerative disorders, including Parkinson's, Alzheimer's, and Huntington's diseases. Inherited mutations leading to inefficient folding and quick degradation of a protein can also be a cause of loss of function as observed in other diseases, such as cystic

Protein Aggregation in Bacteria: Functional and Structural Properties of Inclusion Bodies in Bacterial Cells, First Edition. Edited by Silvia Maria Doglia and Marina Lotti.

fibrosis, Fabry disease, and various cancers. Moreover, during normal aging the capacity of protein homeostasis declines gradually with time, having an impact on cellular integrity and on physiological responses (Balch et al., 2008).

Among unicellular eukaryotes, QC was, and still is, mostly studied in baker's yeast, *Saccharomyces cerevisiae*. Interestingly, such QC systems are highly conserved and the characterization of the pathways regulating these systems is not only intriguing to boost our fundamental insight, but it may also point to potentially new therapies for the aforementioned protein folding diseases. As such, scientists took advantage of this evolutionary conservation and expressed proteins involved in a variety of diseases in yeast (Gitler, 2008; Winderickx et al., 2008; Braun et al., 2009; Khurana and Lindquist, 2010; Van Rossom et al., 2012). By doing so, researchers have created *humanized yeast models* recapitulating basic key aspects of the disorders, which facilitated and will help us to better understand these complex phenomena.

In this chapter we describe how unicellular eukaryotic cells, especially the budding yeast *S. cerevisiae*, deal with the presence of misfolded protein, beginning with their detection [in the endoplamic reticulum (ER), cytosol, or nucleus], their attempted refolding, and eventually, their degradation (proteasome or autophagy). In addition, we describe examples of proteins that naturally assume amyloid-like structures (i.e., class I hydrophobins EAS, HET-s, Sup35), and provide a brief overview of validated humanized yeast models.

5.2 UPR: UNFOLDED PROTEIN RESPONSE IN THE ER

The endoplasmic reticulum is an organelle that coordinates the folding and maturation process of newly synthesized proteins destined for intracellular organelles, membranous structures, or that need to be secreted (Ellgaard et al., 1999). This organelle has its own QC system, which allows only properly processed proteins to be transported to their final destination. The accumulation of unfolded or misfolded proteins in the ER leads to ER stress and activation of the unfolded protein response (UPR). In the next step, proteins that do not pass ER quality control are directed to the cytoplasm, where they are marked for degradation, a process called ER-associated degradation (ERAD) (Stolz and Wolf, 2010).

One of the main actors in the UPR is the ER-localized type I transmembrane protein Ire1, or IreA for filamentous fungi such as *Aspergillus fumigatus*. Ire1 protein is the only ER stress sensor in fungi, but its function is highly conserved among eukaryotes (Cox et al., 1993; Tirasophon et al., 1998).

In metazoans two additional ER stress sensors exist: the protein kinase RNA (PKR)-like ER kinase (PERK) and the activating transcription factor-6 (ATF6). The protein PERK, which resembles Ire1, possesses kinase activity able to phosphorylate the eukaryotic initiation factor 2α (eIF2α) at Ser51, a modification that causes an attenuation of protein synthesis. ATF6 is a type II transmembrane protein that usually resides in the ER. Upon detection of ER stress, ATF6 is transported from the ER to the Golgi. In this compartment ATF6 is subject to digestion by the proteases S1P and S2P. The resulting subunits are then transported to the nucleus, where they will induce expression of chaperones and ERAD-related genes (Kohno, 2010; Walter and Ron, 2011).

At protein homeostasis the N-terminal region of Ire1 interacts with Kar2, the most abundant BiP chaperone in the yeast ER (Normington et al., 1989; Rose et al., 1989; Kohno, 2010). Under normal conditions, Kar2 is also involved in preventing misfolding of secretory and transmembrane protein precursors. However, the recognition of exposed basic and hydrophobic residues of a not properly folded protein by Ire1 (Gardner and Walter, 2011) disturbs this equilibrium and leads to Kar2 dissociation (Kostova and Wolf, 2003). This phenomenon results in subsequent changes to the quaternary structure of the Ire1 protein, which first dimerizes and then further oligomerizes (Lee et al., 2008; Korennykh et al., 2009). This is a key event for full activation of the UPR response (Aragon et al., 2009; Korennykh et al., 2009). Next, Ire1 becomes trans-autophosphorylated (Shamu and Walter, 1996; Welihinda and Kaufman, 1996), with the consequent activation of its ribonuclease activity at the C-terminal region. Interestingly, the presence of the kinase activity of Ire1 or regulation by BiP is not necessary for UPR activation (Kimata et al., 2004; Pincus et al., 2010). Nevertheless, kinase activity seems very important for switching off the UPR signal once ER stress is resolved, since failing to do so diminishes cell survival (Chawla et al., 2011; Rubio et al., 2011). Indeed, ER proteins much as Bxi1 link the UPR to programmed cell death (Cebulski et al., 2011).

The RNAse activity of Ire1 induces the unconventional splicing of the mRNA encoding for Hac1, at the cytosolic side of the ER. During this splicing an intron of 252 nucleotides within *HAC1* is cleaved, and the two remaining fragments are ligated by the yeast tRNA Rlg1p ligase (Sidrauski et al., 1996; Sidrauski and Walter, 1997) to form the messenger coding for a basic domain/leucin zipper (bZIP) transcription factor. A recent study demonstrated that Hac1 induces 7 to 8% of the genes encoded by the yeast genome (Li et al., 2010), including genes coding for chaperones, oxido-reductase, ER-associated degradation (ERAD) machinery, and lipid biosynthesis (Mori et al., 1992). Overall, with this change in expression profile, cells attempt to restore homeostasis by increasing the protein folding capacity of the ER as well as the ER abundance itself, a strategy that is preferred over activation of the degradation pathway (Weibezahn et al., 2004). Interestingly, the size of the intron that is spliced from the *HACA* mRNA in *A. niger*, *A. nidulans*, and *T. reesei* is only 20 bp long, and in *C. albicans*, only 19 bp. In these organisms, full activation of UPR depends on the unconventional splicing of *HACA* and on the truncation of the 5′ UTR of its mRNA. The processed form of *hacA* leads to upregulation of UPR and of the *hacA* gene itself (Saloheimo et al., 2003; Mulder et al., 2004; Mulder and Nikolaev, 2009). Recently, a very divergent bZIP transcription factor named Hxl1 has been described in the human fungal pathogen *Cryptococcus neoformans*. Here, unconventional splicing of HXL1 is also mediated by Ire1, and the production of active Hxl1 is also associated with thermotolerance, cell wall integrity, and virulence. The latter characteristic makes Hxl, and generally the UPR response, possible new targets for the development of novel antifungal targets (Cheon et al., 2011).

As reported previously, activation of UPR increases QC and protein-folding capacity. This phenomenon has been considered to be an advantage for the heterologous expression of proteins in the yeast *Pichia pastoris*. Indeed, in this organism the UPR seems to be constitutively active, and furthermore, overexpression of Hac1 can still increase the amount of proteins secreted (Guerfal et al., 2010). Similar results were also reported for *S. cerevisiae*, *T. reesei, A. fumigates*, and *A. niger* var. *awamori* (Valkonen et al., 2003).

5.3 REMOVING PERSISTENT MISFOLDED PROTEINS WITH THE PROTEASOME

Proteins that fail to fold properly are targeted by compartment-specific ubiquitin ligases to be marked for degradation by the proteasome complex (Tsai et al., 2002; Kostova and Wolf, 2003). It has been estimated that up to 30% of newly synthesized proteins are degraded by the proteasome (Schubert et al., 2000). Misfolded proteins present in the ER, even after induction of the UPR response, need to be retro-translocated across the ER membrane (Figure 5.1). This event is mediated by the translocon, a system able to create a channel within the membrane of the ER. This channel is used both to translocate secretory and membrane proteins to the ER for folding and to retro-translocate misfolded proteins back to the cytosol. The translocon is composed of Sec61, the largest and major subunit of the complex, Sss1, a protein that stabilizes the structure, and Sbh1 (Esnault et al., 1994;

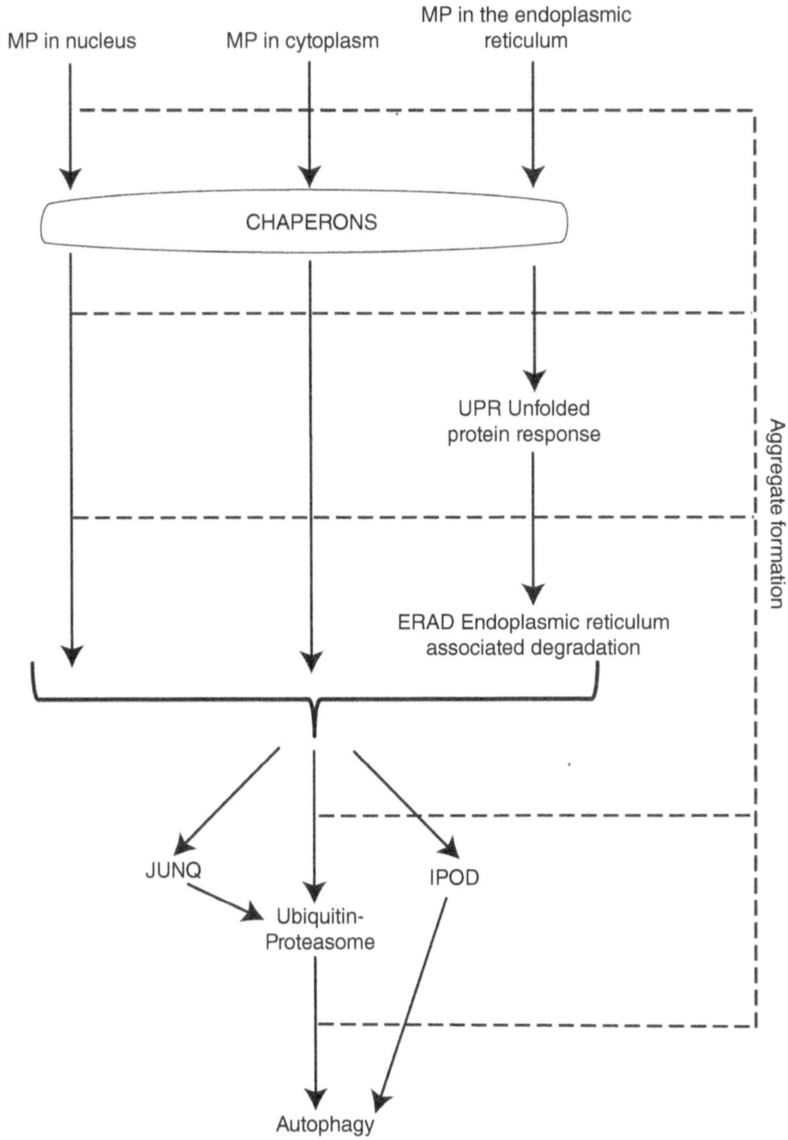

FIGURE 5.1 Schematic view of the journey of misfolded proteins (MPs). MPs can occur in the nucleus, in the cytoplasm, or in the endoplasmic reticulum (ER). In all cases the first line of defense is represented by chaperone proteins, which try to refold proteins to their native state. Failing in doing so in the ER activates the unfolded protein response (UPR) and ER-associated degradation. In a subsequent step, misfolded proteins can be degraded by the ubiquitin–proteasome system or by the autophagic pathway. Denatured proteins can also accumulate in the juxta nuclear quality control compartment (JUNQ) while waiting to be degraded by the ubiquitin–proteasome pathway, or in insoluble protein deposits (IPODs) before being digested by the autophagic machinery. At any step, misfolded proteins can accumulate and generate aggregates potentially toxic to the cell.

Wilkinson et al., 1997; Feng et al., 2007). Once in the cytosol and as soon as the first lysine residue becomes available, the polypeptides are quickly ubiquitinated, a modification that prevents the polypeptide from sliding back into the ER.

Targeting of misfolded protein by ubiquitin occurs in several steps. First, ubiquitin is activated thanks to the attachment of its carboxy-terminal carboxyl group to a cystein residue within the activating E1 enzyme, an event that requires hydrolysis of ATP. In a second step, activated ubiquitin is transferred to the ubiquitin conjugating enzyme E2. In yeast this activity is carried out by a protein complex that is composed of Ubc7, Cue1, Ubc1, and Ubc6 (Tsai et al., 2002). The targets of this complex are misfolded secretory proteins. Misfolded cytoplasmatic proteins are targeted by the E2 enzymes Ubc4 and Ubc5 (Park et al., 2007). These components are then anchored to the E3-ubiquitin ligases via Ubc7 or Ubc1 (Bays et al., 2001; Deak and Wolf, 2001). E3 ligases are involved in transferring ubiquitin to the ε-amino group of a lysine. In yeast the E3 enzymes Hrd1 and Rma1 add ubiquitin to misfolded luminal and membrane proteins localized on the ER membrane (Vembar and Brodsky, 2008). Hrd3, the Hsp70 chaperon Kar2, and the lectin-like Yos9 are all partners of Hrd1 for ER substrate recognition. Hrd3 has a stabilizing role and the link between Hrd1 and Hrd3 is further fortified by Usa1 (Carvalho et al., 2006). Furthermore, this complex interacts with the multipass transmembrane protein Der1 (Knop et al., 1996). Yos9p assists in recognizing specific N-linked glycan moieties in the misfolded targets (Gauss et al., 2006), and it has been reported that Hrd1/Kar2 could help recognition of hydrophobic residues, which would naturally be buried in a properly folded protein (Carvalho et al., 2006). Doa10, Ubr1, Cul5, Hul5, Ltn1, and Ufd4 add ubiquitin to cytosolic polypeptides (Eisele and Wolf, 2008; Ehrlich et al., 2009; Bengtson and Joazeiro, 2010; Fang et al., 2011; Heck et al., 2010), and San1 ubiquitinates abnormal nuclear proteins (Carvalho et al., 2006; Deng and Hochstrasser, 2006). Next, the multiubiquitin chain assembly factors E4, (i.e., Hul5 and Ufd2) elongate the polyubiquitin chain on the target substrate (Koegl et al., 1999; Aviram and Kornitzer, 2010), an activity that seems to facilitate ERAD activity in some cases (Vembar and Brodsky, 2008). This process is opposed by the ubiquitin protease Ubp6, located at the base of the 26S proteasome (Crosas et al., 2006). In the ER, a final extraction step is needed to pull the protein from the ER to the cytosol. This step is facilitated by the Cdc48–Ufd1–Npl4 complex (Jarosch et al., 2002). Cdc48 belongs to the AAA family of ATPase. This exists as two homo-hexamer rings stacked on each other (Ye et al., 2001). Upon ATP hydrolysis, this protein undergoes robust conformational changes, a phenomenon connected with the progressive displacement and pulling of the substrate from the ER into the cytosol (Rosenbaum et al., 2011). Recently, it has been found that the ubiquitin recognition protein Ufd1 links ER stress to cell cycle. Indeed, in the case of prolonged stress, a delay to enter the G1 phase is mediated by Cdh1–Skp2–p27 thanks to regulation through Ufd1 and by the deubiquitinating enzyme USP13 (Chen et al., 2011).

In a final step, ubiquitinated proteins are degraded by the 26S proteasome (Heinemeyer et al., 1991; Baumeister et al., 1998). The proteasome has a cylindrical arrangement composed of two distinct structures: the regulatory 19S, which recognizes, unfolds, and translocates peptides, and the catalytic 20S particles. The 20S is made up of four stacked rings, each containing seven different α- or β-subunits (Groll et al., 1997; Baumeister et al., 1998). The α-subunit regulates the entry of the substrate, while the active sites are located within the β-subunit. The 20S subunits Pre3, Pup1, and Pre2 display peptidyl glutamyl peptide amino acids hydrolyzing activity and trypsin-like and chymotrypsin-like activities (Chen and Hochstrasser, 1996; Arendt and Hochstrasser, 1997; Heinemeyer et al., 1997), which degrade proteins into fragments of 3 to 25 amino acids (Dick et al., 1998; Nussbaum et al., 1998). Correct assembly of the 20S subunit is crucial for the cell and, so far, over 10 factors supporting this event have been identified (Chen and Hochstrasser, 1996; Le Tallec et al., 2007; Kusmierczyk et al., 2008).

The regulatory 19S subunit is composed of a base and a lid. The base consists of a ring of ATPases (Rpt1 to 6) interacting with the α-rings of the 20S subunit and of three non-ATPase proteins (Rpn1, Rpn2, and Rpn10) (Kostova and Wolf, 2003). ATP-dependent interaction promotes opening of the pore and provides access for the substrates. The lid of the 19S subunit is composed of eight subunits (Rpn3p, Rpn5 to Rpn9, Rpn11, Rpn12) whose essential functions are mostly unidentified, but it is generally known that they are able to regulate attachment and removal of ubiquitinated proteins (Kohler et al., 2001; Elsasser et al., 2002; Lam et al., 2002; Demartino and Gillette, 2007). Among those, Rpn11 is the best characterized. This protein is able to remove ubiquitin upon threading of the substrate into the 20S channel, an event that irreversibly commits polypeptides to proteolysis. At the end, polyubiquitin chains are also broken down, an event that releases monomers, which are then reused in a new round of protein degradation. The transcription regulator Rpn4 is responsible for the induction of proteasome genes upon sensing of acute stress or misfolded proteins by the cell. Rpn4 interacts with the proteasome subunit Rpn2 and has a short half-life (around 2 min) (Xie and Varshavsky, 2001; Ju and Xie, 2004). Indeed, an optimal homeostasis requires Rpn4 to be degraded by an ubiquitin-dependent pathway that involves the ubiquitin ligase Ubr2, and by a ubiquitin-independent pathway that has yet to be characterized (Xie and Varshavsky, 2001; Wang et al., 2004). Rpn4 was also found to be necessary for tolerating misfolded membrane proteins (Metzger and Michaelis, 2009).

5.4 LYSOSOMAL/VACUOLAR PROTEOLYSIS (OVERLOAD UPS)

5.4.1 Autophagy

Besides degradation by the proteasome, misfolded and aggregated proteins can also be degraded by autophagy. It has been suggested that the proteasome and the autophagic pathways are interconnected and that the latter plays compensatory roles when the activity of the first is compromised or exceeded (Kruse et al., 2006; Pandey et al., 2007). Furthermore, it has been proposed that Ire1 and Hac1 from the UPR pathway are also required for induction of autophagy, but how this regulation works is still a matter of debate (Bernales et al., 2006). Degradation of misfolded proteins by the vacuole is also present in filamentous fungi (Kimura et al., 2011). During autophagy, cytosolic vesicles with a double membrane, called *autophagosomes*, trap parts of the cytoplasm while they are being formed. At the moment, the source of this membrane is still unclear, but contributions seem to come from the ER membrane, the plasma membrane, the mitochondrial outer membrane, and the Golgi complex (Tooze and Yoshimori, 2010). Autophagy can also be substrate-specific, since the presence of particular cargos can direct selective digestion of cellular components, as discussed below. In the next step, the mature autophagosome moves toward the vacuole. The outer membrane of the autophagosome merges with the membrane of the vacuole, thereby releasing the internal membrane into the lumen of the vacuole, where hydrolases will digest both the membrane and the content of the vesicle (Baba et al., 1994). The catabolic products from this digestive process are then translocated to the cytoplasm, where they will be reused for *de novo* synthesis of macromolecules.

Autophagy has been studied extensively in *S. cerevisiae*, as it plays a crucial role in regulating cell survival under nutrient starvation conditions (Mitchener et al., 1976; Takeshige et al., 1992). In *S. cerevisiae*, many "Atg proteins" are involved in this process, most of which have homologs in other eukaryotes. Under nutrient starvation, low phosphorylation of the Atg13, a key component of the Atg13/Atg1/Atg17 complex, is a prerequisite for autophagy initiation (Cebollero and Reggiori, 2009). This phosphorylation is regulated by the serine/threonine kinase complex TORC1 (Diaz-Troya et al., 2008). In the next step, a pre-autophagosomal structure (PAS), or phagophore, needs to be assembled at the perivacuolar site. The class III phosphatidylinositol-3-kinase complex (PI-3 K) is involved in this step. This complex consists of the Ser/Thr protein kinase Vps15, Vps34, Atg6, and Atg14 (Kihara et al., 2001). In a following step, the mature PI-3 K complex also recruits Atg18, Atg21, Atg20, Atg24, and Atg27 (Nice et al., 2002;

Wurmser and Emr, 2002; Stromhaug et al., 2004). Next, two interrelated ubiquitin-like conjugating systems, Atg12/Atg5/Atg16 and Atg8, are recruited. During vesicle formation, the protease Atg4 processes the ubiquitin-like protein Atg8, which becomes covalently attached to phosphatidylethanolamine (PE) of PAS through an amide bond. This step requires the action of the E1 enzyme Atg7 and of the E2 enzyme Atg3. In parallel, binding of the second ubiquitin-like protein Atg12 to Atg5 needs to occur. This step is promoted by Atg7 and the E2 enzyme Atg10. Finally, for proper maturation of the autophagosome, the complex Atg5/Atg12 needs to recruit Atg16. At present, the source of the autophagosomal membrane is still uncertain, but it has been suggested that Atg9, a protein localized at PAS and at peripheral structures, might act as a membrane carrier (He et al., 2008). In, anycase, once the phagophore is mature, it fuses to the vacuole, thanks to homotypic vacuole membrane fusion. For this fusion several proteins are necessary: the Rab family GTPase Ypt7, Sec18; the SNARE proteins Vam3, Vam7, Vti1, Ykt6; and the class C/HOPS proteins, Ccz1 and Mon1 (Wang and Klionsky, 2003; Wang et al., 2003; Klionsky, 2005). Here, digestion of the phagophore content is carried out by the hydrolases PrA, PrB, CPS, CPY and the lipase Atg15 (Epple et al., 2001, 2003; Kiel, 2010). Finally, the efflux of the molecules is dependent on the putative vacuole permease Atg22 and the vacuolar permeases Avt3 and Avt4 (Russnak et al., 2001; Yang et al., 2006).

Although autophagy is considered to promote survival, its continuous activation due to the presence of cellular aggregates can also be deleterious to a cell (Cherra and Chu, 2008). For example, heterologous expression of α-synuclein, an aggregation-prone protein involved in Parkinson's disease, is in part toxic to *S. cerevisiae*, due to constant activation of autophagy, as demonstrated by a reduction of yeast longevity and participation of autophagy to cell death (Sampaio-Marques et al., 2011).

5.4.2 Selective Types of Autophagy

The Cvt Pathway The cytoplasm-to-vacuole (Cvt) pathway is the best-characterized selective type of autophagy. This pathway delivers precursors of the aminopeptidase I Ape1 and α-mannosidase Ams1 to the vacuole, where they will be activated (Teter and Klionsky, 2000). Cvt requires the presence of Atg19, Atg11, and actin (Monastyrska et al., 2008). To form the Cvt complex, binding of Atg19 to Ape1 and Ams1 is necessary. In a second stage, the adaptor Atg11 links this complex to PAS and the general autophagic machinery. This step requires the actin cytoskeleton and the VFT complex, formed by Vps51, Vps52, Vps53, and Vps54. VFT, in conjunction with Vps45, Tlg1, and Tlg2, facilitates the recruitment of Atg9, a protein involved in delivering membrane to the expanding vesicle (Stromhaug et al., 2004; Lynch-Day and Klionsky, 2010).

ER-phagy As noted previously, the presence of misfolded proteins induces the UPR system and ERAD. The first cellular response to this phenomenon is the increase of the protein-folding capacity of the ER, which is followed by an expansion of the ER itself. Once the UPR system is no longer required, the volume of the ER needs to be brought back to its original size and therefore ER-phagy or reticulophagy will take place (Bernales et al., 2006). This pathway requires Atg8, Atg5, Atg1, and Atg12, and possibly Atg19 and Atg20 (Bernales et al., 2006; Ogata et al., 2006; Yorimitsu et al., 2006). ER degradation can also be induced under starvation conditions, a route known to depend on Atg16 but not on Atg5 and Atg12 (Hamasaki et al., 2005; Mijaljica et al., 2006).

Mitophagy The quality of mitochondria, the cellular energy supplier, needs to be regulated carefully to avoid excessive accumulation of toxic reactive oxygen species. Selective autophagy regulates the vacuolar degradation of damaged mitochondria (Bhatia-Kissova and Camougrand, 2010). In order for this to happen, the mitochondrial receptors Atg32 and Atg33 need to bind to the adapter Atg11, which will then bring the organelles to the PAS (Kanki and Klionsky, 2010). Two more proteins have been identified as possibly important for the process: Aup1 and Uth1. Aup1 is a phosphatase localized in the inner mitochondrial membrane space, where it has an important function for entrance in the stationary phase. Uth1 is an outer membrane protein required during starvation (Kraft et al., 2009).

Pexophagy Peroxysomes are involved in lipid metabolism and the elimination of peroxides. Their concentration is carefully balanced within the cell. To eliminate an excess of peroxysomes, a double membrane will first engulf the organelle, which will then be transported to the vacuole. This requires the pexophagy-specific proteins Pex14 and Pex3, the general adaptor Atg11, and possibly Atg30 (Kraft et al., 2009; Kiel, 2010).

Piecemeal Microautophagy of the Nucleus During starvation, part of the nucleus can also be subject to degradation. When this happens, nuclear–vacuolar junctions are formed, a process that requires the outer nuclear proteins Nvj1, Osh1, and Tsc13; the vacuolar membrane proteins Vac8, and Atg11; and the core Atg machinery (Roberts et al., 2003; Kiel, 2010).

Ribophagy Ribosome degradation also occurs and requires the Atg machinery as well as specific proteins such as the Ub protease Ubp3 and its cofactor Bre5, suggesting that the process involves ubiquitin (Baxter et al., 2005; Kiel, 2010).

5.5 REFOLDING OF PROTEIN AGGREGATES IN CYTOSOL AND NUCLEUS

Misfolded proteins and aggregates can form in the nucleus or within the cytosol. Under these circumstances, the aggregates interact with a cooperative system composed of Ssa1 (Hsp70), the J protein co-chaperon Ydi1 (Hsp40), and the oligomeric chaperone Hsp104 (Glover and Lindquist, 1998). Binding of Hsp40 to Hsp70 stabilizes and stimulates Hsp70 activity. Formation of this complex limits access of proteases to the aggregates and allows their transfer to the processing pore of Hsp104 (Haslberger et al., 2007, 2008). Substrate binding to Hsp70 is regulated by the rate of ATP/ADP exchange, a phenomenon stimulated by the nucleotide exchange factors Fes1 and Snl1 (Kabani et al., 2002; Sondermann et al., 2002). The active form of Hsp104 is a ring-shaped hexamer and, as a member of the Hsp100, it contains two AAA+ modules per subunit. It has been proposed that Hsp104 can employ a threading activity mediated by the aromatic residues located within the pore. Thanks to the energy given by ATP hydrolysis, these residues pull the substrate into the channel, a phenomenon that leads to opening of aggregates and proper folding of the peptides (Weibezahn et al., 2004). Once tightly folded domains are being detected, Hsp104 dissociates from the substrate. Furthermore, Hsp104 was also found to be important in the segregation of protein aggregates during cell division, a topic discussed later in the chapter.

5.6 JUNQ AND IPOD

As explained before, when proteins are beyond repair, they will either be ubiquitinated and degraded by the proteasome, or they can form protein aggregates (Meriin and Sherman, 2005; Ding and Yin, 2008). Recent research has revealed that protein aggregates are not random deposits of conglomerated material, but that they are formed through an active and regulated process that involves transport of small deposits along components of the cytoskeleton. In this respect it has been observed that small inclusions are transported along microtubules, forming larger protein aggregates at the centrosome called *aggresomes* (Kopito, 2000). These are not inert waste disposal sites, but they recruit chaperones and proteasomes to assist in the removal of the misfolded and damaged proteins.

Moreover, two subcellular compartments containing misfolded protein aggregates were recently described (Kaganovich et al., 2008). On the one hand, perivacuolar protein inclusions can act as insoluble protein deposits (IPOD), where damaged protein can be stored awaiting clearance by autophagy. On the other hand, protein inclusions exist in close proximity to the nucleus, where aggregated proteins can still be exchanged with the cytosolic pool. This compartment, called a juxta nuclear quality control compartment (JUNQ), functions as a quality control center that attracts chaperones and proteasomes in order to either refold or degrade misfolded proteins. Partitioning of misfolded proteins between these two compartments seems to depend on their ubiquitination and aggregation state. The Hsp104 chaperon, which normally interacts with misfolded and aggregated proteins, was localized at both compartments. Here this protein could have a role in supporting protein refolding or degradation. Interestingly, disease-like amyloidogenic proteins were found to accumulate primarily within the IPOD, and not in a JUNQ.

5.7 SEGREGATION OF AGGREGATES IN YEAST

Saccharomyces cerevisiae can replicate only a finite number of times, and during this process it undergoes an asymmetrical division. This means that a large mother cell generates smaller daughter cells. This process is accompanied by the uneven distribution of possibly toxic materials. It has been observed that oxidatively damaged proteins accumulate during cell growth, but they are largely retained in the mother cell during cytokinesis, releasing a rejuvenated daughter cell with full replicative potential (Aguilaniu et al., 2003). Furthermore, in aged cells or in cells exposed to oxidative stress or heat shock, damaged proteins form aggregates containing the protein-aggregate remodeling factor Hsp104 (Erjavec et al., 2007). Such Hsp104-containing protein aggregates are also retained selectively in the mother cells during cell division. This process is dependent on a functional actin cytoskeleton and the sirtuin Sir2. A recent study by the group of Nyström identified the polarisome as a crucial determinant for this asymmetric segregation (Liu et al., 2010, 2011). The polarisome is a protein complex localized at the bud tip that mediates the localized assembly of actin monomers into filaments that extend into the mother cell. In this way, continued actin polymerization creates a conveyer belt that transports Hsp104-containing protein aggregates from the growing bud into the mother cell, where they usually end up in one big aggregate. Preliminary evidence links Sir2 to this mechanism via CCT, a chaperonin that is required for folding actin to its native conformation (Brackley and Grantham, 2009). In *sir2Δ* cells, CCT was hyperacetylated and displayed lowered activity, which was also reflected in a reduced level of native actin (Liu et al., 2010). It is thus tempting to speculate that Sir2, which is known to deacetylate histones, can also deacetylate CCT, in this way activating this chaperonin and promoting actin folding. Besides the involvement of Sir2 in the segregation of native Hsp104-containing aggregates, it was also shown that segregation of the aggregation-prone disease proteins Htt103Q and synphilin-1 relies on a sirtuin-dependent process (Büttner et al., 2010; Liu et al., 2010, 2011). Moreover, synphilin-1 inclusions localized to actin cables, suggesting that smaller synphilin-1 aggregates are transported along actin cables, to form one or two larger aggregates (Figure 5.2). Altogether, these findings provide evidence that the compartimentalization and segregation of damaged proteins is a factor-dependent process (Liu et al., 2011).

The phenomenon of asymmetrical division of protein aggregates has also been observed in another microorganism, the bacterium *Escherichia coli*. Cells of an *E. coli* population were shown to be present in two distinct populations: a damage-free population, including cells that remain reproductively competent during the stationary phase, and a damage-enriched population, which loses its life-supporting mechanisms and eventually dies (Desnues et al., 2003). Another study reported that upon division of this organism, aggregated proteins accumulate in cells with the older poles (Lindner et al., 2008). However, the mechanism for this asymmetrical partitioning is still unclear.

5.8 PROTEINS FORMING NONPATHOLOGICAL AMYLOID-LIKE FIBRILS IN UNICELLULAR EUKARYOTES

Interestingly, not all protein aggregates are toxic and deleterious to cells. Indeed, in nature there are some examples of proteins having functions in both their soluble and amyloid-like forms. The fungus *Neurospora crassa* produces, for example, class I hydrophobins EAS. These are small secreted proteins able to self-assemble into strong amphipathic films (Linder et al., 2005). These proteins can reverse the wettability of the surface on which they are deposited, a phenomenon that promotes the formation of aerial structures such as hyphae and spores. Similar proteins are also produced by the fungi *Magnaporthe grisea* and *Asperigillus fumigatus* (Talbot et al., 1996; Aimanianda et al., 2009). In addition, prions, infectious proteins propagating information implanted in their amyloid-like conformation, have been found in at least two fungi: *Podospora anserina* and *S. cerevisiae*. In *P. anserine*, the HET-s prion is involved in heterokaryon incompatibility (Coustou et al., 1997; Wickner, 1997; Glass and Dementhon, 2006). When cells from different individuals fuse, a non-self-recognition process under the control of the

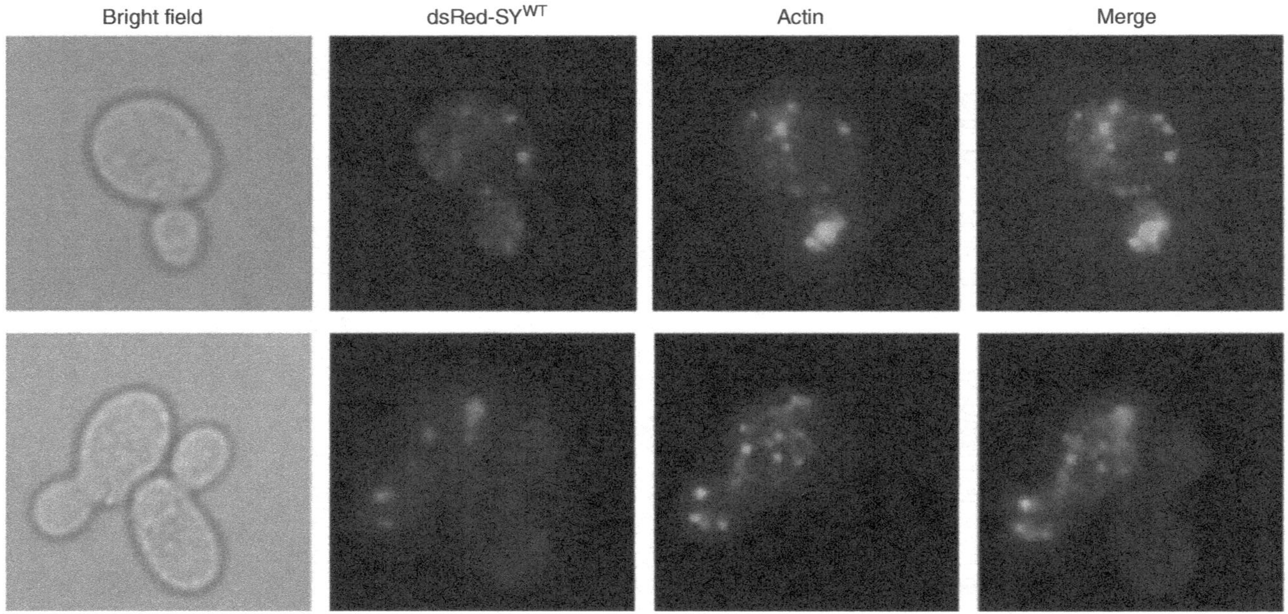

FIGURE 5.2 Transport of synphilin-1 inclusions along actin cables. Fluorescence microscopy images of late exponential yeast cells expressing fluorescently tagged synphilin-1 (dsRed-SY^{WT}) stained with Alexa Fluor 488-phalloidin to visualize actin patches and actin fibers. (*See insert for color representation of the figure.*)

het gene is activated. This process leads to cell death of the mixed population. This system has also been described for the fungus *Fusarium graminearum* (Wasmer et al., 2010). Finally, in *S. cerevisae* at least 10 prions have been identified, among which [PSI$^+$] and [URE3] are the best characterized. [PSI$^+$] is the prion form of Sup35. Sequestration of Sup35 by aggregation allows read-through of premature stop codons (Wickner et al., 1995). [URE3] is the prion form of Ure2, a protein regulating nitrogen metabolism (Wickner, 1994). Aggregation promotes use of poor nitrogen sources even if good sources such as ammonia are still present (Edskes, 2001). Among the other prions identified, [SWI$^+$], [OCT$^+$], and [MOT3] regulate transcription, while [ISP$^+$] and [NSI$^+$] can influence the fidelity of translation. [MCA] aggregation results in a read-through nonsense codon and the possibility of growing on adenine-deficient media. The [GAR$^+$] phenotype supports growth in glycerol in the presence of the nonmetabolizable glucose analog glucosamine (Crow and Li, 2011). A function for [PIN$^+$] has yet to be assigned. Whether the presence of prions in yeast is advantageous or deleterious is still a matter of debate (True and Lindquist, 2000; True et al., 2004; Nakayashiki et al., 2005; Wickner et al., 2007). It is known that these structures can promote evolvability and growth in unusual media or under stress conditions (Tyedmers et al., 2008; Lancaster et al., 2010).

5.9 HUMANIZED YEAST MODELS

S. cerevisiae has been used extensively as a model system over the past decades because of its many advantages. First, it has a short life cycle and can be grown at low cost. It is relatively easy to be genetically manipulated, its genome is completely defined, and strain collections of genome-wide deletions or overexpressions are available. Moreover, many signaling pathways are conserved between yeast and mammalian cells. This also allowed the use of budding yeast to deepen the study of pathological mammalian proteins, whose misfolding cause severe cytotoxic effects. These analyses helped us advance our understanding of protein quality control systems. A number of human diseases, collectively called *protein-misfolding diseases*, are characterized by the chronic abundance of aggregation-prone misfolded proteins. Although the proteins that cause these disorders often have unrelated sequences and functions, they all aggregate in intracellular inclusions that are connected to the disease pathology. The deleterious effects of the misfolded proteins in these diseases can either be the result of loss of protein function or the gain of toxicity of the aggregates.

Examples of protein-misfolding disorders with loss of protein function are cystic fibrosis and retinitis pigmentosa. The cystic fibrosis transmembrane receptor CFTR is an ion channel on the human epithelial cell membrane. Expression of this protein in yeast helped clarifying the connection between degradation by the proteasome and/or by autophagy. The integrity of CFTR is usually maintained by the chaperon Hsp90 (Youker et al., 2004). Only 20% of the CFTR expressed is not degraded immediately. Indeed, cells expressing this receptor accumulate the protein first in the perinuclear and subcortical ER, where it is degraded by the proteasome. Next, the protein accumulates in ER-associated complexes, where it forms aggregates that can be removed by autophagy (Fu and Sztul, 2009). The common deletion ΔF508–CFTR causes the protein to be misfolded and degraded prematurely. When this mutant is expressed in yeast it is degraded completely by the ERAD pathway. For this degradation, the lectin Htm1, the proteins Hrd1 and Doa1, and the complex Cdc48–Ufd1–Npl4 are necessary (Gnann et al., 2004).

The T8993C point mutation in the ATP synthase subunit 6 gene, which encodes a mitochondrial enzyme complex V protein, leads to neurogenic muscle weakness, ataxia, retinitis pigmentosa (NARP), and/or maternally inherited Leigh syndrome. These diseases are caused by a severe impairment of the ATP synthase, but the molecular mechanism leading to this phenotype is still not clear. A study by Morava et al. (2006) proposed that the decrease in activity could be caused by inefficient assembly of the ATP synthase 6 subunit due to instability of the mutated protein. Studies in yeast have been useful to better understand the consequences of this mutation, although results on the assembly efficiency of this variant gave discordant results (Rak et al., 2007; Kucharczyk et al., 2009). Nevertheless, these models have been useful for the identification of chlorhexidine and oleate as compounds able to rescue phenotypes derived from mutations of ATP synthase in yeast as well as in cell lines derived from NARP patients (Couplan et al., 2011).

Protein-misfolding disorders also include several neurodegenerative diseases such as Huntington's, Parkinson's, and Alzheimer's diseases. These diseases are typical examples of how aggregation-prone proteins can cause neurodegeneration by a gain of toxic function. Huntington's disease (HD) is a neurodegenerative disorder that is caused by an abnormal expansion of polyglutamine (polyQ) tracts in the huntingtin (Htt) protein. Most yeast models developed to study the folding and behavior of proteins with an expanded polyQ domain express the N-terminus of Htt, including the polyQ tract (Duennwald, 2011). Several of these yeast models demonstrate that the heterologous expression of Htt results in a polyQ-length-dependent aggregation and toxicity (Krobitsch and Lindquist, 2000; Hughes et al., 2001; Meriin et al., 2002; Sokolov et al., 2006). Moreover, the sequences adjacent to the polyQ tract strongly influence the Htt toxicity (Duennwald et al., 2006). Studying Htt aggregation and toxicity in these models revealed protective functions of chaperones such as Hsp104 and family members of Hsp70 and

Hsp40 (Krobitsch and Lindquist, 2000; Muchowski et al., 2000; Gokhale et al., 2005). In addition to chaperones, aggregation and toxicity of the Htt protein was also connected with the microtubule cytoskeleton (Muchowski et al., 2002), apoptotic cell death, and apoptosis (Meriin et al., 2003; Sokolov et al., 2006; Solans et al., 2006). Furthermore, genetic screenings, as well as the use of these HD yeast models as cell-based high-throughput screening systems, have led to the identification of several modifiers of Htt aggregation and toxicity (Willingham et al., 2003; Giorgini et al., 2005; Zhang et al., 2005).

α-Synuclein is one of the major factors causally linked to the aetiology of Parkinson's disease (PD). Several groups have expressed human α-synuclein in yeast and found that these yeast PD models recapitulate robustly important aspects of synucleinopathy. Heterologous expression of both wild-type α-synuclein and the clinical mutant A53T in yeast results in a dose-dependent growth inhibition and formation of intracellular inclusions (Outeiro and Lindquist, 2003; Zabrocki et al., 2005, 2008). This is accompanied by interference with several events, including vesicular trafficking, endocytosis, vacuolar degradation, mitochondrial dysfunction, apoptosis, and autophagy (Flower et al., 2005, 2007; Zabrocki et al., 2005, 2008; Büttner et al., 2008; Gitler et al., 2008; Soper et al., 2008, 2011; Su et al., 2010; Sampaio-Marques et al., 2012). Furthermore, these yeast PD models have proven to be ideal tools to screen for genes and compounds that modulate α-synuclein-induced toxicity (Griffioen et al., 2006; Zabrocki et al., 2008; Kritzer et al., 2009; Su et al., 2010; Tardiff et al., 2012). In a more recent study, another yeast PD model was developed to investigate the importance of the α-synuclein interaction partner synphilin-1 (Figure 5.2) for toxicity associated with the formation of aggregates (Büttner et al., 2010). In this model, synphilin-1 was expressed alone or together with α-synuclein in yeast cells. Monitoring the exponential growth of these cells revealed that synphilin-1 induced only a minor growth defect. However, synphilin-1 did obtain a toxic character in stationary and aging yeast cells. Interestingly, this toxicity was absent in cells lacking the sirtuin Sir2, indicating an important role for Sir2 in synphilin-1-mediated toxicity. In the majority of the cells, synphilin-1 was dispersed throughout the cytoplasm. However, in approximately one-third of the cells, synphilin-1 formed small aggregates, which transformed into one or a few larger inclusions upon entering the stationary phase (Figure 5.3). Although the mechanism responsible for the formation of these inclusions is still unclear, a role for the cytoskeleton and probably also Sir2 in this process has been suggested (Swinnen et al., 2011). Since the α-synuclein and synphilin-1 yeast models faithfully reproduce one of the essential hallmarks of PD development (i.e., the aggregation of these human proteins into inclusions), these models are valuable tools to decipher the molecular processes involved in the formation of inclusions.

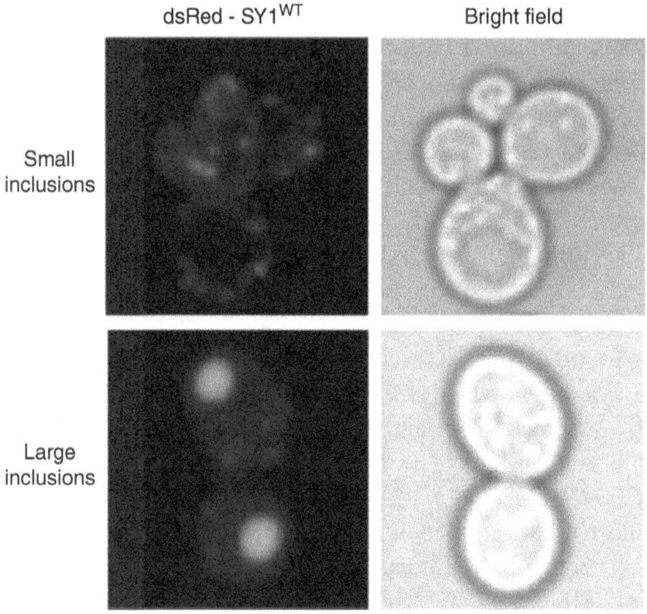

FIGURE 5.3 Aggregation of synphilin-1 in a humanized yeast model for Parkinson's disease. Fluorescence microscopy images of yeast cells expressing fluorescently tagged synphilin-1 (dsRed-SY1^{WT}).

Alzheimer's disease (AD) is characterized by the presence of extra neuronal plaques composed primarily of aggregated Aβ peptide as well as intracellular paired helical filaments and neurofibrillary tangles composed of hyperphosphorylated protein tau. Various yeast models have been developed to study AD pathology. Initial studies in yeast focused on the role of the various secretases in the processing and generation of Aβ peptides from the amyloid precursor protein (APP). These studies demonstrated that expression of human APP in yeast resulted in its cleavage at the C-terminus, indicating that yeast possesses α-secretase activity similar to that of multicellular organisms (Zhang et al., 1994, 1997). Since yeast lacks endogenous β- and γ-secretase activity, strains were developed that express the human β-secretase and reconstitute the γ-secretase complex (Edbauer et al., 2003; Luthi et al., 2003; Middendorp et al., 2004). These models were of great importance to learn more about human APP processing. Moreover, they were used to search for β-secretase inhibitors and to investigate the individual function of each component of the γ-secretase complex. Furthermore, these yeast systems also appeared to be useful tools to study Aβ oligomerization and the molecular pathways that underlie Aβ aggregation as well as to screen for compounds that modulate the aggregation process (Hughes et al., 1996; Bagriantsev and

Liebman, 2006; Caine et al., 2007; von der Haar et al., 2007; Macreadie et al., 2008). Recently, a yeast model were engineered to study Aβ toxicity (Treusch et al., 2011, D'Angelo et al., 2012). Such models allowed for screening of the entire genome to find modifiers of Aβ toxicity (Treusch et al., 2011) as well as for the identification of compounds interfering with Aβ aggregation (Bharadwaj et al., 2012; Lopez et al., 2012).

In contrast to Aβ, only a few studies report on yeast models for protein tau. Nevertheless, these yeast models summarize several fundamental aspects related to the pathogenesis of the protein since human tau in yeast acquired pathological phosho-epitopes, assumed a pathological conformation, and formed aggregates. These processes were found to be dependent on Mds1 and Pho85, the yeast orthologs of the two most important mammalian tau-kinases, GSK-3β and cdk5, respectively (Vandebroek et al., 2005). Modulation of the tau phosphorylation status by expression of the protein in Mds1 and Pho85 deletion strains confirmed that increased tau phosphorylation associates with enhanced tau immunoreactivity to the pathological conformation-dependent antibody MC1 and higher levels of aggregated tau, while it has an adverse effect on the ability of tau to bind and stabilize microtubules (Vandebroek et al., 2005, 2006). Furthermore, detailed analysis of wild-type tau and several clinical tau mutants demonstrated that phosphorylation of ser-409 plays a key role in tau-aggregation and the disengagement of tau to bind microtubules. Moreover, oxidative stress and mitochondrial dysfunction were found to strongly induce tau-insolubility independent of its phosphorylation status (Vanhelmont et al., 2010).

5.10 CONCLUDING REMARKS

In healthy cells, QC mechanisms monitor the proper folding of native proteins and respond to damaged and misfolded proteins. Defects in these mechanisms of proteostasis result in the accumulation of misfolded proteins and the formation of protein aggregates. In addition to the research on protein aggregation in higher eukaryotic organisms, the study of protein misfolding in eukaryotic unicellular model organisms has been highly instructive in understanding the molecular mechanisms driving the cellular buildup of protein aggregates. The strong conservation of the basic cellular machineries has allowed unicellular eukaryotes to confirm and further extend our knowledge about how cells handle the misfolded protein load. The budding yeast *S. cerevisiae* especially has been very instrumental in this research. Several "humanized" yeast models have been developed by heterologous expression of principal disease-causing proteins such as Huntingtin, tau, and α-synuclein, allowing a study of the biological processes leading to the accumulation of toxic protein aggregates in neurodegenerative diseases such as Huntington's, Alzheimer's, and Parkinson's diseases. In addition, humanized yeast models have also proven to be convenient screening tools for the discovery of novel drugs for treating corresponding protein-misfolding diseases. Collectively, yeast and other unicellular eukaryotes provide an excellent model for understanding the molecular mechanisms of protein misfolding and aggregation and an effective tool to tackle protein-misfolding diseases found in humans.

ACKNOWLEDGMENTS

This work was supported by the Flemish Society for Alzheimer Research and by the IWT-Vlaanderen, the KU Leuven Research Fund, and the Fund for Scientific Research of Flanders.

REFERENCES

Aguilaniu H, Gustafsson L, Rigoulet M, Nystrom T (2003) Asymmetric inheritance of oxidatively damaged proteins during cytokinesis. Science 299:1751–1753.

Aimanianda V, Bayry J, Bozza S, Kniemeyer O, Perruccio K, Elluru SR, Clavaud C, Paris S, Brakhage AA, Kaveri SV, Romani L, Latge JP (2009) Surface hydrophobin prevents immune recognition of airborne fungal spores. Nature 460:1117–1121.

Aragon T, van Anken E, Pincus D, Serafimova IM, Korennykh AV, Rubio CA, Walter P (2009) Messenger RNA targeting to endoplasmic reticulum stress signalling sites. Nature 457:736–740.

Arendt CS, Hochstrasser M (1997) Identification of the yeast 20S proteasome catalytic centers and subunit interactions required for active-site formation. Proc Natl Acad Sci U S A 94:7156–7161.

Aviram S, Kornitzer D (2010) The ubiquitin ligase Hul5 promotes proteasomal processivity. Mol Cell Biol 30:985–994.

Baba M, Takeshige K, Baba N, Ohsumi Y (1994) Ultrastructural analysis of the autophagic process in yeast: detection of autophagosomes and their characterization. J Cell Biol 124:903–913.

Bagriantsev S, Liebman S (2006) Modulation of Abeta42 low-*n* oligomerization using a novel yeast reporter system. BMC Biol 4:32.

Balch WE, Morimoto RI, Dillin A, Kelly JW (2008) Adapting proteostasis for disease intervention. Science 319:916–919.

Baumeister W, Walz J, Zuhl F, Seemuller E (1998) The proteasome: paradigm of a self-compartmentalizing protease. Cell 92:367–380.

Baxter B K, Abeliovich H, Zhang X, Stirling AG, Burlingame AL, Goldfarb DS (2005) Atg19p ubiquitination and the cytoplasm to vacuole trafficking pathway in yeast. J Biol Chem 280:39067–39076.

Bays NW, Gardner RG, Seelig LP, Joazeiro CA, Hampton RY (2001) Hrd1p/Der3p is a membrane-anchored ubiquitin ligase required for ER-associated degradation. Nat Cell Biol 3:24–29.

Bengtson MH, Joazeiro CA (2010) Role of a ribosome-associated E3 ubiquitin ligase in protein quality control. Nature 467:470–473.

Bernales S, McDonald KL, Walter P (2006) Autophagy counterbalances endoplasmic reticulum expansion during the unfolded protein response. PLoS Biol 4:e423.

Bharadwaj PR, Verdile G, Barr RK, Gupta V, Steele JW, Lachenmayer ML, Yue Z, Ehrlich ME, Petsko G, Ju S, Ringe D, Sankovich SE, Caine JM, Macreadie IG, Gandy S, Martins RN (2012) Latrepirdine (Dimebon) enhances autophagy and reduces intracellular GFP-Abeta42 levels in yeast. J Alzheimers Dis 32:949–967.

Bhatia-Kissova I, Camougrand N (2010) Mitophagy in yeast: actors and physiological roles. FEMS Yeast Res 10:1023–1034.

Brackley KI, Grantham J (2009) Activities of the chaperonin containing TCP-1 (CCT): implications for cell cycle progression and cytoskeletal organisation. Cell Stress Chaperones 14:23–31.

Braun RJ, Büttner S, Ring J, Kroemer G, Madeo F (2009) Nervous yeast: modeling neurotoxic cell death. Trends Biochem Sci 35:135–144.

Büttner S, Bitto A, Ring J, Augsten M, Zabrocki P, Eisenberg Y, Jungwirth H, Hutter S, Carmona-Gutierrez D, Kroemer G, Winderickx J, Madeo F (2008) Functional mitochondria are required for alpha-synuclein toxicity in aging yeast. J Biol Chem 283:7554–7560.

Büttner S, Delay C, Franssens V, Bammens T, Ruli D, Zaunschirm S, de Oliveira RM, Outeiro TF, Madeo F, Buee L, Galas MC, Winderickx J (2010) Synphilin-1 enhances alpha-synuclein aggregation in yeast and contributes to cellular stress and cell death in a Sir2-dependent manner. PLoS One 5:e13700.

Caine J, Sankovich S, Antony H, Waddington L, Macreadie P, Varghese J, Macreadie I (2007) Alzheimer's Abeta fused to green fluorescent protein induces growth stress and a heat shock response. FEMS Yeast Res 7:1230–1236.

Carvalho P, Goder V, Rapoport TA (2006) Distinct ubiquitin-ligase complexes define convergent pathways for the degradation of ER proteins. Cell 126:361–373.

Cebollero E, Reggiori F (2009) Regulation of autophagy in yeast *Saccharomyces cerevisiae*. Biochim Biophys Acta 1793:1413–1421.

Cebulski J, Malouin J, Pinches N, Cascio V, Austriaco N (2011) Yeast Bax inhibitor, Bxi1p, is an ER-localized protein that links the unfolded protein response and programmed cell death in *Saccharomyces cerevisiae*. PLoS One 6:e20882.

Chawla A, Chakrabarti S, Ghosh G, Niwa M (2011) Attenuation of yeast UPR is essential for survival and is mediated by IRE1 kinase. J Cell Biol 193:41–50.

Chen M, Gutierrez GJ, Ronai ZA (2011) Ubiquitin-recognition protein Ufd1 couples the endoplasmic reticulum (ER) stress response to cell cycle control. Proc Natl Acad Sci U S A 108:9119–9124.

Chen P, Hochstrasser M (1996) Autocatalytic subunit processing couples active site formation in the 20S proteasome to completion of assembly. Cell 86:961–972.

Cheon SA, Jung KW, Chen YL, Heitman J, Bahn YS, Kang HA (2011) Unique evolution of the UPR pathway with a novel bZIP transcription factor, Hxl1, for controlling pathogenicity of *Cryptococcus neoformans*. PLoS Pathog 7:e1002177.

Cherra SJ, 3rd, Chu CT (2008) Autophagy in neuroprotection and neurodegeneration: a question of balance. Future Neurol 3:309–323.

Couplan E, Aiyar RS, Kucharczyk R, Kabala A, Ezkurdia N, Gagneur J, St Onge RP, Salin B, Soubigou F, Le Cann M, Steinmetz LM, di Rago JP, Blondel M (2011) A yeast-based assay identifies drugs active against human mitochondrial disorders. Proc Natl Acad Sci U S A 108:11989–11994.

Coustou V, Deleu C, Saupe S, Begueret J (1997) The protein product of the HET-s heterokaryon incompatibility gene of the fungus *Podospora anserina* behaves as a prion analog. Proc Natl Acad Sci U S A 94:9773–9778.

Cox JS, Shamu CE, Walter P (1993) Transcriptional induction of genes encoding endoplasmic reticulum resident proteins requires a transmembrane protein kinase. Cell 73:1197–1206.

Crosas B, Hanna J, Kirkpatrick DS, Zhang DP, Tone Y, Hathaway NA, Buecker C, Leggett DS, Schmidt M, King RW, Gygi SP, Finley D (2006) Ubiquitin chains are remodeled at the proteasome by opposing ubiquitin ligase and deubiquitinating activities. Cell 127:1401–1413.

Crow ET, Li L (2011) Newly identified prions in budding yeast, and their possible functions. Semin Cell Dev Biol 22:452–459.

D'Angelo F, Vignaud H, Di Martino J, Salin B, Devin A, Cullin C, Marchal C (2012) A yeast model for amyloid-beta aggregation exemplifies the role of membrane trafficking and PICALM in cytotoxicity. Dis Model Mech 6:206–216.

Deak PM, Wolf DH (2001) Membrane topology and function of Der3/Hrd1p as a ubiquitin-protein ligase (E3) involved in endoplasmic reticulum degradation. J Biol Chem 276:10663–10669.

Demartino GN, Gillette TG (2007) Proteasomes: machines for all reasons. Cell 129:659–662.

Deng M, Hochstrasser M (2006) Spatially regulated ubiquitin ligation by an ER/nuclear membrane ligase. Nature 443:827–831.

Desnues B, Cuny C, Grégori G, Dukan S, Aguilaniu H, Nyström T (2003) Differential oxidative damage and expression of stress defence regulons in culturable and non-culturable *Escherichia coli* cells. EMBO Rep 4:400–404.

Diaz-Troya S, Perez-Perez ME, Florencio FJ, Crespo JL (2008) The role of TOR in autophagy regulation from yeast to plants and mammals. Autophagy 4:851–865.

Dick TP, Nussbaum AK, Deeg M, Heinemeyer W, Groll M, Schirle M, Keilholz W, Stevanovic S, Wolf DH, Huber R, Rammensee HG, Schild H (1998) Contribution of proteasomal beta-subunits to the cleavage of peptide substrates analyzed with yeast mutants. J Biol Chem 273:25637–25646.

Ding WX, Yin XM (2008) Sorting, recognition and activation of the misfolded protein degradation pathways through macroautophagy and the proteasome. Autophagy 4:141–150.

Duennwald ML (2011) Monitoring polyglutamine toxicity in yeast. Methods 53:232–237.

Duennwald ML, Jagadish S, Muchowski PJ, Lindquist S (2006) Flanking sequences profoundly alter polyglutamine toxicity in yeast. Proc Natl Acad Sci U S A 103:11045–11050.

Edbauer D, Winkler E, Regula JT, Pesold B, Steiner H, Haass C (2003) Reconstitution of gamma-secretase activity. Nat Cell Biol 5:486–488.

Edskes HK (2001) Protein-based inheritance in *Saccharomyces cerevisiae*: [URE3] as a prion form of the nitrogen regulatory protein Ure2. Res Microbiol 152:605–612.

Ehrlich ES, Wang T, Luo K, Xiao Z, Niewiadomska AM, Martinez T, Xu W, Neckers L, Yu XF (2009) Regulation of Hsp90 client proteins by a Cullin5-RING E3 ubiquitin ligase. Proc Natl Acad Sci U S A 106:20330–20335.

Eisele F, Wolf DH (2008) Degradation of misfolded protein in the cytoplasm is mediated by the ubiquitin ligase Ubr1. FEBS Lett 582:4143–4146.

Ellgaard L, Molinari M, Helenius A (1999) Setting the standards: quality control in the secretory pathway. Science 286:1882–1888.

Elsasser S, Gali RR, Schwickart M, Larsen, Leggett DS, Muller B, Feng MT, Tubing F, Dittmar GA, Finley D (2002) Proteasome subunit Rpn1 binds ubiquitin-like protein domains. Nat Cell Biol 4:725–730.

Epple UD, Suriapranata I, Eskelinen EL, Thumm M (2001) Aut5/Cvt17p, a putative lipase essential for disintegration of autophagic bodies inside the vacuole. J Bacteriol 183:5942–5955.

Epple UD, Eskelinen EL, Thumm M (2003) Intravacuolar membrane lysis in *Saccharomyces cerevisiae*: Does vacuolar targeting of Cvt17/Aut5p affect its function? J Biol Chem 278:7810–7821.

Erjavec N, Larsson L, Grantham J, Nystrom T (2007) Accelerated aging and failure to segregate damaged proteins in Sir2 mutants can be suppressed by overproducing the protein aggregation-remodeling factor Hsp104p. Genes Dev 21: 2410–2421.

Esnault Y, Feldheim D, Blondel MO, Schekman R, Kepes F (1994) SSS1 encodes a stabilizing component of the Sec61 subcomplex of the yeast protein translocation apparatus. J Biol Chem 269:27478–27485.

Fang NN, Ng AH, Measday V, Mayor T (2011) Hul5 HECT ubiquitin ligase plays a major role in the ubiquitylation and turnover of cytosolic misfolded proteins. Nat Cell Biol 13:1344–1352.

Feng D, Zhao X, Soromani C, Toikkanen J, Romisch K, Vembar SS, Brodsky JL, Keranen S, Jantti J (2007) The transmembrane domain is sufficient for Sbh1p function, its association with the Sec61 complex, and interaction with Rtn1p. J Biol Chem 282:30618–30628.

Flower TR, Chesnokova LS, Froelich CA, Dixon C, Witt SN (2005) Heat shock prevents alpha-synuclein-induced apoptosis in a yeast model of Parkinson's disease. J Mol Biol 351:1081–1100.

Flower TR, Clark-Dixon C, Metoyer C, Yang H, Shi R, Zhang Z, Witt SN (2007) YGR198w (YPP1) targets A30P alpha-synuclein to the vacuole for degradation. J Cell Biol 177:1091–1104.

Fu L, Sztul E (2009) ER-associated complexes (ERACs) containing aggregated cystic fibrosis transmembrane conductance regulator (CFTR) are degraded by autophagy. Eur J Cell Biol 88:215–226.

Gardner BM, Walter P (2011) Unfolded proteins are Ire1-activating ligands that directly induce the unfolded protein response. Science 333:1891–1894.

Gauss R, Jarosch E, Sommer T, Hirsch C (2006) A complex of Yos9p and the HRD ligase integrates endoplasmic reticulum quality control into the degradation machinery. Nat Cell Biol 8:849–854.

Giorgini F, Guidetti P, Nguyen Q, Bennett SC, Muchowski PJ (2005) A genomic screen in yeast implicates kynurenine 3-monooxygenase as a therapeutic target for Huntington disease. Nat Genet 37:526–531.

Gitler AD (2008) Beer and bread to brains and beyond: Can yeast cells teach us about neurodegenerative disease? Neurosignals 16:52–62.

Gitler AD, Bevis BJ, Shorter J, Strathearn KE, Hamamichi S, Su LJ, Caldwell KA, Caldwell GA, Rochet JC, McCaffery JM, Barlowe C, Lindquist S (2008) The Parkinson's disease protein alpha-synuclein disrupts cellular Rab homeostasis. Proc Natl Acad Sci U S A 105:145–150.

Glass NL, Dementhon K (2006) Non-self recognition and programmed cell death in filamentous fungi. Curr Opin Microbiol 9:553–558.

Glover JR, Lindquist S (1998) Hsp104, Hsp70, and Hsp40: a novel chaperone system that rescues previously aggregated proteins. Cell 94:73–82.

Gnann A, Riordan JR, Wolf DH (2004) Cystic fibrosis transmembrane conductance regulator degradation depends on the lectins Htm1p/EDEM and the Cdc48 protein complex in yeast. Mol Biol Cell 15:4125–4135.

Gokhale KC, Newnam GP, Sherman MY, Chernoff YO (2005) Modulation of prion-dependent polyglutamine aggregation and toxicity by chaperone proteins in the yeast model. J Biol Chem 280:22809–22818.

Griffioen G, Duhamel H, Van Damme N, Pellens K, Zabrocki P, Pannecouque C, van Leuven F, Winderickx J, Wera S (2006) A yeast-based model of alpha-synucleinopathy identifies compounds with therapeutic potential. Biochim Biophys Acta 1762:312–318.

Groll M, Ditzel L, Lowe J, Stock D, Bochtler M, Bartunik HD, Huber R (1997) Structure of 20S proteasome from yeast at 2.4Å resolution. Nature 386: 463–471.

Guerfal M, Ryckaert S, Jacobs PP, Ameloot P, Van Craenenbroeck K, Derycke R, Callewaert N (2010) The HAC1 gene from *Pichia pastoris*: characterization and effect of its overexpression on the production of secreted, surface displayed and membrane proteins. Microb Cell Fact 9:49.

Hamasaki M, Noda T, Baba M, Ohsumi Y (2005) Starvation triggers the delivery of the endoplasmic reticulum to the vacuole via autophagy in yeast. Traffic 6:56–65.

Haslberger T, Weibezahn J, Zahn R, Lee S, Tsai FT, Bukau B, Mogk A (2007) M domains couple the ClpB threading motor with the DnaK chaperone activity. Mol Cell 25:247–260.

Haslberger T, Zdanowicz A, Brand I, Kirstein J, Turgay K, Mogk A, Bukau B (2008) Protein disaggregation by the AAA^+ chaperone ClpB involves partial threading of looped polypeptide segments. Nat Struct Mol Biol 15:641–650.

He C, Baba M, Cao Y, Klionsky DJ (2008) Self-interaction is critical for Atg9 transport and function at the phagophore assembly site during autophagy. Mol Biol Cell 19:5506–5516.

Heck JW, Cheung SK, Hampton RY (2010) Cytoplasmic protein quality control degradation mediated by parallel actions of the E3 ubiquitin ligases Ubr1 and San1. Proc Natl Acad Sci U S A 107:1106–1111.

Heinemeyer W, Kleinschmidt JA, Saidowsky J, Escher C, Wolf DH (1991) Proteinase yscE, the yeast proteasome/multicatalytic-multifunctional proteinase: mutants unravel its function in stress induced proteolysis and uncover its necessity for cell survival. EMBO J 10:555–562.

Heinemeyer W, Fischer M, Krimmer T, Stachon U, Wolf DH (1997) The active sites of the eukaryotic 20S proteasome and their involvement in subunit precursor processing. J Biol Chem 272:25200–25209.

Hughes RE, Lo RS, Davis C, Strand AD, Neal CL, Olson JM, Fields S (2001) Altered transcription in yeast expressing expanded polyglutamine. Proc Natl Acad Sci U S A 98:13201–13206.

Hughes SR, Goyal S, Sun JE, Gonzalez-DeWhitt P, Fortes MA, Riedel NG, Sahasrabudhe SR (1996) Two-hybrid system as a model to study the interaction of beta-amyloid peptide monomers. Proc Natl Acad Sci U S A 93:2065–2070.

Jarosch E, Taxis C, Volkwein C, Bordallo J, Finley D, Wolf DH, Sommer T (2002) Protein dislocation from the ER requires polyubiquitination and the AAA–ATPase Cdc48. Nat Cell Biol 4:134–139.

Ju D, Xie Y (2004) Proteasomal degradation of RPN4 via two distinct mechanisms, ubiquitin-dependent and -independent. J Biol Chem 279:23851–23854.

Kabani M, Beckerich JM, Brodsky JL (2002) Nucleotide exchange factor for the yeast Hsp70 molecular chaperone Ssa1p. Mol Cell Biol 22:4677–4689.

Kaganovich D, Kopito R, Frydman J (2008) Misfolded proteins partition between two distinct quality control compartments. Nature 454:1088–1095.

Kanki T, Klionsky DJ (2010) The molecular mechanism of mitochondria autophagy in yeast. Mol Microbiol 75:795–800.

Khurana V, Lindquist S (2010) Modelling neurodegeneration in *Saccharomyces cerevisiae*: Why cook with baker's yeast? Nat Rev Neurosci 11:436–449.

Kiel JA (2010) Autophagy in unicellular eukaryotes. Philos Trans R Soc Lond B 365:819–830.

Kihara A, Noda T, Ishihara N, Ohsumi Y (2001) Two distinct Vps34 phosphatidylinositol 3-kinase complexes function in autophagy and carboxypeptidase Y sorting in *Saccharomyces cerevisiae*. J Cell Biol 152:519–530.

Kimata Y, Oikawa D, Shimizu Y, Ishiwata-Kimata Y, Kohno K (2004) A role for BiP as an adjustor for the endoplasmic reticulum stress-sensing protein Ire1. J Cell Biol 167:445–456.

Kimura S, Maruyama J, Kikuma T, Arioka M, Kitamoto K (2011) Autophagy delivers misfolded secretory proteins accumulated in endoplasmic reticulum to vacuoles in the filamentous fungus *Aspergillus oryzae*. Biochem Biophys Res Commun 406:464–470.

Klionsky D J (2005) The molecular machinery of autophagy: unanswered questions. J Cell Sci 118:7–18.

Knop M, Finger A, Braun T, Hellmuth K, Wolf DH (1996) Der1, a novel protein specifically required for endoplasmic reticulum degradation in yeast. EMBO J 15:753–763.

Koegl M, Hoppe T, Schlenker S, Ulrich HD, Mayer TU, Jentsch S (1999) A novel ubiquitination factor, E4, is involved in multiubiquitin chain assembly. Cell 96:635–644.

Kohler A, Cascio P, Leggett DS, Woo KM, Goldberg AL, Finley D (2001) The axial channel of the proteasome core particle is gated by the Rpt2 ATPase and controls both substrate entry and product release. Mol Cell 7:1143–1152.

Kohno K (2010) Stress-sensing mechanisms in the unfolded protein response: similarities and differences between yeast and mammals. J Biochem 147: 27–33.

Kopito RR (2000) Aggresomes, inclusion bodies and protein aggregation. Trends Cell Biol 10:524–530.

Korennykh AV, Egea PF, Korostelev AA, Finer-Moore J, Zhang C, Shokat KM, Stroud RM, Walter P (2009) The unfolded protein response signals through high-order assembly of Ire1. Nature 457:687–693.

Kostova Z, Wolf DH (2003) For whom the bell tolls: protein quality control of the endoplasmic reticulum and the ubiquitin-proteasome connection. EMBO J 22:2309–2317.

Kraft C, Reggiori F, Peter M (2009) Selective types of autophagy in yeast. Biochim Biophys Acta 1793:1404–1412.

Kritzer JA, Hamamichi S, McCaffery JM, Santagata S, Naumann TA, Caldwell KA, Caldwell GA, Lindquist S (2009) Rapid selection of cyclic peptides that reduce alpha-synuclein toxicity in yeast and animal models. Nat Chem Biol 5:655–663.

Krobitsch S, Lindquist S (2000) Aggregation of huntingtin in yeast varies with the length of the polyglutamine expansion and the expression of chaperone proteins. Proc Natl Acad Sci U S A 97:1589–1594.

Kruse KB, Brodsky JL, McCracken AA (2006) Characterization of an ERAD gene as VPS30/ATG6 reveals two alternative and functionally distinct protein quality control pathways: one for soluble Z variant of human alpha-1 proteinase inhibitor (A1PiZ) and another for aggregates of A1PiZ. Mol Biol Cell 17:203–212.

Kucharczyk R, Rak M, di Rago JP (2009) Biochemical consequences in yeast of the human mitochondrial DNA 8993T>C mutation in the ATPase6 gene found in NARP/MILS patients. Biochim Biophys Acta 1793:817–824.

Kusmierczyk AR, Kunjappu MJ, Funakoshi M, Hochstrasser M (2008) A multimeric assembly factor controls the formation of alternative 20S proteasomes. Nat Struct Mol Biol 15:237–244.

Lam YA, Lawson TG, Velayutham M, Zweier JL, Pickart CM (2002) A proteasomal ATPase subunit recognizes the polyubiquitin degradation signal. Nature 416:763–767.

Lancaster AK, Bardill JP, True HL, Masel J (2010) The spontaneous appearance rate of the yeast prion [PSI+] and its implications for the evolution of the evolvability properties of the [PSI+] system. Genetics 184:393–400.

Le Tallec B, Barrault MB, Courbeyrette R, Guerois R, Marsolier-Kergoat MC, Peyroche A (2007) 20S proteasome assembly is orchestrated by two distinct pairs of chaperones in yeast and in mammals. Mol Cell 27:660–674.

Lee KP, Dey M, Neculai D, Cao C, Dever TE, Sicheri F (2008) Structure of the dual enzyme Ire1 reveals the basis for catalysis and regulation in nonconventional RNA splicing. Cell 132:89–100.

Li H, Korennykh AV, Behrman SL, Walter P (2010) Mammalian endoplasmic reticulum stress sensor IRE1 signals by dynamic clustering. Proc Natl Acad Sci U S A 107:16113–16118.

Linder MB, Szilvay GR, Nakari-Setala T, Penttila ME (2005) Hydrophobins: the protein-amphiphiles of filamentous fungi. FEMS Microbiol Rev 29:877–896.

Lindner AB, Madden R, Demarez A, Stewart EJ, Taddei F (2008) Asymmetric segregation of protein aggregates is associated with cellular aging and rejuvenation. Proc Natl Acad Sci U S A 105:3076–3081.

Liu B, Larsson L, Caballero A, Hao X, Oling D, Grantham J, Nystrom T (2010) The polarisome is required for segregation and retrograde transport of protein aggregates. Cell 140:257–267.

Liu B, Larsson L, Franssens V, Hao X, Hill SM, Andersson V, Hoglund D, Song J, Yang X, Oling D, Grantham J, Winderickx J, Nystrom T (2011) Segregation of protein aggregates involves actin and the polarity machinery. Cell 147:959–961.

Lopez LC, Dos-Reis S, Espargaro A, Carrodeguas JA, Maddelein ML, Ventura S, Sancho J (2012) Discovery of novel inhibitors of amyloid beta-peptide 1–42 aggregation. J Med Chem 55:9521–9530.

Luthi U, Schaerer-Brodbeck C, Tanner S, Middendorp O, Edler K, Barberis A (2003) Human beta-secretase activity in yeast detected by a novel cellular growth selection system. Biochim Biophys Acta 1620:167–178.

Lynch-Day MA, Klionsky DJ (2010) The Cvt pathway as a model for selective autophagy. FEBS Lett 584:1359–1366.

Macreadie I, Lotfi-Miri M, Mohotti S, Shapira D, Bennett L, Varghese J (2008) Validation of folate in a convenient yeast assay suited for identification of inhibitors of Alzheimer's amyloid-beta aggregation. J Alzheimers Dis 15:391–396.

Meriin AB, Sherman MY (2005) Role of molecular chaperones in neurodegenerative disorders. Int J Hyperthermia 21:403–419.

Meriin AB, Zhang X, He X, Newnam GP, Chernoff YO, Sherman MY (2002) Huntington toxicity in yeast model depends on polyglutamine aggregation mediated by a prion-like protein Rnq1. J Cell Biol 157:997–1004.

Meriin AB, Zhang X, Miliaras NB, Kazantsev A, Chernoff YO, McCaffery JM, Wendland B, Sherman MY (2003) Aggregation of expanded polyglutamine domain in yeast leads to defects in endocytosis. Mol Cell Biol 23: 7554–7565.

Metzger MB, Michaelis S (2009) Analysis of quality control substrates in distinct cellular compartments reveals a unique role for Rpn4p in tolerating misfolded membrane proteins. Mol Biol Cell 20:1006–1019.

Middendorp O, Ortler C, Neumann U, Paganetti P, Luthi, Barberis A (2004) Yeast growth selection system for the identification of cell-active inhibitors of beta-secretase. Biochim Biophys Acta 1674:29–39.

Mijaljica D, Prescott M, Devenish RJ (2006) Endoplasmic reticulum and Golgi complex: contributions to, and turnover by, autophagy. Traffic 7:1590–1595.

Mitchener JS, Shelburne JD, Bradford WD, Hawkins HK (1976) Cellular autophagocytosis induced by deprivation of serum and amino acids in HeLa cells. Am J Pathol 83:485–492.

Monastyrska I, Reggiori F, Klionsky DJ (2008) Harpooning the Cvt complex to the phagophore assembly site. Autophagy 4:914–916.

Mori K, Sant A, Kohno K, Normington K, Gething MJ, Sambrook JF (1992) A 22 bp *cis*-acting element is necessary and sufficient for the induction of the yeast KAR2 (BiP) gene by unfolded proteins. EMBO J 11:2583–2593.

Muchowski PJ, Schaffar G, Sittler A, Wanker EE, Hayer-Hartl MK, Hartl FU (2000) Hsp70 and hsp40 chaperones can inhibit self-assembly of polyglutamine proteins into amyloid-like fibrils. Proc Natl Acad Sci U S A 97:7841–7846.

Muchowski PJ, Ning K, D'Souza-Schorey C, Fields S (2002) Requirement of an intact microtubule cytoskeleton for aggregation and inclusion body formation by a mutant huntingtin fragment. Proc Natl Acad Sci U S A 99:727–732.

Mulder HJ, Nikolaev I (2009) HacA-dependent transcriptional switch releases hacA mRNA from a translational block upon endoplasmic reticulum stress. Eukaryot Cell 8:665–675.

Mulder HJ, Saloheimo M, Penttila M, Madrid SM (2004) The transcription factor HACA mediates the unfolded protein response in *Aspergillus niger*, and up-regulates its own transcription. Mol Genet Genomics 271:130–140.

Nakayashiki T, Kurtzman CP, Edskes HK, Wickner RB (2005) Yeast prions [URE3] and [PSI+] are diseases. Proc Natl Acad Sci U S A 102:10575–10580.

Nice DC, Sato TK, Stromhaug PE, Emr SD, Klionsky DJ (2002) Cooperative binding of the cytoplasm to vacuole targeting pathway proteins, Cvt13 and Cvt20, to phosphatidylinositol 3-phosphate at the pre-autophagosomal structure is required for selective autophagy. J Biol Chem 277:30198–30207.

Normington K, Kohno K, Kozutsumi Y, Gething MJ, Sambrook J (1989) *S. cerevisiae* encodes an essential protein homologous in sequence and function to mammalian BiP. Cell 57:1223–1236.

Nussbaum AK, Dick TP, Keilholz W, Schirle M, Stevanovic S, Dietz K, Heinemeyer W, Groll M, Wolf DH, Huber R, Rammensee HG, Schild H (1998) Cleavage motifs of the yeast 20S proteasome beta subunits deduced from digests of enolase 1. Proc Natl Acad Sci U S A 95:12504–12509.

Ogata M, Hino S, Saito A, Morikawa K, Kondo S, Kanemoto S, Murakami T, Taniguchi M, Tanii I, Yoshinaga K, Shiosaka S, Hammarback JA, Urano F, Imaizumi K (2006) Autophagy is activated for cell survival after endoplasmic reticulum stress. Mol Cell Biol 26:9220–9231.

Outeiro TF, Lindquist S (2003) Yeast cells provide insight into alpha-synuclein biology and pathobiology. Science 302:1772–1775.

Pandey UB, Nie Z, Batlevi Y, McCray BA, Ritson GP, Nedelsky NB, Schwartz SL, DiProspero NA, Knight MA, Schuldiner O, Padmanabhan R, Hild M, Berry DL, Garza D, Hubbert CC, Yao TP, Baehrecke EH, Taylor JP (2007) HDAC6 rescues neurodegeneration and provides an essential link between autophagy and the UPS. Nature 447:859–863.

Park SH, Bolender N, Eisele F, Kostova Z, Takeuchi J, Coffino P, Wolf DH (2007) The cytoplasmic Hsp70 chaperone machinery subjects misfolded and endoplasmic reticulum import-incompetent proteins to degradation via the ubiquitin-proteasome system. Mol Biol Cell 18:153–165.

Pincus D, Chevalier MW, Aragon T, van Anken E, Vidal SE, El-Samad H, Walter P (2010) BiP binding to the ER-stress sensor Ire1 tunes the homeostatic behavior of the unfolded protein response. PLoS Biol 8:e1000415.

Rak M, Tetaud E, Duvezin-Caubet S, Ezkurdia N, Bietenhader M, Rytka J, di Rago JP (2007) A yeast model of the neurogenic ataxia retinitis pigmentosa (NARP) T8993G mutation in the mitochondrial ATP synthase-6 gene. J Biol Chem 282:34039–34047.

Roberts P, Moshitch-Moshkovitz S, Kvam E, O'Toole E, Winey M, Goldfarb DS (2003) Piecemeal microautophagy of nucleus in *Saccharomyces cerevisiae*. Mol Biol Cell 14:129–141.

Rose MD, Misra LM, Vogel JP (1989) KAR2, a karyogamy gene, is the yeast homolog of the mammalian BiP/GRP78 gene. Cell 57:1211–1221.

Rosenbaum JC, Fredrickson EK, Oeser ML, Garrett-Engele CM, Locke MN, Richardson LA, Nelson ZW, Hetrick ED, Milac TI, Gottschling DE, Gardner RG (2011) Disorder targets misorder in nuclear quality control degradation: a disordered ubiquitin ligase directly recognizes its misfolded substrates. Mol Cell 41:93–106.

Rubio C, Pincus D, Korennykh A, Schuck S, El-Samad H, Walter P (2011) Homeostatic adaptation to endoplasmic reticulum stress depends on Ire1 kinase activity. J Cell Biol 193:171–184.

Russnak R, Konczal D, McIntire SL (2001) A family of yeast proteins mediating bidirectional vacuolar amino acid transport. J Biol Chem 276:23849–23857.

Saloheimo M, Valkonen M, Penttila M (2003) Activation mechanisms of the HAC1-mediated unfolded protein response in filamentous fungi. Mol Microbiol 47:1149–1161.

Sampaio-Marques B, Felgueiras C, Silva A, Rodrigues F, Ludovico P (2011) Yeast chronological lifespan and proteotoxic stress: Is autophagy good or bad? Biochem Soc Trans 39:1466–1470.

Sampaio-Marques B, Felgueiras C, Silva A, Rodrigues M, Tenreiro S, Franssens V, Reichert AS, Outeiro TF, Winderickx J, Ludovico P (2012) SNCA (alpha-synuclein)-induced toxicity in yeast cells is dependent on sirtuin 2 (Sir2)-mediated mitophagy. Autophagy 8:1494–1509.

Schubert U, Anton LC, Gibbs J, Norbury CC, Yewdell JW, Bennink JR (2000) Rapid degradation of a large fraction of newly synthesized proteins by proteasomes. Nature 404:770–774.

Shamu CE, Walter P (1996) Oligomerization and phosphorylation of the Ire1p kinase during intracellular signaling from the endoplasmic reticulum to the nucleus. EMBO J 15:3028–3039.

Sidrauski C, Walter P (1997) The transmembrane kinase Ire1p is a site-specific endonuclease that initiates mRNA splicing in the unfolded protein response. Cell 90:1031–1039.

Sidrauski C, Cox JS, Walter P (1996) tRNA ligase is required for regulated mRNA splicing in the unfolded protein response. Cell 87:405–413.

Sokolov S, Pozniakovsky A, Bocharova N, Knorre D, Severin F (2006) Expression of an expanded polyglutamine domain in yeast causes death with apoptotic markers. Biochim Biophys Acta 1757:660–666.

Solans A, Zambrano A, Rodriguez M, Barrientos A (2006) Cytotoxicity of a mutant huntingtin fragment in yeast involves early alterations in mitochondrial OXPHOS complexes II and III. Hum Mol Genet 15:3063–3081.

Sondermann H, Ho AK, Listenberger LL, Siegers K, Moarefi I, Wente SR, Hartl FU, Young JC, (2002) Prediction of novel Bag-1 homologs based on structure/function analysis identifies Snl1p as an Hsp70 co-chaperone in *Saccharomyces cerevisiae*. J Biol Chem 277:33220–33227.

Soper JH, Roy S, Stieber A, Lee E, Wilson RB, Trojanowski JQ, Burd CG, Lee VM (2008) Alpha-synuclein-induced aggregation of cytoplasmic vesicles in *Saccharomyces cerevisiae*. Mol Biol Cell 19:1093–1103.

Soper JH, Kehm V, Burd CG, Bankaitis VA, Lee VM (2011) Aggregation of alpha-synuclein in *S. cerevisiae* is associated with defects in endosomal trafficking and phospholipid biosynthesis. J Mol Neurosci 43:391–405.

Stolz A, Wolf DH (2010) Endoplasmic reticulum associated protein degradation: a chaperone assisted journey to hell. Biochim Biophys Acta 1803:694–705.

Stromhaug PE, Reggiori F, Guan J, Wang CW, Klionsky DJ (2004) Atg21 is a phosphoinositide binding protein required for efficient lipidation and localization of Atg8 during uptake of aminopeptidase I by selective autophagy. Mol Biol Cell 15:3553–3566.

Su LJ, Auluck PK, Outeiro TF, Yeger-Lotem E, Kritzer JA, Tardiff DF, Strathearn KE, Liu F, Cao S, Hamamichi S, Hill KJ, Caldwell KA, Bell GW, Fraenkel E, Cooper AA, Caldwell GA, McCaffery JM, Rochet JC, Lindquist S (2010) Compounds from an unbiased chemical screen reverse both ER-to-Golgi trafficking defects and mitochondrial dysfunction in Parkinson's disease models. Dis Model Mech 3:194–208.

Swinnen E, Büttner S, Outeiro TF, Galas MC, Madeo F, Winderickx J, Franssens V (2011) Aggresome formation and segregation of inclusions influence toxicity of alpha-synuclein and synphilin-1 in yeast. Biochem Soc Trans 39:1476–1481.

Takeshige K, Baba M, Tsuboi S, Noda T, Ohsumi Y (1992) Autophagy in yeast demonstrated with proteinase-deficient mutants and conditions for its induction. J Cell Biol 119:301–311.

Talbot NJ, Kershaw MJ, Wakley GE, De Vries O, Wessels J, Hamer JE (1996) MPG1 encodes a fungal hydrophobin involved in surface interactions during infection-related development of *Magnaporthe grisea*. Plant Cell 8:985–999.

Tardiff DF, Tucci ML, Caldwell KA, Caldwell GA, Lindquist S (2012) Different 8-hydroxyquinolines protect models of TDP-43 protein, alpha-synuclein, and polyglutamine proteotoxicity through distinct mechanisms. J Biol Chem 287:4107–4120.

Teter SA, Klionsky DJ (2000) Transport of proteins to the yeast vacuole: autophagy, cytoplasm-to-vacuole targeting, and role of the vacuole in degradation. Semin Cell Dev Biol 11:173–179.

Tirasophon W, Welihinda AA, Kaufman RJ (1998) A stress response pathway from the endoplasmic reticulum to the nucleus requires a novel bifunctional protein kinase/endoribonuclease (Ire1p) in mammalian cells. Genes Dev 12:1812–1824.

Tooze SA, Yoshimori T (2010) The origin of the autophagosomal membrane. Nat Cell Biol 12:831–835.

Treusch S, Hamamichi S, Goodman JL, Matlack KE, Chung CY, Baru V, Shulman JM, Parrado A, Bevis BJ, Valastyan JS, Han H, Lindhagen-Persson M, Reiman EM, Evans DA, Bennett DA, Olofsson A, DeJager PL, Tanzi RE, Caldwell KA, Caldwell GA, Lindquist S (2011) Functional links between Abeta toxicity, endocytic trafficking, and Alzheimer's disease risk factors in yeast. Science 334:1241–1245.

True HL, Lindquist SL (2000) A yeast prion provides a mechanism for genetic variation and phenotypic diversity. Nature 407:477–483.

True HL, Berlin I, Lindquist SL (2004) Epigenetic regulation of translation reveals hidden genetic variation to produce complex traits. Nature 431:184–187.

Tsai B, Ye Y, Rapoport TA (2002) Retro-translocation of proteins from the endoplasmic reticulum into the cytosol. Nat Rev Mol Cell Biol 3:246–255.

Tyedmers J, Madariaga ML, Lindquist S (2008) Prion switching in response to environmental stress. PLoS Biol 6:e294.

Tyedmers J, Mogk A, Bukau B (2010) Cellular strategies for controlling protein aggregation. Nat Rev Mol Cell Biol 11:777–788.

Valkonen M, Ward M, Wang H, Penttila M, Saloheimo M (2003) Improvement of foreign-protein production in *Aspergillus niger* var. *awamori* by constitutive induction of the unfolded-protein response. Appl Environ Microbiol 69: 6979–6986.

Van Rossom S, Op de Beeck K, Franssens V, Swinnen E, Schepers A, Ghillebert R, Caldara M, Van Camp G, Winderickx J (2012) The splicing mutant of the human tumor suppressor protein DFNA5 induces programmed cell death when expressed in the yeast *Saccharomyces cerevisiae*. Front Oncol 2:77.

Vandebroek T, Vanhelmont T, Terwel D, Borghgraef P, Lemaire K, Snauwaert J, Wera S, Van Leuven F, Winderickx J (2005) Identification and isolation of a hyperphosphorylated, conformationally changed intermediate of human protein tau expressed in yeast. Biochemistry 44:11466–11475.

Vandebroek T, Terwel D, Vanhelmont T, Gysemans M, Van Haesendonck C, Engelborghs Y, Winderickx J, Van Leuven F (2006) Microtubule binding and clustering of human Tau-4R and Tau-P301L proteins isolated from yeast deficient in orthologues of glycogen synthase kinase-3beta or cdk5. J Biol Chem 281: 25388–25397.

Vanhelmont T, Vandebroek T, De Vos A, Terwel D, Lemaire K, Anandhakumar J, Franssens V, Swinnen E, Van Leuven F, Winderickx J (2010) Serine-409 phosphorylation and oxidative damage define aggregation of human protein tau in yeast. FEMS Yeast Res 10:992–1005.

Vembar SS, Brodsky JL (2008) One step at a time: endoplasmic reticulum-associated degradation. Nat Rev Mol Cell Biol 9:944–957.

Verghese J, Abrams J, Wang Y, Morano KA (2012) Biology of the heat shock response and protein chaperones: budding yeast (*Saccharomyces cerevisiae*) as a model system. Microbiol Mol Biol Rev 76:115–158.

von der Haar T, Josse L, Wright P, Zenthon J, Tuite MF (2007) Development of a novel yeast cell-based system for studying the aggregation of Alzheimer's disease-associated Abeta peptides in vivo. Neurodegener Dis 4:136–147.

Walter P, Ron D (2011) The unfolded protein response: from stress pathway to homeostatic regulation. Science 334:1081–1086.

Wang CW, Klionsky DJ (2003) The molecular mechanism of autophagy. Mol Med 9:65–76.

Wang CW, Stromhaug PE, Kauffman EJ, Weisman LS, Klionsky DJ (2003) Yeast homotypic vacuole fusion requires the Ccz1–Mon1 complex during the tethering/docking stage. J Cell Biol 163:973–985.

Wang L, Mao X, Ju D, Xie Y (2004) Rpn4 is a physiological substrate of the Ubr2 ubiquitin ligase. J Biol Chem 279:55218–55223.

Wasmer C, Zimmer A, Sabate R, Soragni A, Saupe SJ, Ritter C, Meier BH (2010) Structural similarity between the prion domain of HET-s and a homologue can explain amyloid cross-seeding in spite of limited sequence identity. J Mol Biol 402:311–325.

Weibezahn J, Tessarz P, Schlieker C, Zahn R, Maglica Z, Lee S, Zentgraf H, Weber-Ban EU, Dougan DA, Tsai FT, Mogk A, Bukau B (2004) Thermotolerance requires refolding of aggregated proteins by substrate translocation through the central pore of ClpB. Cell 119:653–665.

Welihinda AA, Kaufman RJ (1996) The unfolded protein response pathway in *Saccharomyces cerevisiae*: oligomerization and trans-phosphorylation of Ire1p (Ern1p) are required for kinase activation. J Biol Chem 271:18181–18187.

Wickner RB (1994) [URE3] as an altered URE2 protein: evidence for a prion analog in *Saccharomyces cerevisiae*. Science 264:566–569.

Wickner RB (1997) A new prion controls fungal cell fusion incompatibility. Proc Natl Acad Sci U S A 94:10012–10014.

Wickner RB, Masison DC, Edskes HK (1995) [PSI] and [URE3] as yeast prions. Yeast 11:1671–1685.

Wickner RB, Edskes HK, Shewmaker F, Nakayashiki T, Engel A, McCann L, Kryndushkin D (2007) Yeast prions: evolution of the prion concept. Prion 1:94–100.

Wilkinson BM, Esnault Y, Craven RA, Skiba F, Fieschi J, K'Epes F, Stirling CJ (1997) Molecular architecture of the ER translocase probed by chemical crosslinking of Sss1p to complementary fragments of Sec61p. EMBO J 16:4549–4559.

Willingham S, Outeiro TF, DeVit MJ, Lindquist SL, Muchowski PJ (2003) Yeast genes that enhance the toxicity of a mutant huntingtin fragment or alpha-synuclein. Science 302:1769–1772.

Winderickx J, Delay C, De Vos A, Klinger H, Pellens K, Vanhelmont T, Van Leuven F, Zabrocki P (2008) Protein folding diseases and neurodegeneration: lessons learned from yeast. Biochim Biophys Acta 1783:1381–1395.

Wurmser AE, Emr SD (2002) Novel PtdIns(3)P-binding protein Etf1 functions as an effector of the Vps34 PtdIns 3-kinase in autophagy. J Cell Biol 158:761–772.

Xie Y, Varshavsky A (2001) RPN4 is a ligand, substrate, and transcriptional regulator of the 26S proteasome: a negative feedback circuit. Proc Natl Acad Sci U S A 98:3056–3061.

Yang Z, Huang J, Geng J, Nair U, Klionsky DJ (2006) Atg22 recycles amino acids to link the degradative and recycling functions of autophagy. Mol Biol Cell 17:5094–5104.

Ye Y, Meyer HH, Rapoport TA (2001) The AAA ATPase Cdc48/p97 and its partners transport proteins from ER into the cytosol. Nature 414:652–656.

Yorimitsu T, Nair U, Yang Z, Klionsky DJ (2006) Endoplasmic reticulum stress triggers autophagy. J Biol Chem 281:30299–30304.

Youker RT, Walsh P, Beilharz T, Lithgow T, Brodsky JL (2004) Distinct roles for the Hsp40 and Hsp90 molecular chaperones during cystic fibrosis transmembrane conductance regulator degradation in yeast. Mol Biol Cell 15:4787–4797.

Zabrocki P, Pellens K, Vanhelmont T, Vandebroek T, Griffioen G, Wera S, Van Leuven F, Winderickx J (2005) Characterization of alpha-synuclein aggregation and synergistic toxicity with protein tau in yeast. FEBS J 272:1386–1400.

Zabrocki P, Bastiaens I, Delay C, Bammens T, Ghillebert R, Pellens K, De Virgilio C, Van Leuven F, Winderickx J (2008) Phosphorylation, lipid raft interaction and traffic of alpha-synuclein in a yeast model for Parkinson. Biochim Biophys Acta 1783:1767–1780.

Zhang H, Komano H, Fuller RS, Gandy SE, Frail DE (1994) Proteolytic processing and secretion of human beta-amyloid precursor protein in yeast: evidence for a yeast secretase activity. J Biol Chem 269:27799–27802.

Zhang W, Espinoza D, Hines V, Innis M, Mehta P, Miller DL (1997) Characterization of beta-amyloid peptide precursor processing by the yeast Yap3 and Mkc7 proteases. Biochim Biophys Acta 1359:110–122.

Zhang X, Smith DL, Meriin AB, Engemann S, Russel DE, Roark M, Washington SL, Maxwell MM, Marsh JL, Thompson LM, Wanker EE, Young AB, Housman DE, Bates GP, Sherman MY, Kazantsev AG (2005) A potent small molecule inhibits polyglutamine aggregation in Huntington's disease neurons and suppresses neurodegeneration in vivo. Proc Natl Acad Sci U S A 102:892–897.

6

STRUCTURAL PROPERTIES OF BACTERIAL INCLUSION BODIES

ANTONINO NATALELLO, DILETTA AMI, AND SILVIA MARIA DOGLIA

6.1 INTRODUCTION

The overexpression of a recombinant protein in bacteria often leads to its deposition in cytoplasmic or periplasmic aggregates, called *inclusion bodies* (IBs). Many factors contribute to IB formation, such as the inability of bacteria to perform the specific post-translational modifications required by the heterologous protein and the saturation of the cellular folding modulators due to the recombinant protein overexpression (Baneyx and Mujacic, 2004). Since IB formation is obviously associated with a reduced yield of the soluble protein, IBs have been considered the main bottleneck of recombinant protein production. Moreover, for a long time IBs were believed to be unstructured aggregates made up of only unfolded/misfolded and nonfunctional polypeptides, and thus regarded as a waste by-product of protein production (García-Fruitós et al., 2012). Therefore, while many efforts have been devoted to inhibit IB formation and to develop new protocols for the *in vitro* refolding and recovery of functional proteins from these aggregates, only a few functional and structural characterizations of IBs have been performed in the past.

Protein Aggregation in Bacteria: Functional and Structural Properties of Inclusion Bodies in Bacterial Cells, First Edition. Edited by Silvia Maria Doglia and Marina Lotti.

Interestingly, in recent years, the observation that proteins can also be active when embedded in insoluble IBs, and that inactive proteins can be found in the cell soluble fraction, raised the question about the correlation between solubility and functionality (González-Montalbán et al., 2007). As a consequence, bacterial IBs have recently gained the attention (García-Fruitós, 2010; Gatti-Lafranconi et al., 2011) of academic and industrial research groups aimed at investigating the peculiar IB structural properties and exploiting their possible unexpected uses in biotechnology and biomedicine. In particular, IBs can be considered as new promising nanomaterials to be employed in biocatalysis (García-Fruitós et al., 2005), regenerative medicine (García-Fruitós et al., 2010), and in the controlled delivery of therapeutic polypeptides and drugs (Liovic et al., 2012; Villaverde et al., 2012). Moreover, the finding that IBs contain amyloid-like structures makes them a powerful model system in biomedicine for amyloid studies (Carrió et al., 2005; de Groot et al., 2009; García-Fruitós et al., 2011). Indeed, amyloid aggregation is the hallmark of a number of pathological conditions associated with the formation of highly organized protein assemblies, such as in the case of Alzheimer's, prion, and Parkinson's diseases (Chiti and Dobson 2006). In this context, IB formation offers a possibility to investigate protein aggregation *in vivo*—in the presence of cellular molecular crowding, folding modulators, and proteases—therefore using bacteria as model systems easier to be studied and modified than eukaryotic cells (Ami et al., 2013). However, the further development of these IB applications requires a deep knowledge of their structural properties, which are the subject of a few chapters of this book. In particular, we present here the results of recent works that highlight the peculiar structural features of IBs as well as the methods employed to monitor their aggregation process and structure. Interestingly, the possibility of tuning this process and, consequently, the structural and functional features of the final assemblies paves the way for improving IB applications and disclosing new ones.

6.2 INTERMEDIATE SPECIES IN INCLUSION BODY FORMATION

A simplified model of IB formation is reported in Figure 6.1. The polypeptide newly synthetized by the ribosome reaches its native conformation through folding intermediates that can be more prone to aggregation compared to the native form. Under heterologous overexpression conditions, the saturation of the cellular folding modulators, together with other factors such as the absence of specific post-translational modifications, will increase the concentration of aggregation-prone folding intermediates, which in turn will assemble into

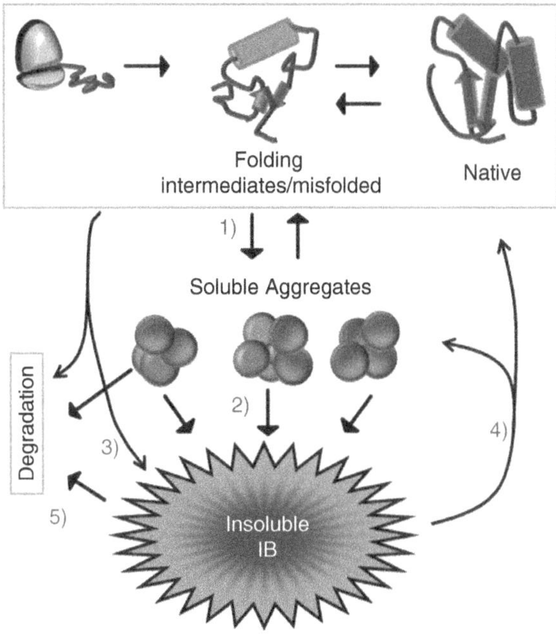

FIGURE 6.1 Scheme of inclusion body formation. See text for details. (*See insert for color representation of the figure.*)

soluble and insoluble aggregates. Partially folded and misfolded species can also originate from the native protein as a consequence of local or global conformational changes induced by normal thermal fluctuation and by stress conditions (step 1 in Figure 6.1). In addition to the expression conditions and to the action of cellular folding modulators, the intrinsic properties of the overexpressed protein are key factors in determining the above events. Concerning this point, mutational analyses of model proteins expressed in *Escherichia coli* showed that both the conformational stability of the native structure and the intrinsic aggregation propensity of the fully or partially unfolded states are able to modulate the distribution of the mutated variants between the soluble and the insoluble cell fractions (Calloni et al., 2005; Espargaro et al., 2008; Winkelmann et al., 2010). Interestingly, these studies also showed that the same physicochemical properties determining the protein aggregation propensity *in vitro* (such as hydrophobicity, β-sheet propensity, and charges) are also crucial *in vivo*. Therefore, the aggregation of heterologous proteins can be inhibited by mutations able not only to stabilize the native structure, but also to tune the aggregation rate of the fully or partially unfolded states.

IB formation from the folded/misfolded intermediates seems to occur starting from multiple soluble small assemblies—with random cellular distribution—that subsequently merge into one or two large aggregates localized at the cell poles (step 2 in Figure 6.1).

In an alternative model, a single or a few nucleation sites trigger the deposition of newly synthetized polypeptides into the growing IBs (step 3 in Figure 6.1). Interestingly, analyses of the aggregate size distribution during IB formation of the human growth hormone (hGH) and of asparaginase suggested that the two proteins follow different aggregation mechanisms: the formation of several small assemblies that merge into larger aggregates in the case of asparaginase IBs, and the formation of a nucleation seed followed by the addition of monomers to the existing aggregates in the case of hGH IBs (Upadhyay et al., 2012). It is noteworthy that the IBs obtained by the two mechanisms showed different structural properties; in particular, hGH IBs displayed higher density, final size, and stability to proteinase K and urea treatments compared to the asparaginase IBs (Upadhyay et al., 2012).

Since IBs are highly dynamic structures, proteins can be released from them and refolded or degraded by the bacterial protein quality control system (Carrió and Villaverde, 2002) (steps 4 and 5 in the Figure 6.1). Therefore, a continuum of protein assemblies and conformations can occur, ranging from the native species to soluble and insoluble aggregates, both of which might contain active and inactive proteins (González-Montalbán et al., 2007). Accordingly, it has been found that the specific activity or fluorescence of the soluble protein fractions can be lower than that of IBs (García-Fruitós et al., 2005); on the other hand, in the soluble fraction of recombinant proteins, soluble aggregates—with different morphologies, ranging from particulate material to fibril-like structures—can also be observed (Martinez-Alonso et al., 2008).

It should be noted that a better comprehension of the mechanisms leading to IB formation has been achieved monitoring protein aggregation *in vivo*, thanks to the availability of different and complementary approaches. The main spectroscopic methods are illustrated in Table 6.1. For instance, the spatial distribution of protein aggregates in *E. coli* can be studied by fluorescence microscopy employing model proteins fused to a fluorescent tag (Ami et al., 2013). Even if the spatial resolution of optical microscopies (see Section 6.4) is not sufficient for a fine structural characterization of aggregates, it enables to monitor the behavior of small fluorescence-labeled aggregation foci. In particular, it has been observed that thermolabile proteins (*Photinus pyralis* luciferase and *E. coli* MetA) fused with the yellow and green fluorescent proteins (YFP, GFP) formed polar foci after temperature upshift, and that the main determinant of polar localization of protein aggregates is the nucleoid occlusion, namely the aggregates formed at the nucleoid-free space (Winkler et al., 2010). It is still an open question whether the delivery of randomly distributed aggregates to the cell poles, with the consequent formation of larger protein aggregates, is an

TABLE 6.1 Spectroscopic Methods for the Study of Protein Aggregation in Intact Cells

Method	Measurements	Application Examples
Fluorescence		
Fusion of the target polypeptide with a fluorescent protein	Bulk cell fluorescence; Fluorescence microscopy and laser scanning confocal microscopy	These approaches can be used: 1. to monitor protein aggregation (Ignatova and Gierasch, 2004)
Fusion of the target polypeptide with the tetra-Cys tag	Förster resonance energy transfer (FRET) experiments	2. for the screening of aggregation inhibitors (Kim et al., 2006)
Thioflavin-S	Flow cytometry	3. for the detection and localization of functional polypeptides within IBs (García-Fruitós et al., 2007a) 4. to follow the spatial distribution of fluorescence-labeled aggregation foci (Rokney et al., 2009; Winkler et al., 2010) 5. to study IB formation at the single cell level (Zhao et al., 2012) 6. for the detection of amyloid-like aggregates within intact cells (Espargaro et al., 2012)
FTIR		
	Label-free (micro) spectroscopy of intact cells	Monitoring the formation of intermolecular β-sheet structures in protein aggregates within intact cells (Doglia et al., 2008); membrane lipid rearrangement induced by heterologous protein expression (Ami et al., 2009)
NMR		
	Solid-state NMR of whole cells	Detailed structural information of the protein embedded within IBs (Curtis-Fisk et al., 2008)

Source: Adapted from Ami et al. (2013).

active and energy-dependent process (Rokney et al., 2009; Winkler et al., 2010; Kirkpatrick and Viollier, 2011). Soluble aggregates have also been observed and characterized during the recombinant expression of *Brucella abortus* PdhS histidine kinase fused to the mCherry fluorescent protein. In particular, the fusion protein forms soluble aggregates where both polypeptides are functional, as indicated by the fact that these soluble aggregates are fluorescent and able to recruit known PdhS partners (Van der Henst et al., 2010).

We should add that another very important property of IBs concerns the possible co-aggregation of the embedded proteins. Interestingly, this issue was investigated in prokaryotic cells by Morell and colleagues (2008) by means of Förster resonance energy transfer (FRET, see Section 6.4), which made it possible to evaluate the degree of specificity of the aggregation process *in vivo*. To this end, the self-aggregating Aβ42 amyloid peptide and the VP1 capsid protein were labeled with proper fluorescent protein variants, the blue (BFP) and the green (GFP). Interestingly, when the two dyes were fused to the same polypeptide, the FRET efficiency was found to be higher than in the case of BFP and GFP fused to the two different polypeptides, indicating that the *in vivo* aggregation of the two proteins was highly specific (Morell et al., 2008).

Overall, these studies indicate that different conformations and aggregation states of the recombinant proteins can be found in both the soluble and the insoluble cell fractions. This is in contrast to what is proposed by the conventional model, stating that native functional polypeptides are present only in the soluble fraction, while only disordered nonfunctional polypeptide are embedded in the insoluble IBs. Interestingly, a clear identification and precise characterization of subclasses of insoluble aggregates have been reported in the work of Schrodel and de Marco (2005). Indeed, these authors showed that following the expression at 20°C of a recombinant GFP–GST fusion protein in *E. coli*, the insoluble aggregates can be separated—by a sucrose step gradient—in several subclasses characterized by different density and specific conformational properties. In particular: (1) in all classes the fusion protein retained at least part of its GFP fluorescence; (2) an increasing Thioflavin T (ThT) binding and larger aggregates were observed, going from the upper to the lower gradient fractions; and (3) aggregates in the upper fractions were disaggregated more efficiently *in vitro* by a combination of molecular chaperones. Moreover, a preferential binding of specific chaperones was observed for each aggregate subclass, and interestingly, the distribution of the recombinant protein among the different subclasses was found to depend on DnaK availability (Schrodel and de Marco, 2005).

All these results emphasize the complexity of heterologous protein aggregation in bacterial cells where, under the cell quality control machinery of chaperones and proteases, the recombinant protein can assume several conformational states embedded into different soluble and insoluble supramolecular assemblies.

6.3 STRUCTURAL CHARACTERIZATION OF INCLUSION BODIES

6.3.1 Composition, Overall Structure, Shape, and Morphology

IBs consist primarily of the recombinant protein, which can represent up to 80 to 100% of the total cell content and whose percent fraction usually increases at longer expression times. Moreover, IBs might contain proteolytic fragments of the recombinant protein, ribosomal components, and traces of phospholipids and nucleic acids, as well as a small fraction of polypeptides unrelated to the target protein (Carrió et al., 1998; Carrió and Villaverde, 2002). We should note that potentially, the inactivation of these cellular proteins represents a cytotoxic event, as reported recently in the case of amyloid-like aggregate toxicity in human cells (Olzscha et al., 2011). Moreover, cellular folding modulators such as DnaK, GroEL, and the small heat-shock proteins IbpA and IbpB, have been found in IBs, where they could also have a functional role. For example, IbpA and IbpB might be involved in the release of proteins from IBs, while DnaK in the processing of the IB surface (Carrió et al., 1998; Carrió and Villaverde, 2002). Note of worthy is the fact that the heterogeneity of IB composition can be modulated not only by the expression time but also by temperature.

In transmission electron microscopy (TEM) imaging of ultrathin cell sections (see Table 6.2 and Section 6.4), IBs appeared as electron-dense material with electron-transparent regions of 20 to 30Å, probably due to internal pores (Bowden et al., 1991). The electron microscopy (EM) analyses of IBs, both purified or within the cell, revealed different shapes (ovoid, cylindrical, and spherical), with size ranging from 0.05 to more than 1.6 μm. An IB density of the order of about 1.3 g/mL has been found combining EM and sedimentation field-flow fractionation (sedFFF), a technique that can be employed to characterize the size distribution of bacterial IBs (Margreiter et al., 2008).

When imaged by scanning electron microscopy (SEM), IBs appear as porous structures characterized by surfaces with very different properties: namely, rough or, on the contrary, smooth. A highly porous structure is in agreement with IB density characterizations by centrifugal sedimentation analysis. Indeed, it has been found that the IB density increases with that of the suspending solvent, indicating that the solvent can penetrate inside IBs. These results suggested the presence of accessible void substructures which can reach 85% of the total volume, as in the case of prochymosin IBs (Taylor et al., 1986). This porous architecture can be exploited for biocatalysis applications when the protein embedded in the IBs retains its activity. Indeed, thanks to this property, substrates and reaction products can, respectively, go in and out from the biocatalyst.

TABLE 6.2 Structural Properties of IBs[a]

Functional and Structural Properties	Methods of Investigation	Biotechnological Implications
Overall stability		
Stability is dependent on the specific protein and on the expression conditions. For instance, hGH IBs increased their stability at higher induction time [1].	Proteinase K and chemical denaturation by turbidity [1–3].	Mild solubilization for improved recovery of functional proteins. Nonclassical IBs can be used as a delivery system of functional proteins [4].
Size, shape, morphology		
Ovoid, cylindrical, and spherical shape; tear-shaped in cells deficient in the ClpP protease [5]. Size ranging from 0.05 to >1.6 μm. Porous morphology with rough or smooth surface. IBs appeared as electron-dense material with regions electron transparent of 20–30 Å, probably due to internal pores [6].	Electron microscopy [5,6]. AFM [7]. Size analyzer [8]. Sedimentation field-flow fractionation [9]. Dynamic light scattering [7].	The IB shape and volume are mainly dependent on the host genetic background and on the conditions and time of growth [5,9–12]. The possibility to tune IB geometry and mechanical properties has a great potential for the use of IBs as biomaterial in tissue engineering [4,5,7].
Native-Like		
The proteins embedded in IB can retain the native structure and function.	FTIR [13]. Fluorescence and enzymatic activity [14]. NMR [15].	Active IBs can be employed as efficient, stable, and reusable biocatalysts. Release of functional proteins [4].
Amyloid-Like		
High presence of intermolecular β-sheets. Binding to amyloid dyes: Congo Red, Thioflavin T and S. Seeding of protein fibrillogenesis *in vitro*.	FTIR [13,18]. CD. Binding to amyloid-specific dye. X-ray. Electron microscopy [17].	IB as a model system to study amyloid aggregation in cells and for the screening of antiamyloidogenic compounds [17,18,22–25].

(*Continued*)

TABLE 6.2 (Continued)

Functional and Structural Properties	Methods of Investigation	Biotechnological Implications
Amyloid-Like		
Specific protein–protein interaction in the aggregation process. Presence of prefibrillar and fibrillar structures within IBs [16,17].	AFM [17,18]. NMR [19–21].	

[a]References cited: [1] Upadhyay et al., 2012. [2] Espargaro et al., 2008. [3] Cano-Garrido et al., 2013. [4] García-Fruitós et al., 2012. [5] García-Fruitós et al., 2010. [6] Bowden et al., 1991. [7] García-Fruitós et al., 2009. [8] Taylor et al., 1986. [9] Margreiter et al., 2008. [10] Carrió et al., 1998. [11] Carrió and Villaverde 2003. [12] García-Fruitós et al., 2007b. [13] Doglia et al., 2008. [14] García-Fruitós et al., 2005. [15] Curtis-Fisk et al., 2008. [16] Carrió et al., 2005. [17] de Groot et al., 2009. [18] Invernizzi et al., 2012. [19] Wasmer et al., 2009, [20] Wang et al., 2008. [21] Taylor et al., 2011. [22] García-Fruitós et al., 2011. [23] Espargaro et al., 2012. [24] Villar-Piqué et al., 2012. [25] Villar-Piqué and Ventura, 2012.

The IB size is mainly dependent on the host genetic background, on the conditions and time of growth, and on the specific recombinant proteins (Carrió et al., 1998; Carrió and Villaverde, 2003; García-Fruitós et al., 2007b, 2010; Margreiter et al., 2008). For example, an opposite effect of GroEL and DnaK on the size of VP1LAC IB has been observed. While the absence of GroEL resulted in very small IBs and in a larger number, the absence of DnaK was associated with larger IBs compared to wild-type cells (Carrió and Villaverde, 2003). An additional interesting example is that of asparaginase and hGH IBs; indeed, while the first protein formed small aggregates of about 200 nm in size, constant during the expression time, hGH IBs increased in size from 200 to 800 nm (Upadhyay et al., 2012). In addition, not only IB size, but also their shape can be tuned by the host genetic background. In particular, it has been found that inactivation of the cytosolic protease ClpP led to the formation of tear-shaped IBs of the fusion protein VP1-GFP, as observed by scanning electron microscopy and confocal microscopy (García-Fruitós et al., 2010). It is noteworthy that these IBs displayed a high level of fluorescence, indicative of the presence of native-like polypeptides, and were also able to promote the proliferation of PC112 cells when employed for surface decoration. This last result underlines the potential use of IBs as substrate materials for tissue engineering (García-Fruitós et al., 2010).

It has been proposed that the sponge-like architecture of IBs is supported by a proteinase K-resistant and amyloid-like scaffold whose gap can be filled with more native-like and proteinase K-sensitive polypeptides (Cano-Garrido et al., 2013). Indeed, it has been found that the treatment of IBs of a fluorescent protein with proteinase K leads to a progressive reduction of IB fluorescence and density without detectable alterations of IB size and morphology, as studied by DLS, confocal fluorescence microscopy, and Cryo-TEM (Cano-Garrido et al., 2013). Similarly, a cotton-like matrix for the inner IB architecture has been proposed for other proteins, examining the effect on IB morphology and dimension of low pH, low detergent concentrations, or protease treatment (Peternel and Komel, 2011).

6.3.2 Native-like and Amyloid-like Structures

For a long time, IBs have been considered amorphous deposits consisting of completely denatured proteins. Interestingly, this view changed significantly in recent years, and several pilot studies highlighted the IB molecular architecture, also shedding new light on their physiological role (Carrió et al., 2000). In particular, it has been shown that IBs are heterogeneous and dynamic ensembles in which the embedded proteins can retain a considerable amount of native-like structures and therefore biological activity, thus making IBs important sources of active proteins (García-Fruitós et al., 2007a). In this perspective, IBs can be used for the release of functional proteins under mild denaturation conditions or as biocatalysts (García-Fruitós et al., 2012). Indeed, several enzymes have already been investigated for IB-based catalysis, including aldolases, oxidases, and kinases. These studies pointed out several practical advances of functional IBs, including efficient *in vivo* immobilization of the protein at high yields as IBs are enriched in recombinant proteins, efficient mass transfer thanks to their porous nature, and long-term stability under storage and reaction conditions that allow the reuse of the enzyme. We should add that IB-based catalysts can be used as naked materials, cross-linked or matrix-embedded (García-Fruitós et al., 2012).

The presence of native-like structures in proteins embedded in IBs has been explored by different optical spectroscopies (Table 6.2) at different spatial resolution, ranging from polypeptide backbone conformations to single residues. Among these approaches, fluorescence is one of the most widely used. Indeed, different fluorescence-based techniques can take advantage, for example, of the use of a fluorescent partner bound to the target protein. In particular, the discovery of naturally fluorescent proteins such as the GFP, and the development of some of its derivatives, including yellow, blue, and red ones, were shown to be very useful for this goal. For example, García-Fruitós and colleagues (2007a) were able to localize functional polypeptides within IBs, monitoring the fluorescence of GFP that was used as a model system. In particular, the authors found that the fluorescence emission was higher in the core of IBs than in their external shell, an observation indicating that the active and therefore native-like protein was present primarily in the inner part of the protein deposits. Interestingly, this result indicated that proteins are not trapped passively in IBs; rather, IBs are the result of a dynamic process that consists of protein deposition and release. The monitoring of GFP fluorescence also enabled (Garcia-Fruitós et al., 2007b) to investigate the influence of a protein quality control system in *E. coli*, a key factor in managing aggregation. The authors showed in particular that it is possible to modulate chaperone and protease activities to obtain aggregates with different properties, making it possible in this way to obtain IBs containing a high level of functional proteins, useful in a variety of applications.

Among the spectroscopic methods that enable to obtain structural information on the proteins within IBs, we should mention Fourier transform infrared (FTIR) spectroscopy (see Section 6.4). FTIR is a vibrational technique that provides information on the protein secondary structures in a label-free way, through the analysis of the amide I absorption band, due to the CO stretching vibration of the peptide bond (Barth, 2007). Moreover, direct information on protein aggregates can also be achieved through the monitoring of a specific IR response due to intermolecular β-sheets, whose intensity and peak position—approximately between 1620 and 1630 cm^{-1} when measurements are performed in H_2O—are affected by the extent and compactness of the aggregates. We should add that the study of protein secondary structures requires the use of resolution-enhancement procedures, such as second derivatives, that allow to resolve the various components overlapping under the broad amide I band (Susi and Byler, 1986). In this way it is therefore possible to better assign each spectral component to a specific protein secondary structure.

An interesting application of the FTIR approach is the in situ monitoring of the formation of IBs directly within intact bacterial cells that makes it possible, for example, to follow the kinetics of protein aggregation in real time. As an example, in this way we monitored the expression of *P. fragi* lipase (PFL) in intact *E. coli* cells grown at two temperatures, 27 and 37°C (Ami et al., 2005). In particular, the aggregation occurring at the two temperatures was found to produce IBs with a different extent of native-like protein: at 27°C the intensity of the α-helix component at approximately 1656 cm^{-1} was higher than that due to aggregates at about 1627 cm^{-1}. The opposite behavior was observed for IBs produced at 37°C. Moreover, to better explore the structural differences of IBs formed at the two temperatures, we also measured IBs extracted from cells (see Figure 6.2). We again observed significant differences between the two samples, which confirmed the results found in situ. Indeed, a different residual secondary native-like structure of the protein was found to characterize the IBs obtained by cells grown at the two temperatures, indicating that PFL assumes different conformational states within IBs, depending on the expression conditions.

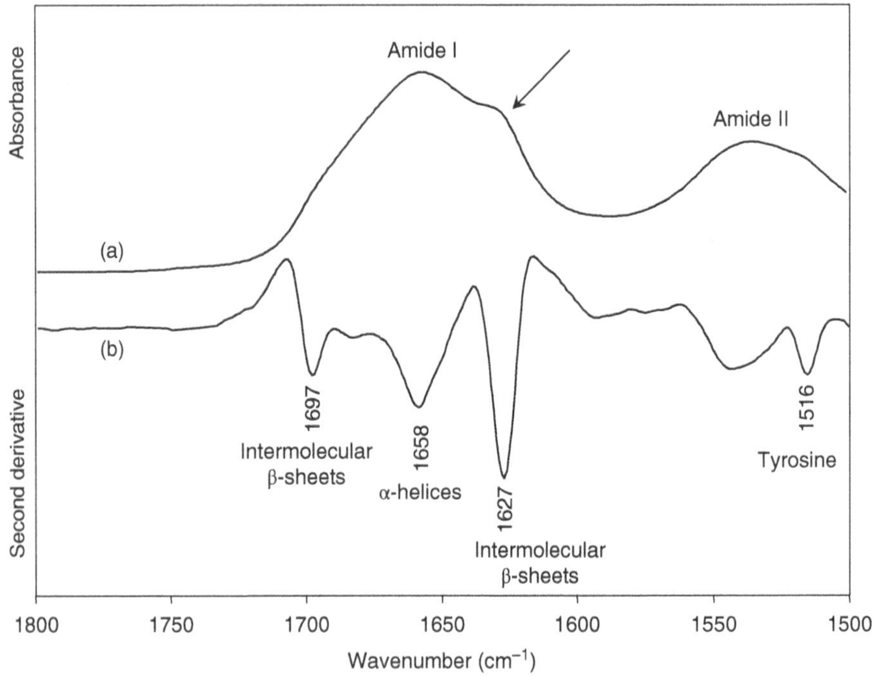

FIGURE 6.2 Infrared spectroscopy of IBs. The FTIR absorption spectrum (a) of PFL IBs, extracted at 8 h of induction at 37°C, and its second derivative (b) are reported in the amide I and II regions. The peak positions of the main components and their assignment to the protein secondary structures are also indicated. The arrow points to the absorption band due to the intermolecular β-sheet structures in aggregates.

The relationship between the retention of residual native-like structure and IB formation was also investigated by FTIR in the case of two recombinant *E. coli strains* expressing different levels of human interferon-alpha-2b (IFN-α-2b) (Ami et al., 2003, 2006). In particular, the marker band of aggregates, around 1628 cm^{-1}, was found to increase in intensity with the level of protein expression. Moreover, the high producer strain turned out to be characterized by a minor extent of the α-helical native-like component compared to the low producer strain.

Also in this case, the in situ results were confirmed by the comparison with the IR response of IBs extracted from low and high producer strains (Ami et al., 2006). Indeed, while a similar extent of residual native-like structure component was observed in the two strains, a significant difference in intensity was found for the aggregate component, which was higher in the case of the high producer strain, as expected. Thus, the ratio between the aggregate band and that of the residual native-like structures in IBs was found to be dramatically different in the two cases, depending on the protein expression level.

A vibrational approach complementary to FTIR, able to provide information on IB structural properties, is Raman spectroscopy. It should be noted that whereas FTIR is particularly sensitive to the backbone amide I vibrations, Raman is more sensitive to vibrations due to the amino acid side chains. Moreover, this approach could give information on disulfide bond formation, highlighting the role of cysteines in the aggregation process. Indeed, among the Raman studies of IBs, one of the leading works is that of Przybycien and colleagues (1994); here the authors reported the analysis of the secondary structure of β-lactamase embedded in bacterial IBs, formed at different culture conditions, 37 and 42°C. In particular, they observed an increasing extent of β-sheet structures at higher temperature, even if all the IBs studied were found to retain a significant amount of native-like structures.

The presence of β-sheet structures in IBs has also been detected by x-ray diffraction. Indeed, the x-ray diffraction spectra of IBs displayed two diffraction rings, at about 4.7 and 10Å, respectively assigned to the spacing between strands in a β-sheet and to the spacing between β-sheets in a cross-β-sheet amyloid-like conformation (Wang et al., 2008). Indeed, this diffraction pattern has also been observed for amyloid fibrils, where the two reflections typically appeared in an orthogonal position compared to the circular profile observed for IBs, suggesting not strongly aligned β-sheets within IBs (Wang et al., 2008).

To obtain information on protein conformations at the residue level, nuclear magnetic resonance (NMR) spectroscopy is the technique of choice, as described in Chapter 7 of this book. In particular, solid-state NMR has been used to study the structural properties of proteins embedded in IBs, not only extracted but also within intact cells. For example, in the pioneering work of Curtis-Fisk and colleagues, the authors characterized—in whole cells—the functional domain of the influenza virus hemagglutinin protein HA2 subunit, in the form of IBs (Curtis-Fisk et al., 2008). Noteworthy, they identified the presence of residual native α-helices at some specific residues, also revealing the presence of conformational heterogeneity of the protein within IBs.

Another interesting application of solid-state NMR to the in situ investigation of the conformational properties of a protein within IBs concerned the Fgp41 fusion protein, which contains the integral membrane protein of the HIV ectodomain. In particular, in this case the authors were able to disclose the presence of a six-helix bundle stable fold for the protein embedded within IBs (Vogel et al., 2011). We should emphasize that an important advantage of NMR-based approaches is that they make it possible to correlate the structural properties of aggregates—identified at specific residues—with their function, such as cellular toxicity or enzymatic activity (Wang et al., 2010).

Altogether, these studies indicate that the same protein, when expressed as IBs, can retain a residual native-like structure to a varying extent, leading to

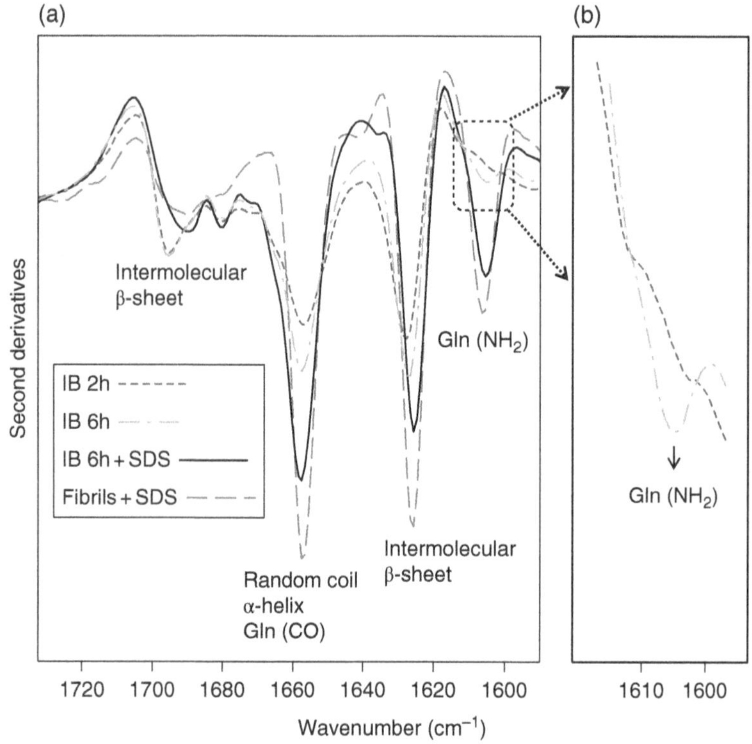

FIGURE 6.3 Ataxin-3 aggregation in form of IBs. (a) Second derivative infrared spectra in the amide I region of extracted IBs of AT3-Q55, collected after 2 h of induction in *E. coli* and after 6 h. The second derivatives spectra of AT3-Q55 fibrils, obtained *in vitro*, and of IBs at 6 h of induction are also reported after SDS treatments. (b) Second derivative spectra of IBs, extracted 2 h and 6 h after induction, are reported on an enlarged wavenumber scale in the spectral region of the glutamine NH_2 deformation modes. The band assignment to the secondary structure components is indicated. The second derivative spectra are normalized at the tyrosine peak at about 1515 cm^{-1}, to take into account possible differences in protein content. (*See insert for color representation of the figure.*)

aggregates of different complexity. Indeed, in addition to proteins with native-like structures and function, it has been found that IBs also contain a variable amount of amyloid-like structures (see also Chapter 4 in this book by Villar-Piqué and Ventura for a discussion of the amyloid-like properties of IBs). In particular, IBs are the result of a highly protein-specific aggregation process and are characterized by the presence of intermolecular β-sheets. They can also bind the typical amyloid dyes (i.e., Congo Red, Thioflavin T, and Thioflavin S) and can seed protein fibrillogenesis *in vitro* (Carrió et al., 2005; Morell et al., 2008; Wang et al., 2008; de Groot et al., 2009; García-Fruitós et al., 2011). Although it is still not clear how the native- and amyloid-like conformations of the same protein coexist in IBs, several experimental data underline the possibility of tuning the relative fractions of these structures by the genetic background of the host cells and by the expression conditions.

As already discussed, an interesting model has recently been proposed, where the IB amyloid-like components formed a proteinase K-resistant scaffold responsible for the mechanical properties of the entire IB structure, whereas proteins with a more native-like conformation and sensitive to proteinase K were embedded in the matrix. These latter components are eventually responsible for the IB activity (Cano-Garrido et al., 2013). It is worthy to note that the presence of amyloid-like features makes IBs a useful model system for the study of amyloid aggregation in a cellular environment and for the screening of anti-amyloid compounds (de Groot et al., 2009; García-Fruitós et al., 2011; Espargaro et al., 2012; Villar-Piqué et al., 2012; Villar-Piqué and Ventura, 2012; Ami et al., 2013). For example, the heterologous expression in *E. coli* of the amyloid protein ataxin-3 made it possible to study the multistep aggregation mechanism of an authentic polyglutamine (polyQ) protein in a cellular milieu (Invernizzi et al., 2012).

In particular, proteins containing a stretch of repeated glutamine (the polyQ region) undergo aggregation when the number of glutamines exceeds a critical threshold, resulting in an amyloid disease (Machado–Joseph disease or spinocerebellar ataxia type 3 in the case of ataxin-3). In Figure 6.3 we reported the second derivative infrared spectrum of IBs extracted after 2 h of induction of the pathological expanded ataxin-3 variants containing 55 glutamines in the polyQ tract (AT3-Q55). In the amide I region, the spectrum displayed two marker bands due to the formation of intermolecular β-sheets and a component at around 1657 cm^{-1} that can be assigned to the unordered and/or α-helical structures of the proteins. The spectrum of AT3-Q55 IBs collected after 6 h of induction showed an additional band at around 1604 cm^{-1} (Invernizzi et al., 2012), which we demonstrated in a previous work (Natalello et al., 2011) to be due to the NH_2 deformation mode of the glutamine side chains involved in an ordered network of hydrogen bonds. We also showed that this band is a marker of the irreversible formation of SDS-insoluble amyloid aggregates (Natalello et al., 2011). Indeed, the 1604-cm^{-1} peak is preserved and displayed a higher relative intensity after SDS treatment of the AT3-Q55 IBs extracted at 6 h of induction time (Figure 6.3) (Invernizzi et al., 2012). Moreover, the IR spectrum of these SDS-insoluble materials is very similar to that of AT3-Q55 fibrils measured after SDS treatment (Natalello et al., 2011), as shown in Figure 6.3. In particular, the spectrum is characterized by components due to intermolecular β-sheet interactions and by peaks at about 1657 and 1604 cm^{-1}, respectively, due to the C=O stretching and NH_2 deformation modes of the glutamine side chains, which are involved in a network of strong hydrogen bonds in the final aggregates (Natalello et al., 2011). These results disclosed that in a cellular environment, AT3-Q55 undergoes a multistep aggregation, as reported previously *in vitro* (Ellisdon et al., 2006; Natalello et al., 2011). In this process, the formation of intermolecular β-sheets is independent on the polyQ stretch and precedes the rearrangement of the glutamine side chains into an ordered side-chain hydrogen bonding. This, in turn, results in the generation of irreversible and SDS-insoluble aggregates (Ellisdon et al., 2006; Natalello et al., 2011). Moreover, these data indicated the presence of amyloid-like materials in AT3-Q55 IBs, as confirmed by the fibrillar morphology of these aggregates after SDS treatment (Invernizzi et al., 2012).

Direct structural comparison at the residue-specific resolution of IBs and amyloid fibrils of the same proteins has also been reported in the literature by NMR spectroscopy (Wasmer et al., 2009; Wang et al., 2010; Taylor et al., 2011). In particular, it has been found that IBs of the C-terminal region (residues 218 to 289) of the *Podospora anserina* prion [HET-s(218–289)] display an identical chemical shift of HET-s(218–289) fibrils in the $^{13}C–^{13}C$ proton-driven spin-diffusion solid-state NMR spectra. Interestingly, HET-s(218–289) IBs and fibrils were found to be characterized by similar infectivity, chemical stability, seeding competency, and by a similar H/D exchange pattern as studied by liquid-state NMR (Wasmer et al., 2009). All these results (Sabate et al., 2009; Wasmer et al., 2009) clearly indicated that IBs of the HET-s fungal prion have the same molecular structures of HET-s fibrils formed *in vitro*. It is noteworthy that the opposite has been found in the case of β2microglobulin (b2m). Indeed, b2m IBs and fibrils (obtained *in vitro* at pH 2.5) displayed important differences in ThT binding and molecular structure and dynamics, as observed by solid-state NMR (Taylor et al., 2011). These results underlined that the presence of amyloid-like structures in IBs also reflects the intrinsic amyloid propensity of a given protein, strengthening the use of IBs as model systems for the study of protein aggregation.

Finally, we should note that IBs are characterized by an intrinsic dynamic nature beyond the structural and compositional changes induced and modulated by the cellular machineries. As an example, IBs are not only able to release functional polypeptides, but the proteins embedded within IBs can undergo conformational transitions in situ as well as in a cell-free environment (González-Montalbán et al., 2008). Indeed, it has been found that IBs of the misfolding-prone β-galactosidase protein VP1LAC displayed increased native-like secondary structures after arrest of protein synthesis, as detected by FTIR spectroscopy. Interestingly, the in situ refolding was associated with an increase in IB enzymatic activity both *in vitro* (i.e., without soluble cellular factors) and *in vivo*, where this in situ activation is regulated negatively by the chaperone DnaK (González-Montalbán et al., 2008). These data suggest that protein-folding intermediates can also complete their folding process after the deposition in IBs, which are highly hydrated and porous aggregates characterized by an unexpected conformational plasticity. Note that the structural reorganization of proteins observed within bacterial IBs (González-Montalbán et al., 2008; Invernizzi et al., 2012) appeared to be a feature common to other aggregates of different origins, such as in the case of amyloid formation from native-like precursor aggregates (Plakoutsi et al., 2005).

In conclusion, IBs appeared to be highly dynamic protein deposits in which the embedded polypeptides can assume a continuum of conformations ranging from the native to the amyloid structures, whose relative fractions can be tuned by the genetic background of the cell and by the expression conditions. In this way, IBs can ensure their biological role *in vivo* and can have very promising biotechnological applications. Indeed, IBs are characterized by a stable architecture able to release *in vivo* and *in vitro* functional proteins, therefore being a source of functional proteins, even if in other cases their enzymatic activity is associated with a very low release of proteinaceous materials. In the latter case, IBs can be employed directly as reusable biocatalysts. Moreover, they can be considered a model system for amyloid studies as well as biocompatible materials that can stimulate mammalian cell proliferation, with possible uses in tissue engineering. We should emphasize that the use of several biophysical approaches made it possible to disclose, also *in vivo*, unexpected properties of IBs. Indeed, it would be desirable to further improve new experimental approaches to better understand the mechanisms of protein aggregation and the factors that can modulate aggregation. To this end, methods allowing in situ studies of aggregation should be developed further, since they could also provide important information for the application of IBs in recombinant protein production and in biomedicine.

6.4 APPENDIX: EXPERIMENTAL METHODS USED IN IB STRUCTURAL CHARACTERIZATION

We give here a brief overview of the main experimental methods applied to IB structural characterization: fluorescence spectroscopy and microscopy, flow cytometry, circular dichroism, Fourier transform infrared spectroscopy, electron microscopies, atomic force microscopy, and x-ray fiber diffraction.

Fluorescence Spectroscopy and Microscopy Fluorescence relies on the light absorption of a fluorophore at a given excitation wavelength and on the subsequent emission of fluorescence at a longer wavelength. After labeling the cell components with fluorophores with different fluorescence excitation and emission spectra, it is possible to obtain information on biomolecules in solution and on cell suspension (Cantor and Schimmel, 1980; Michalet et al., 2003; Lakowicz 2006; Sabate and Ventura, 2013). In addition, as reported widely in the literature in recent years, fluorescence microscopies have been very successful in determining the intracellular localization of fluorescently labeled biomolecules and obtaining the fluorescence images of different cell components within single living cells, at a spatial resolution limited by diffraction. In the case of laser scanning confocal fluorescence microscopy (Paddock, 2000), the spatial resolution can be of the order of 0.1 μm in the image X,Y-plane and about 0.5 μm in the axial direction. This resolution can be improved by one order of magnitude when working with new super-resolution microscopies such as STED (Hell, 2007). However, using fluorescence microscopy, the localization of two interacting fluorescent molecules can be determined at higher resolution, exploiting the techniques of fluorescence resonance energy transfer (FRET) (Sun et al., 2011). FRET is a mechanism in which a transfer of excitation energy occurs between two separated fluorophores, a donor and an acceptor. In this process, the excitation energy of the donor in an excited electronic state is transferred nonradiatively to the acceptor, which in turn emits a sensitized fluorescence in the absence of any radiative excitation process.

This transfer requires a long-range dipole–dipole interaction between the electronic systems of donor and acceptor and occurs over intermolecular distances of about 10 nm. Its efficiency, depending on the dipole–dipole interaction between the two partners, is therefore sensitive to their distance, to the alignment of their electric transition dipoles, and to the resonance condition between the donor and acceptor electronic transitions. This condition can be evaluated experimentally through the overlapping of the acceptor absorption spectrum with the emission spectrum of the donor. Therefore, when FRET occurs between two fluorescently labeled cell components (properly chosen to satisfy the requirements to be FRET partners), one can say that their separation is less than 10 nm, a distance that cannot be evaluated by confocal microscopy, which would, instead, see the two components as overlapped. To verify experimentally if FRET takes place between donor and acceptor, a quenching of the donor fluorescence is expected to occur, accompanied almost simultaneously by an increase in the fluorescence emission of the acceptor. This behavior could therefore be monitored upon excitation of the donor fluorescence and detection of the acceptor emission fluorescence.

To exploit FRET microscopy to study intracellular protein interactions, fusion of the target polypeptide with a fluorescent tag protein has been developed. The two fusion proteins GFP and BFP can behave as donor (BFP) and acceptor (GFP) partners, respectively. When FRET takes place, the acceptor emission at 510 nm increases upon excitation of the donor in its absorption band at 380 nm.

Under this excitation, if no acceptor fluorescence emission is observed, no FRET interaction occurs between the two labeled proteins.

Flow Cytometry It is a rapid technique that can count and examine microscopic particles and cells in suspension, in a very short time of only a few seconds for thousands of particles (Jacquemin-Sablon, 1993; Espargaro et al., 2012; Zhao et al., 2012). Multiple parameters can be detected at a single cell level, allowing to identify different cell populations on the basis of their parameters (Jacquemin-Sablon, 1993). The flow cytometer is composed of a fluid stream transporting cells properly stained with specific dyes or fluorescent labels. When, by hydrodynamic focusing, single cells in suspension enter into the focus of a collimated laser beam, their fluorescence emission at different wavelengths is excited, while the incident light is scattered in different directions. The intensity of fluorescence emission and scattered light in the forward [forward-angle light scattering (FALS)] and right-angle directions [right-angle light scattering (RALS)] are collected simultaneously. Light scattering data can be used to determine cell number, size, and morphology in order to discriminate among the various cell populations present in the sample. Interestingly, an important advance in instrumentation has been reached by adding the facility of cell sorting to separate and identify the cell populations (Herzenberg et al., 2002).

Circular Dichroism Among optical spectroscopies, circular dichroism (CD) has been widely employed to characterize the structural properties of chiral molecules in solution, such as nucleic acids and proteins (Kelly et al., 2005; Greenfield, 2006). The CD spectrum is the difference in absorption of a left- and right-circularly polarized light $\Delta A = A_L - A_R$ (where A_L and A_R are, respectively, the absorbance of the sample for left- and right-circularly polarized light), occurring in the spectral range where the sample absorbs. The difference in absorption ΔA is about 3×10^{-4}, that is 0.03% of the total absorption. When the two left- and right-circularly polarized lights pass through a dichroic sample, a differential decrease of their amplitude occurs and the emerging beam is, as a consequence, elliptically polarized. The ellipticity θ (in degrees) of the outcoming beam is related to the difference in absorption ΔA by the equation $\theta = 32.98\ \Delta A$.

Among chiral biomolecules, proteins have been successfully studied using CD. In this case the data are usually (Greenfield, 2006) converted to mean residue ellipticity [θ,] given in deg cm^2 dmol^{-1} as $[\theta] = \theta/LCN$, where θ is the ellipticity in millidegrees, L is the pathlength in millimeters, C is the protein molar concentration, and N is the number of protein residues. Interestingly, the CD spectrum of a protein enables to obtain information on its secondary structure, since the different protein secondary structures give rise to characteristic CD signatures in the far-ultraviolet (UV) range. The α-helix displays a positive band below 193 nm and a negative band with "double minima" at about 208 nm and 222 nm; random coil is, instead, characterized by a large negative band with a minimum around 195 nm and a low positive band above about 210 nm. A negative band with a minimum at about 218 nm and a positive band around 195 nm is found for β-sheet structures. CD spectroscopy therefore became a very useful and easy diagnostic technique for the determination of protein secondary structures in solution (Kelly et al., 2005; Greenfield, 2006). Interestingly, additional structural information can be obtained for proteins by CD in different spectral regions. In particular, the CD spectrum in the near UV (320 to 260 nm) gives information on the protein tertiary structures. In addition, an induced dichroism can occur on a molecule with no intrinsic chirality upon binding or interaction with a chiral molecule, a phenomenon known as induced Cotton effect. An example is the Cotton effect for the ThT dye bound to an amyloid fibril (Dzwolak et al., 2013).

Fourier Transform Infrared Spectroscopy The absorption of light from a molecule in the infrared range of long wavelengths from 2 to 50 μm (corresponding to wavenumbers from 5000 to 200 cm^{-1}) is due to the transitions between the vibrational levels of the molecule in its ground electronic state (Barth and Zscherp, 2002). As for the UV and visible light absorption, the vibrational band intensities are determined by the light-induced transition dipoles, but with an extinction coefficient lower by more than one order of magnitude than that of the electronic transitions.

In the last decades, Fourier transform infrared (FTIR) spectroscopy emerged as a powerful technique to provide a rapid determination of protein secondary structure, conformational changes, and aggregation, with no restriction on their molecular weight. This has been made possible thanks to important technical advances, such as the use of interferometric spectroscopy, of highly sensitive infrared detectors, and of the availability of powerful data analysis approaches, enabling spectral corrections and resolution-enhancement procedures. In addition, the great potential of this technique relies on the possibility of investigating samples in different physical states, not only in solutions, but also in solid states, including highly scattering systems, such as complex biological systems, due to the fact that scattering effects in the infrared spectral range are limited compared to those in the UV and visible ranges.

The FTIR absorption spectrum of a protein is dominated by the absorption bands due to the vibrational modes of the peptide bond, called amide bands. The most important is the amide I band, due primarily to the stretching vibration of the C=O peptide group that occurs in the spectral range from 1700 to

1600 cm^{-1}. The frequency of the amide I band depends on the environment of the peptide C=O group in the polypeptide chain, and it is therefore sensitive to the secondary structures of the protein. Infrared studies on model peptides and on proteins of known structure indicated that the amide I band of the various secondary structures occurs at specific wavenumbers. For α-helices in water, the amide I band was found around 1654 cm^{-1}, within the range 1648 to 1660 cm^{-1}; for β-sheets in water two amide I components were observed around 1633 and 1686 cm^{-1} (extremes: 1640 to 1623 and 1695 to 1674 cm^{-1}, respectively); for β-turns the amide I absorption typically is seen between 1686 and 1662 cm^{-1}; for unordered structures in water the amide I band is found to be around 1654 cm^{-1} (extremes: 1657 to 1642 cm^{-1}), very close to those of α-helices. In D_2O the unordered band shifts to lower wavenumbers, down to 1640 cm^{-1}, and it is therefore possible to discriminate between α-helix and unordered structure through hydrogen/deuterium exchange experiments.

Interestingly, intermolecular β-sheets in protein aggregates display amide I absorption in the region from 1630 to 1620 cm^{-1} (Seshadri et al., 1999; Zandomeneghi et al., 2004; Natalello and Doglia, 2010). The amide I band of a protein is therefore the result of the overlapping of the C=O absorption bands due to the various secondary structures and aggregates (Arrondo and Goni, 1999; Barth, 2007; Natalello et al., 2012). To extract from the FTIR data information on the protein secondary structures and on aggregates, it is therefore necessary to resolve the amide I band into its spectral components using a deconvolution procedure. A simple approach is offered by the second derivative mathematical analysis of the absorption spectrum, since the derivative spectrum is characterized by negative peaks corresponding to the absorption maxima in the measured spectrum (Susi and Byler, 1986). This procedure therefore allows identification of the amide I components and their assignment to various protein secondary structures on the basis of their peak position. To obtain the percentage of secondary structures in a protein, a fitting of the amide I band has to be performed by iterative procedures. For example, a fitting of the amide I experimental band profile into a linear combination of bell functions, representing the different band components, can be performed (Arrondo et al., 1993; Arrondo and Goni, 1999; Natalello et al., 2005).

Interestingly, in recent years, advances in infrared microscopy (Levin and Bhargava, 2005) were successfully employed in the infrared study of complex biological systems. Despite a limited spatial resolution (of the order of a few tens of micrometers), it enabled collection of the response of single intact cells, tissues, and even whole intact organisms (e.g., the nematode *C. elegans*) in a noninvasive and label-free way (Ami et al., 2012). Recently, the use of synchrotron light sources has further improved infrared microscopy, thanks to the gain in signal-to-noise ratio and in spatial resolution, down to 5 μm (Marcelli et al., 2012).

Electron Microscopies The structure of thin biological samples can be determined by electron diffraction using an electron microscope. In this microscope a beam of electrons is focused on the sample by a system of electrostatic and magnetic lenses. Low irradiation intensities are employed to limit damage induced by the colliding electrons, even if this procedure reduces the image contrast. The sample image on a photographic plate can reach a resolution of a few angstroms, much higher than that of a light microscope, thanks to the short wavelength of the electrons, which is of the order of atomic dimensions. This microscopy became, therefore, one of the most informative techniques for high-resolution structural studies of molecules, protein aggregates such as IBs and amyloid fibrils, and complex systems such as cells and tissues. Two types of microscopes were developed. The transmission electron microscope (TEM), limited to the study of samples with thicknesses of a few hundreds of nanometeres, has the highest resolution, about 0.2 nm, but it is affected by the difficulties of sample preparation and also by the impossibility of obtaining three-dimensional structures. These limitations can be overcome using a scanning electron microscopy (SEM), even if at the cost of a much lower resolution (of about 10 nm) than that of a TEM. In the scanning microscope, the electron beam is projected on a small spot on the sample and scanned successively throughout the entire sample area. In both microscopes, the secondary electrons emitted by the sample are collected and projected in a cathode ray tube, where the sample image is formed on the tube fluorescent screen (Cantor and Schimmel, 1980).

Atomic Force Microscopy Atomic force microscopy (AFM) is a high-resolution scanning microscopy that makes it possible to image a surface with nanometer resolution, through scanning of a mechanical probe (Santos and Castanho, 2004; Alessandrini and Facci, 2005; Adamcik and Mezzenga, 2012). Almost any type of sample surface of different materials can be examined, including biological samples, from biopolymers to protein aggregates, as well as cells and tissues.

The microscope consists of a cantilever bearing a sharp microfabricated probe with a tip of a few to tens of nanometers in size. When the tip is brought close to the sample surface, forces between the tip and the sample lead to a deflection of the cantilever according to Hooke's law. The forces involved in the tip–surface interaction are mainly mechanical contact forces, van der Waals forces, electrostatic forces, and solvation forces. These forces are not measured directly but are calculated using Hooke's law after measuring the deflection of the lever and knowing the stiffness of the cantilever. Typically, the deflection is measured using a laser beam reflected from the top surface of the cantilever.

Several modes of operation can be performed in AFM. In the contact mode the tip is adjusted constantly to maintain a constant deflection, and therefore a constant height above the surface. In this mode, drag effects of the cantilever across the sample surface are often introduced, disturbing the surface image of soft biological samples.

In noncontact mode, the cantilever is oscillating externally to the surface sample with its tip close to the sample but not touching it, and consequently, the forces between the tip and the sample are low (10^{-12} N). In this case, changes in the amplitude of the oscillating cantilever induced by the tip–surface interaction provide information on the sample surface.

In the tapping mode, the cantilever oscillates closer to the sample than in the noncontact mode, and the tip intermittently touches or "taps" the surface. This mode is often employed to examine biological soft samples, since in these conditions the drag effects of the cantilever present in the contact mode are almost absent (Santos and Castanho, 2004; Alessandrini and Facci, 2005). In recent years, important information have been obtained by AFM on IBs (de Groot et al., 2009; García-Fruitós et al., 2009; Invernizzi et al., 2012) and amyloid fibril structures (Canale et al., 2011; Natalello et al., 2011; Adamcik and Mezzenga, 2012).

X-ray Fiber Diffraction X-ray scattering of noncrystalline materials can give useful structural information even if no detailed structures can be determined at atomic resolution, such as in the case of crystals. The first x-rays studies on DNA fibers were performed a long time ago.

Recently, important structural information have been obtained by x-ray diffraction on amyloid fibrils, often confirmed by solid-state NMR studies (Makin and Serpell, 2005; Morris and Serpell, 2010; Wang et al., 2010). The fibril diffraction patterns are in general very similar for the various amyloid fibrils, suggesting a similar cross-β-sheet structure. When fibril axis is aligned perpendicularly to the x-ray beam, their diffraction pattern is characterized by two common signals: a sharp reflection at 4.7 Å along the direction of the fiber axes (meridional reflection) and a more diffuse reflection at about 10 Å, which is perpendicular to the fibril direction (equatorial reflection). The x-ray data then indicated that the β-strands are perpendicular to the fiber axis and are hydrogen-bonded with a separation of 4.7 Å. From the equatorial reflection, the packing distance between two contiguous β-sheets was found to be about 10 Å, a value varying with the size of the protein side-chain groups that can change the packing distance between the sheets (Fandrich and Dobson, 2002).

REFERENCES

Adamcik J, Mezzenga R (2012) Study of amyloid fibrils via atomic force microscopy. Curr Opin Colloid Interface Sci 17:369–376.

Alessandrini A, Facci P (2005) AFM: a versatile tool in biophysics. Meas Sci Technol 16:R65–R92.

Ami D, Bonecchi L, Cali S, Orsini G, Tonon G, Doglia SM (2003) FT-IR study of heterologous protein expression in recombinant *Escherichia coli* strains. Biochim Biophys Acta 1624:6–10.

Ami D, Natalello A, Taylor G, Tonon G, Doglia SM (2006) Structural analysis of protein inclusion bodies by Fourier transform infrared microspectroscopy. Biochim Biophys Acta 1764:793–799.

Ami D, Natalello A, Lotti M, Doglia S (2013) Why and how protein aggregation has to be studied in vivo. Microb Cell Fact 12:17.

Ami D, Natalello A, Gatti-Lafranconi P, Lotti M, Doglia SM (2005) Kinetics of inclusion body formation studied in intact cells by FT-IR spectroscopy. FEBS Lett 579:3433–3436.

Ami D, Natalello A, Schultz T, Gatti-Lafnanconi P, Lotti M, Doglia SM, de Marco A (2009) Effects of recombinant protein misfolding and aggregation on bacterial membranes. Biochim Biophys Acta 1794:263–269.

Ami D, Natalello A, Doglia SM (2012) Fourier transform infrared microspectroscopy of complex biological systems: from intact cells to whole organisms. *In*: Intrinsically Disordered Protein Analysis: Vol. 1, Methods and Experimental Tools. VN Uversky and AK Dunker. Methods Mol Biol 895:85–100.

Arrondo JLR, Goni FM (1999) Structure and dynamics of membrane proteins as studied by infrared spectroscopy. Prog Biophys Mol Biol 72:367–405.

Arrondo JLR, Muga A, Castresana J, Goni FM (1993) Quantitative studies of the structure of proteins in solution by Fourier-transform infrared-spectroscopy. Prog Biophys Mol Biol 59:23–56.

Baneyx F, Mujacic M (2004) Recombinant protein folding and misfolding in *Escherichia coli*. Nature Biotechnol 22:1399–1408.

Barth A (2007) Infrared spectroscopy of proteins. Biochim Biophys Acta 1767:1073–1101.

Barth A, Zscherp C (2002) What vibrations tell us about proteins. Q Rev Biophys 35:369–430.

Bowden GA, Paredes AM, Georgiou G (1991) Structure and morphology of protein inclusion-bodies in *Escherichia-coli*. Bio-Technology 9:725–730.

Calloni G, Zoffoli S, Stefani M, Dobson CM, Chiti F (2005) Investigating the effects of mutations on protein aggregation in the cell. J Biol Chem 280:10607–10613.

Canale C, Relini A, Gliozzi A (2011) Atomic force microscopy of ex vivo amyloid fibrils. Methods Mol Bio 736:81–95.

Cano-Garrido O, Rodriguez-Carmona E, Diez-Gil C, Vazquez E, Elizondo E, Cubarsi R, Seras-Franzoso J, Corchero JL, Rinas U, Ratera I, Ventosa N, Veciana J, Villaverde A, García-Fruitós E (2013) Supramolecular organization of protein-releasing functional amyloids solved in bacterial inclusion bodies. Acta Biomater 9:6134–6142.

Cantor CR, Schimmel PR (1980) Biophysical Chemistry, Part II, Techniques for the Study of Biological Structure and Function. Oxford: W. H. Freeman.

Carrió MM, Villaverde A (2002) Construction and deconstruction of bacterial inclusion bodies. J Biotechnol 96:3–12.

Carrió MM, Villaverde A (2003) Role of molecular chaperones in inclusion body formation. FEBS Lett 537:215–221.

Carrió MM, Corchero JL, Villaverde A (1998) Dynamics of in vivo protein aggregation: building inclusion bodies in recombinant bacteria. FEMS Microbiol Lett 169:9–15.

Carrió MM, Cubarsi R, Villaverde A (2000) Fine architecture of bacterial inclusion bodies. FEBS Lett 471:7–11.

Carrió M, González-Montalbán N, Vera A, Villaverde A, Ventura S (2005) Amyloid-like properties of bacterial inclusion bodies. J Mol Biol 347:1025–1037.

Chiti F, Dobson CM (2006) Protein misfolding, functional amyloid, and human disease. Annu Rev Biochem 75:333–366.

Curtis-Fisk J, Spencer RM, Weliky DP (2008) Native conformation at specific residues in recombinant inclusion body protein in whole cells determined with solid-state NMR spectroscopy. J Am Chem Soc 130:12568–12569.

de Groot NS, Sabate R, Ventura S (2009) Amyloids in bacterial inclusion bodies. Trends Biochem Sci 34:408–416.

Doglia SM, Ami D, Natalello A, Gatti-Lafranconi P, Lotti M (2008) Fourier transform infrared spectroscopy analysis of the conformational quality of recombinant proteins within inclusion bodies. Biotechnol J 3:193–201.

Dzwolak W, Surmacz-Chwedoruk W, Babenko V (2013) Conformational memory effect reverses chirality of vortex-induced insulin amyloid superstructures. Langmuir 29:365–370.

Ellisdon AM, Thomas B, Bottomley SP (2006) The two-stage pathway of ataxin-3 fibrillogenesis involves a polyglutamine-independent step. J Biol Chem 281:16888–16896.

Espargaro A, Castillo V, de Groot NS, Ventura S (2008) The in vivo and in vitro aggregation properties of globular proteins correlate with their conformational stability: the SH3 case. J Mol Biol 378:1116–1131.

Espargaro A, Sabate R, Ventura S (2012) Thioflavin-S staining coupled to flow cytometry: a screening tool to detect in vivo protein aggregation. Mol Biosyst 8:2839–2844.

Fandrich M, Dobson CM (2002) The behaviour of polyamino acids reveals an inverse side chain effect in amyloid structure formation. EMBO J 21:5682–5690.

García-Fruitós E (2010) Inclusion bodies: a new concept. Microb Cell Fact 9:80.

García-Fruitós E, González-Montalbán N, Morell M, Vera A, Ferraz R, Aris A, Ventura S, Villaverde A (2005) Aggregation as bacterial inclusion bodies does not imply inactivation of enzymes and fluorescent proteins. Microb Cell Fact 4:27.

García-Fruitós E, Aris A, Villaverde A (2007a) Localization of functional polypeptides in bacterial inclusion bodies Appl Environ Microbiol 73:289–294.

García-Fruitós E, Martinez-Alonso M, González-Montalbán N, Valli M, Mattanovich D, Villaverde A (2007b) Divergent genetic control of protein solubility and conformational quality in *Escherichia coli*. J Mol Biol 374:195–205.

García-Fruitós E, Rodriguez-Carmona E, Diez-Gil C, Ferraz RM, Vazquez E, Corchero JL, Cano-Sarabia M, Ratera I, Ventosa N, Veciana J, Villaverde A (2009) Surface cell growth engineering assisted by a novel bacterial nanomaterial. Adv Mater 21:4249–4253.

García-Fruitós E, Seras-Franzoso J, Vazquez E, Villaverde A (2010) Tunable geometry of bacterial inclusion bodies as substrate materials for tissue engineering. Nanotechnology 21:205101.

García-Fruitós E, Sabate R, de Groot NS, Villaverde A, Ventura S (2011) Biological role of bacterial inclusion bodies: a model for amyloid aggregation. FEBS J 278:2419–2427.

García-Fruitós E, Vazquez E, Diez-Gil C, Corchero JL, Seras-Franzoso J, Ratera I, Veciana J, Villaverde A (2012) Bacterial inclusion bodies: making gold from waste. Trends Biotechnol 30:65–70.

Gatti-Lafranconi P, Natalello A, Ami D, Doglia SM, Lotti M (2011) Concepts and tools to exploit the potential of bacterial inclusion bodies in protein science and biotechnology. FEBS J 278:2408–2418.

González-Montalbán N, García-Fruitós E, Villaverde A (2007) Recombinant protein solubility: Does more mean better? Nat Biotechnol 25:718–720.

González-Montalbán N, Natalello A, García-Fruitós E, Villaverde A, Doglia SM (2008) In situ protein folding and activation in bacterial inclusion bodies. Biotechnol Bioeng 100:797–802.

Greenfield NJ (2006) Using circular dichroism spectra to estimate protein secondary structure. Nat Protoc 1:2876–2890.

Hell SW (2007) Far-field optical nanoscopy. Science 316:1153–1158.

Herzenberg LA, Parks D, Sahaf B, Perez O, Roederer M, Herzenberg LA (2002) The history and future of the fluorescence activated cell sorter and flow cytometry: a view from Stanford. Clin Chem 48:1819–1827.

Ignatova Z, Gierasch LM (2004) Monitoring protein stability and aggregation in vivo by real-time fluorescent labeling. Proc Natl Acad Sci U S A 101:523–528.

Invernizzi G, Aprile FA, Natalello A, Ghisleni A, Penco A, Relini A, Doglia SM, Tortora P, Regonesi ME (2012) The relationship between aggregation and toxicity of polyglutamine-containing ataxin-3 in the intracellular environment of *Escherichia coli*. PLoS ONE 7:e51890.

Jacquemin-Sablon A (1993) Flow cytometry: new developments. New York: Springer-Verlag.

Kelly SM, Jess TJ, Price NC (2005) How to study proteins by circular dichroism. Biochim Biophys Acta 1751:119–139.

Kim W, Kim Y, Min J, Kim DJ, Chang YT, Hecht MH (2006) A high-throughput screen for compounds that inhibit aggregation of the Alzheimer's peptide. ACS Chem Biol 1:461–469.

Kirkpatrick CL, Viollier PH (2011) Poles apart: prokaryotic polar organelles and their spatial regulation. Cold Spring Harbor Persp Biol 3.

Lakowicz JR (2006) Principles of Fluorescence Spectroscopy. New York: Springer-Verlag.

Levin IW, Bhargava R (2005) Fourier transform infrared vibrational spectroscopic imaging: integrating microscopy and molecular recognition. Annu Rev Phys Chem 56:429–474.

Liovic M, Ozir M, Zavec A, Peternel S, Komel R, Zupancic T (2012) Inclusion bodies as potential vehicles for recombinant protein delivery into epithelial cells. Microb Cell Fact 11:67.

Makin OS, Serpell LC (2005) Structures for amyloid fibrils. FEBS J 272:5950–5961.

Marcelli A, Cricenti A, Kwiatek WM, Petibois C (2012) Biological applications of synchrotron radiation infrared spectromicroscopy. Biotechnol Adv 30:1390–1404.

Margreiter G, Messner P, Caldwell KD, Bayer K (2008) Size characterization of inclusion bodies by sedimentation field-flow fractionation. J Biotechnol 138:67–73.

Martinez-Alonso M, González-Montalbán N, García-Fruitós E, Villaverde A (2008) The functional quality of soluble recombinant polypeptides produced in *Escherichia coli* is defined by a wide conformational spectrum. Appl Environ Microbiol 74:7431–7433.

Michalet X, Kapanidis AN, Laurence T, Pinaud F, Doose S, Pflughoefft M, Weiss S (2003) The power and prospects of fluorescence microscopies and spectroscopies. Annu Rev Biophys Biomol Struct 32:161–182.

Morell M, Bravo R, Espargaro A, Sisquella X, Aviles FX, Fernandez-Busquets X, Ventura S (2008) Inclusion bodies: specificity in their aggregation process and amyloid-like structure. Biochim Biophys Acta 1783:1815–1825.

Morris K, Serpell L (2010) From natural to designer self-assembling biopolymers, the structural characterisation of fibrous proteins and peptides using fibre diffraction. Chem Soc Rev 39:3445–3453.

Natalello A, Ami D, Brocca S, Lotti M, Doglia SM (2005) Secondary structure, conformational stability and glycosylation of a recombinant *Candida rugosa* lipase studied by Fourier-transform infrared spectroscopy. Biochem J 385:511–517.

Natalello A, Doglia SM (2010) Intrinsically disordered proteins and induced folding studied by Fourier transform infrared spectroscopy. *In*: Instrumental Analysis of Intrinsically Disordered Proteins: Assessing Structure and Conformation V Uversky and S Longhi. Hoboken, NJ: Wiley, pp. 225–252.

Natalello A, Frana AM, Relini A, Apicella A, Invernizzi G, Casari C, Gliozzi A, Doglia SM, Tortora P, Regonesi ME (2011) A major role for side-chain polyglutamine hydrogen bonding in irreversible ataxin-3 aggregation. PLoS ONE 6:e18789.

Natalello A, Ami D, Doglia S (2012) Fourier transform infrared spectroscopy of intrinsically disordered proteins: measurement procedures and data analyses. *In*: Intrinsically Disordered Protein Analysis. VN Uversky and AK Dunker. Methods Mol Biol 895:229–244.

Olzscha H, Schermann SM, Woerner AC, Pinkert S, Hecht MH, Tartaglia GG, Vendruscolo M, Hayer-Hartl M, Hartl FU, Vabulas RM (2011) Amyloid-like aggregates sequester numerous metastable proteins with essential cellular functions. Cell 144:67–78.

Paddock SW (2000) Principles and practices of laser scanning confocal microscopy. Mol Biotechnol 16:127–149.

Peternel S, Komel R (2011) Active protein aggregates produced in *Escherichia coli*. Int J Mol Sci 12:8275–8287.

Plakoutsi G, Bemporad F, Calamai M, Taddei N, Dobson CM, Chiti F (2005) Evidence for a mechanism of amyloid formation involving molecular reorganisation within native-like precursor aggregates. J Mol Biol 351:910–922.

Przybycien TM, Dunn JP, Valax P, Georgiou G (1994) Secondary structure characterization of beta-lactamase inclusion-bodies. Protein Eng 7:131–136.

Rokney A, Shagan M, Kessel M, Smith Y, Rosenshine I, Oppenheim AB (2009) *E. coli* transports aggregated proteins to the poles by a specific and energy-dependent process. J Mol Biol 392:589–601.

Sabate R, Ventura S (2013) Cross-beta-sheet supersecondary structure in amyloid folds: techniques for detection and characterization. Methods Mol Biol 932:237–257.

Sabate R, Espargaro A, Saupe SJ, Ventura S (2009) Characterization of the amyloid bacterial inclusion bodies of the HET-s fungal prion. Microb Cell Fact 8:56.

Santos NC, Castanho MARB (2004) An overview of the biophysical applications of atomic force microscopy. Biophys Chem 107:133–149.

Schrodel A, de Marco A (2005) Characterization of the aggregates formed during recombinant protein expression in bacteria. BMC Biochem 6:10.

Seshadri S, Khurana R, Fink AL (1999) Fourier transform infrared spectroscopy in analysis of protein deposits. Methods Enzymol 309:559–576.

Sun YS, Wallrabe H, Seo SA, Periasamy A (2011) FRET microscopy in 2010: the legacy of Theodor Forster on the 100th anniversary of his birth. ChemPhysChem 12:462–474.

Susi H, Byler DM (1986) Resolution-enhanced Fourier-transform infrared-spectroscopy of enzymes. Methods Enzymol 130:290–311.

Taylor G, Hoare M, Gray DR, Marston FAO (1986) Size and density of protein inclusion-bodies. bio-Technology 4:553–557.

Taylor GF, Wood SP, Mors K, Glaubitz C, Werner JM, Williamson PTF (2011) Morphological differences between beta(2)-microglobulin in fibrils and inclusion bodies. ChemBioChem 12:556–558.

Upadhyay AK, Murmu A, Singh A, Panda AK (2012) Kinetics of inclusion body formation and its correlation with the characteristics of protein aggregates in *Escherichia coli*. PLoS ONE 7:e33951.

Van der Henst C, Charlier C, Deghelt M, Wouters J, Matroule JY, Letesson JJ, De Bolle X (2010) Overproduced *Brucella abortus* PdhS-mCherry forms soluble aggregates in *Escherichia coli*, partially associating with mobile foci of IbpA-YFP. BMC Microbiol 10:248.

Villar-Piqué A, Ventura S (2012) Modeling amyloids in bacteria. Microb Cell Fact 11:166.

Villar-Piqué A, Espargaro A, Sabate R, de Groot N, Ventura S (2012) Using bacterial inclusion bodies to screen for amyloid aggregation inhibitors. Microb Cell Fact 11:55.

Villaverde A, García-Fruitós E, Rinas U, Seras-Franzoso J, Kosoy A, Corchero JL, Vazquez E (2012) Packaging protein drugs as bacterial inclusion bodies for therapeutic applications. Microb Cell Fact 11:76.

Vogel EP, Curtis-Fisk J, Young KM, Weliky DP (2011) Solid-state nuclear magnetic resonance (NMR) spectroscopy of human immunodeficiency virus gp41 protein that includes the fusion peptide: NMR detection of recombinant Fgp41 in inclusion bodies in whole bacterial cells and structural characterization of purified and membrane-associated Fgp41. Biochemistry 50:10013–10026.

Wang L, Maji SK, Sawaya MR, Eisenberg D, Riek R (2008) Bacterial inclusion bodies contain amyloid-like structure. PLoS Biol. 6:e195.

Wang L, Schubert D, Sawaya MR, Eisenberg D, Riek R (2010) Multidimensional structure–activity relationship of a protein in its aggregated states. Angew Chem Int Ed 49:3904–3908.

Wasmer C, Benkemoun L, Sabate R, Steinmetz MO, Coulary-Salin B, Wang L, Riek R, Saupe SJ, Meier BH (2009) Solid-state NMR spectroscopy reveals that *E. coli* inclusion bodies of HET-s(218–289) are amyloids. Angew Chem Int Ed 48:4858–4860.

Winkelmann J, Calloni G, Campioni S, Mannini B, Taddei N, Chiti F (2010) Low-level expression of a folding-incompetent protein in *Escherichia coli*: search for the molecular determinants of protein aggregation in vivo. J Mol Biol 398: 600–613.

Winkler J, Seybert A, Konig L, Pruggnaller S, Haselmann U, Sourjik V, Weiss M, Frangakis AS, Mogk A, Bukau B (2010) Quantitative and spatio-temporal features of protein aggregation in *Escherichia coli* and consequences on protein quality control and cellular ageing. EMBO J 29:910–923.

Zandomeneghi G, Krebs MRH, McCammon MG, Fandrich M (2004) FTIR reveals structural differences between native beta-sheet proteins and amyloid fibrils. Protein Sci 13:3314–3321.

Zhao Y, He W, Liu WF, Liu CC, Feng LK, Sun L, Yan YB, Hang HY (2012) Two distinct states of *Escherichia coli* cells that overexpress recombinant heterogeneous beta-galactosidase. J Biol Chem 287:9259–9268.

7

RESIDUE-SPECIFIC STRUCTURAL STUDIES OF INCLUSION BODIES

CHRISTIAN WASMER, MARIELLE WÄLTI, YONGLI CHEN, AND LEI WANG

7.1 INTRODUCTION

Nascent protein chains normally fold into their natural three-dimensional conformations under physiological conditions with the facilitation of chaperones. However, various cellular perturbations may trigger their misfolding and cause aggregation. In the human body, protein aggregation into amyloid fibrils is associated with more than a dozen serious diseases, such as Alzheimer's, Parkinson's, and prion disease. Protein aggregation can also present problems in biotechnology: for example, in the formation of bacterial inclusion bodies during recombinant protein production (Dobson, 2003; Selkoe, 2003; Kelly, 2005; Ventura, 2005).

Amyloid fibrils have been studied extensively, due to their close association with human diseases. Under an electron microscope, amyloid fibrils appear as highly ordered fibrillar protein aggregates with average fibril width ranges between 6 and 20 nm (Figure 7.1a) (Sunde and Blake, 1997; Stromer and Serpell, 2005; Sawaya et al., 2007). They can bind to specific dyes, such as Thioflavin T and Congo Red (Klunk et al., 1999; LeVine, 1999), and their x-ray diffraction pattern normally shows a sharp meridional reflection at 4.7 Å and a diffused equatorial reflection at approximately 10 Å

Protein Aggregation in Bacteria: Functional and Structural Properties of Inclusion Bodies in Bacterial Cells, First Edition. Edited by Silvia Maria Doglia and Marina Lotti.

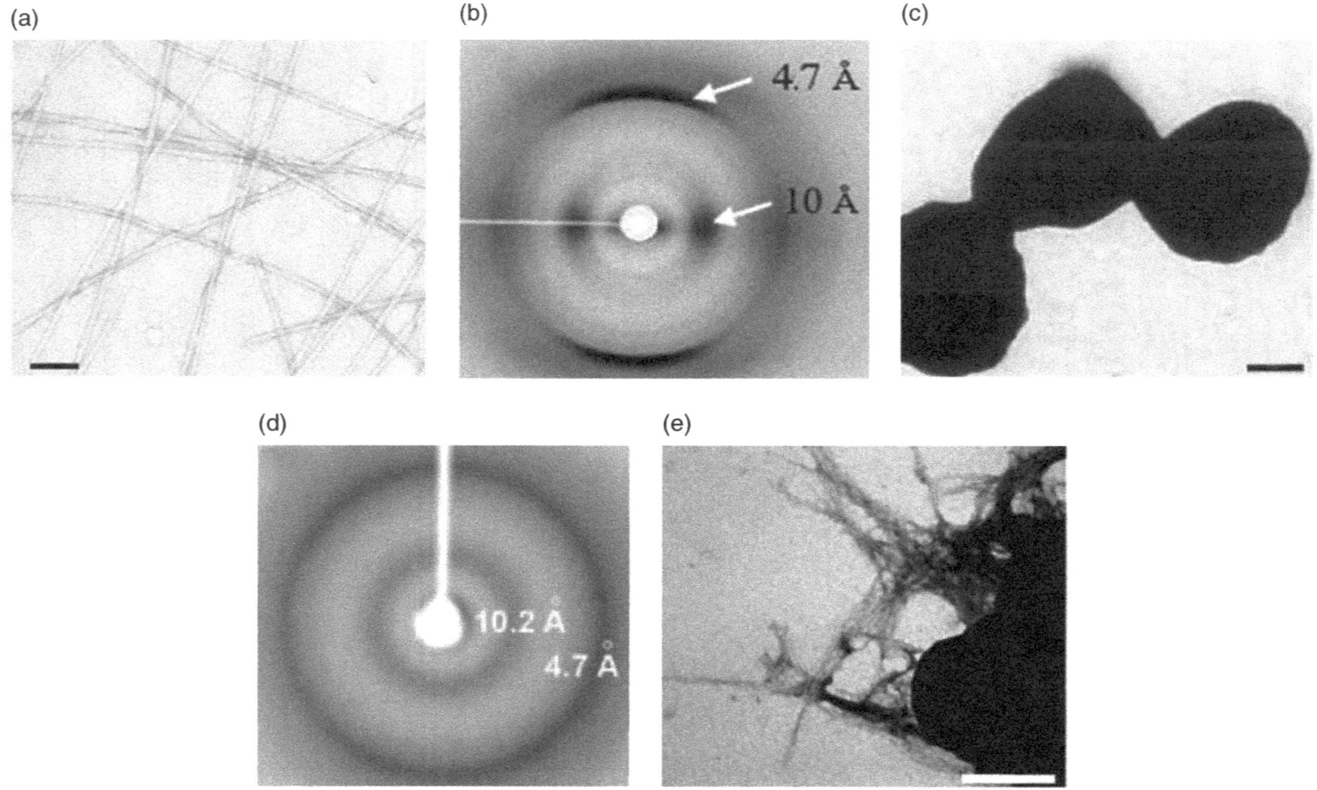

FIGURE 7.1 (a) Electron micrograph of amyloid fibrils of the GGVVIA peptide from amyloid-β protein; (b) x-ray diffraction of amyloid fibrils; (c) electron micrograph of bacterial inclusion bodies of HypF-N protein; (d) x-ray diffraction of HypF-N inclusion bodies; (e) electron micrograph of fibrillar structure of amyloid-β42 inclusion bodies after proteinase K digestion. The scale bar represents 200 nm. [Part (a) reproduced with permission from the Nature Publishing Group; part (b) with permission from Wiley-Liss, Inc; parts (c) and (d) with permission from Wiley-VCH Verlag GmbH; part (e) with permission from Elsevier.]

(Sunde et al., 1997; Stromer and Serpell, 2005) (Figure 7.1b). Infrared spectroscopy and circular dichroism spectroscopy show that when proteins aggregate into amyloid fibrils, the latter contains an enriched β-strand structure compared to the native conformation (Maji et al., 2008, 2009a; Wang et al., 2010). Sequence analysis using software algorithms such as TANGO, 3DPROFILE, AGGRESCAN, and ZYGGREGATOR suggests that proteins forming amyloid fibrils usually contain short sequences that are vulnerable and prone to aggregation (Fernandez-Escamilla et al., 2004; Thompson et al., 2006; Tartaglia and Vendruscolo, 2008; de Groot et al., 2012).

Inclusion bodies are formed during recombinant protein production. Since their presence suggests reduced production of their soluble counterpart, the formation of inclusion bodies was traditionally undesirable in the biotech industry until the discovery of their functional aspects in recent years (García-Fruitós et al., 2010, 2012; Gatti-Lafranconi et al., 2011; Sans et al., 2012; Seras-Franzoso et al., 2012; Vázquez et al., 2012; Villaverde, 2012; Villaverde et al., 2012). Under an electron microscope, inclusion bodies appear to be very different protein aggregates than amyloid fibrils. They normally have amorphous shapes with estimated diameters ranging from 0.2 to 1.2 μm if they are regarded roughly as balls (Carrió et al., 2000, 2005; Wang et al., 2008, 2010) (Figure 7.1c). Despite the morphology difference, inclusion bodies actually have many properties resembling amyloid fibrils (Ventura and Villaverde, 2006). For example, inclusion bodies can also bind to Thioflavin T and Congo Red, the dyes used specifically to recognize amyloid fibrils (Carrió et al., 2005; Wang et al., 2008). Their x-ray diffraction pattern also shows two reflections, at 4.7 Å and approximately 10 Å (Figure 7.1d) (Wang et al., 2008, 2010). Similar to amyloid fibrils, inclusion bodies are found to contain enriched β-strand structure (Carrió et al., 2005; Morell et al., 2008), and proteins that form inclusion bodies are predicted to contain short sequences that are vulnerable and prone to aggregation by using the samealgorithms (Hamodrakas, 2011). Recently, fibril-like entities are observed in inclusion bodies of amyloid-β42, HET-s(218–289), and several other proteins (Morell et al., 2008; Wang et al., 2008; Wasmer et al., 2009) (Figure 7.1e). Besides the amyloid-like properties, studies have also shown that some inclusion bodies have activities similar to those of their native counterparts (García-Fruitós et al., 2005, 2007; Lotti, 2011; Sans et al., 2012), suggesting that their inhomogeneous composition is more complex than that of amyloid fibrils (Ventura and Villaverde, 2006).

Since a protein's functions correlate closely with its structure, the similarity of properties between inclusion bodies and amyloid fibrils suggest possible similarity of their molecular structures. Therefore, understanding the molecular structure of amyloid fibrils may help us to study the structure of inclusion bodies.

7.2 MOLECULAR STRUCTURE OF AMYLOID FIBRILS

Different from other protein aggregates, highly ordered amyloid fibrils are crystalline in nature, which makes their molecular structures more accessible to high-resolution structural methods, such as x-ray crystallography and solid- or liquid-state nuclear magnetic resonance (NMR) spectroscopy. These studies have shown that amyloid fibrils formed from different proteins or peptides have a common structural property: the cross-β-sheet structure.

7.2.1 Structure of Peptide Amyloid Fibrils as Determined by X-ray Crystallography

X-ray crystallography is a structural technique that has been used to determine the majority of protein structures deposited in the Protein Data Bank. To employ this technique, the sample has to be crystallized first. It has been found that amyloid fibrils formed from proteins are difficult to crystallize (Maji et al., 2009b; Greenwald and Riek, 2010). So far, only amyloid fibrils of peptides that are six to eight residues long have been successfully crystallized and studied with x-ray crystallography. These peptides are from fragments of fibril-forming proteins: for example, tau, amyloid-β, α-synuclein, and β2-microglobulin (Ivanova et al., 2004, 2009; Nelson et al.,2005; Sawaya et al., 2007; Wiltzius et al., 2008).

Figure 7.2 a to c shows the structure of amyloid fibrils formed from the VQIVYK fragment of the tau protein (Sawaya et al., 2007). Within the amyloid fibrils, each tau peptide forms a β-strand (the dark or light gray arrow bar); these β-strands parallel-stack on top of each other to form β-sheets, perpendicularly orient to the fibril growing direction (the black line), and form the so-called cross-β-sheet structure (Figure 7.2a). In a β-sheet (e.g., the light gray β-sheet), β-strands interconnect with each other by forming backbone hydrogen bonds (Figure 7.2b); for two β-sheets, they arrange their relative orientation so that peptide sidechains between them can tightly interdigitate to form a steric zipper that connects β-sheets together (Figure 7.2c). The distance between two adjacent β-strands in a β-sheet is 4.7Å and the distance between two adjacent β-sheets is approximately 10Å, which give rise to the characteristic x-ray diffraction pattern in Figure 7.1b (Maji et al., 2009b). To form a steric zipper between β-sheets, self-complementary of the peptide side chains is necessary and makes the formation of amyloid fibrils highly

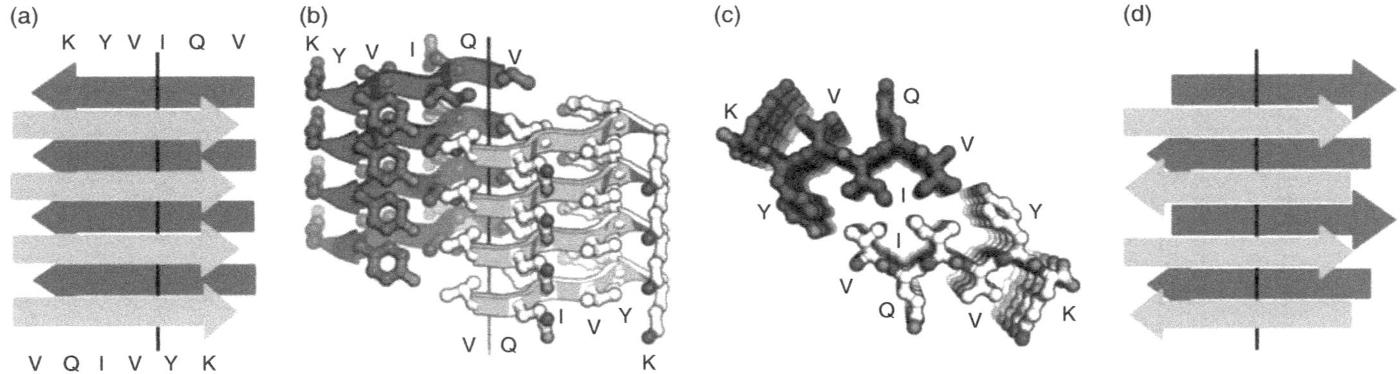

FIGURE 7.2 (a) Parallel cross-β-sheet structure of amyloid fibrils of the VQIVYK peptide from tau protein; (b) side view of VQIVYK amyloid fibrils with side chains; (c) top view of VQIVYK amyloid fibrils with sidechains; (d) antiparallel cross-β-sheet structure. [Parts (b) and (c) are reproduced with permission from the Nature Publishing Group.] (*See insert for color representation of the figure.*)

sequence-dependent. In addition to parallel β-strands, some peptides can stack antiparallel to form a cross-β-sheet (Figure 7.2d).

7.2.2 Structure of HET-s Amyloid Fibrils as Determined by Solid-State NMR Spectroscopy

NMR spectroscopy is another major technique used to determine protein structure. Solid-state NMR spectroscopy focuses particularly on solid samples, and it has the unique ability to measure structural constraints (e.g., internuclear distances, dihedral angles) directly in protein aggregates (Maji et al., 2009). A nice example of amyloid structure determined using solid-state NMR is the prion protein HET-s(218–289) in filamentous fungus *Podospora anserine* (Wasmer et al., 2008a, b). Similar to the tau peptide amyloid fibrils, HET-s(218–289) amyloid fibrils also have a cross-β-sheet structure (Figure 7.3a). In the amyloid fibril, each HET-s(218–289) molecule forms eight β-strands (β1a to β4b), and the rest of the molecule is in a random coil structure (Figure 7.3b: ssNMR). Three β-strands (β1a, β1b, and β2a) form a β-triangle at the bottom; three β-strands (β3a, β3b, and β4a) form another β-triangle at the top (Figure 7.3c); and the entire molecule forms two helical windings. In amyloid fibrils, the β-triangles of all HET-s(218–289) molecules parallel-stack on top of each other in a sandwich manner to form a cross-β-sheet structure with a densely packed hydrophobic core (Figure 7.3a); adjacent β-strands (1a and 3a, 1b and 3b, 2a and 4a) in the same β-sheet are interconnected by forming backbone hydrogen bonds.

7.2.3 Cross-β-Sheet Core of HET-s Amyloid Fibrils as Determined by Solution-State NMR and the H/D-Exchange Method

Solution-state NMR spectroscopy is used to determine the structure of soluble protein samples, and it cannot directly measure the molecular structure of insoluble protein aggregates. However, when combined with the hydrogen/deuterium-exchange procedure (H/D-exchange), solution-state NMR provides a convenient method to identify protein sequences that are involved in the cross-β-sheet structure of amyloid fibrils (Lee and Goto, 2012; Vilar et al., 2012). The mechanism of this method is explained below, and its validity has been verified in solid-state NMR studies. For example, it was identified with the method that residues 226–234, 236–246, 262–270, and 272–282 of HET-s(218–289) are involved in the formation of cross-β-sheet structure of amyloid fibrils (Ritter et al., 2005) (Figure 7.3b, H/D). This result agrees well with the structure of HET-s(218–289) amyloid fibrils determined by solid-state NMR (Wasmer et al., 2008a) (Figure 7.3b: ssNMR).

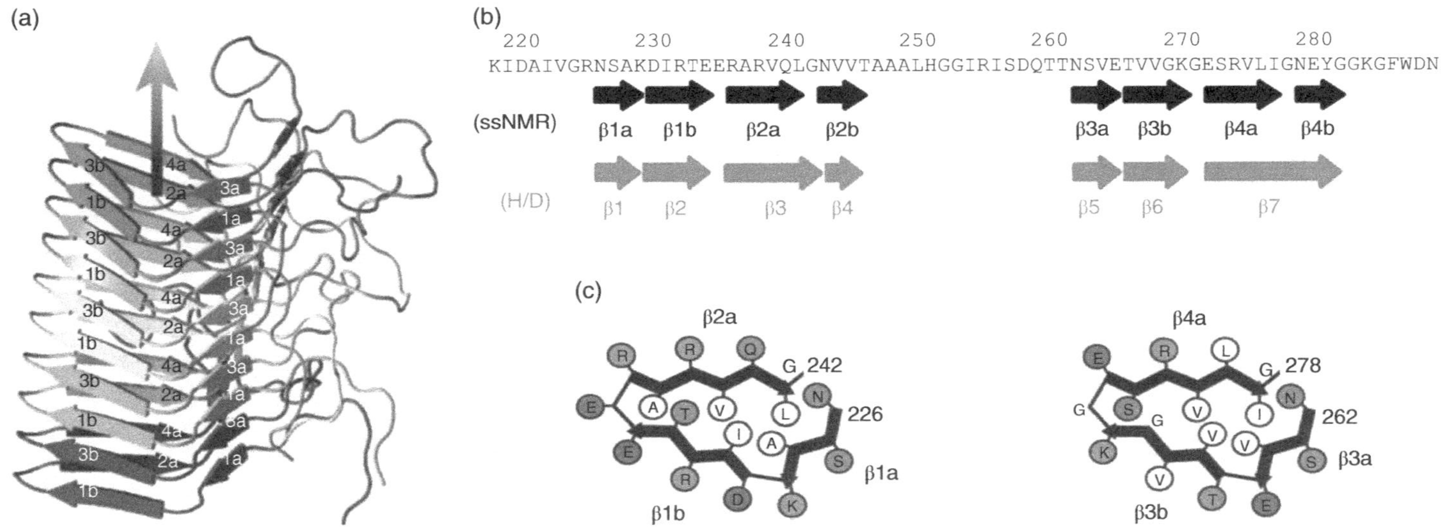

FIGURE 7.3 (a) Cross-β-sheet structure of HET-s(218–289) amyloid fibrils. Each color represents one HET-s(218–289) molecule, and its β-strands are labeled; (b) amino acid sequence of HET-s(218–289). Black arrow bars represent β-strands formed in the amyloid fibrils, as determined by solid-state NMR (ssNMR). Blue arrow bars represent β-strands determined with solution-state NMR and the H/D-exchange method (H/D); (c) residue arrangement around the cross-β-sheet structure: hydrophobic residues are colored white, acidic residues red, basic residues blue, and others green. [Parts (a) and (c) reproduced with permission from *Science*.] (*See insert for color representation of the figure*.)

7.3 STRUCTURAL STUDY OF INCLUSION BODIES WITH SOLUTION-STATE NMR AND THE H/D-EXCHANGE METHOD

Since studies of inclusion bodies suggest that they may have a molecular structure similar to that of amyloid fibrils, and a cross-β-sheet is the structural characteristic of amyloid fibrils, it is reasonable to hypothesize that inclusion bodies may also contain cross-β-sheet structure. The hypothesis is strongly supported by the x-ray diffraction pattern of inclusion bodies (Figure 7.1d; see also Section 2.1) and the fibrillar entities found in inclusion bodies (Figure 7.1e).

To study the structure of inclusion bodies, it would be ideal that if x-ray crystallography or solid-state NMR spectroscopy can be used to determine their molecular structure directly. However, the noncrystalline nature of inclusion bodies prevents the use of x-ray crystallography, and their inhomogeneity presents a challenge for the spectral analysis of solid-state NMR. As an alternative, the solution-state NMR combined with the H/D-exchange method has become a handy tool to obtain residue-specific structural information on inclusion bodies. Use of this method to study protein aggregates was developed by Goto and Alesandrescu to study amyloid fibrils (Alexandrescu, 2001; Hoshino et al., 2002), and it is suitable to study any protein aggregates that may contain a cross-β-sheet structure (Luhrs et al., 2005; Ritter et al., 2005; Vilar et al., 2008; Wang et al., 2008, 2010).

7.3.1 Experimental Mechanism of Solution-State NMR and the H/D-Exchange Method

Preparation To apply the method, the protein sample needs to be produced recombinantly with ^{15}N–^{13}C isotopic labeling. The labeled protein is then allowed to form aggregates at its specific condition and is stored in an appropriate buffer. Figure 7.4a shows protein aggregates that contain a cross-β-sheet structure. Notice that in the cross-β-sheet, β-strands are interconnected through backbone hydrogen bonds (represented by red dots in the figure).

Reference Spectrum To collect a reference NMR spectrum to evaluate an H/D-exchange result, a sample of protein aggregates needs to be dehydrated and then dissolved in deuterated dimethyl sulfoxide (DMSO-d6) that contains a trace amount of deuterated trifluoroacetic acid (TFA-d). DMSO is an excellent organic solvent that can dissolve protein aggregates into monomers for solution-state NMR measurement (Hirota-Nakaoka et al., 2003). An NMR ^{1}H–^{15}N heteronuclear single-quantum coherence spectrum (HSQC) is collected on the dissolved sample as the reference spectrum (Figure 7.4b). Each peak in the spectrum corresponds to one backbone N–H pair of the protein.

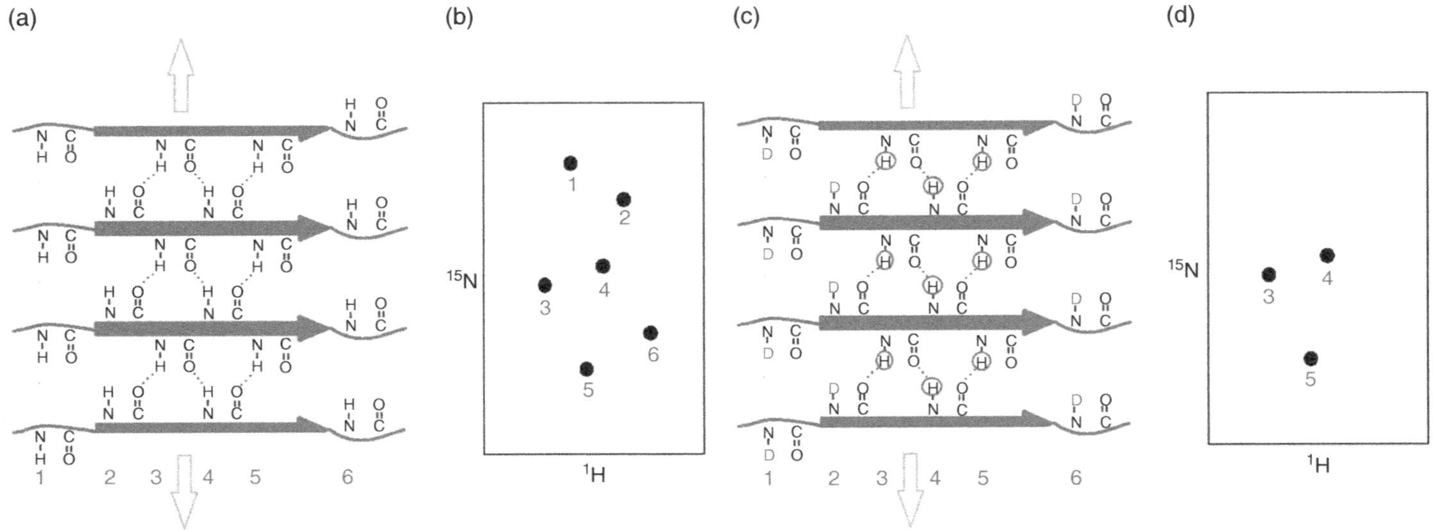

FIGURE 7.4 (a) Protein aggregates that contain cross-β-sheet structure. Each blue arrow bar and two end curves represent one protein molecule; the arrow bar represents β-strand in the cross-β-sheet; and curves represent the rest of the molecule. Backbone N–H and C=O pairs are labeled, and N–H pairs are numbered 1 through 6. Red dots represent hydrogen bonds, and yellow arrows indicate the growing direction of β-sheet. (b) The reference spectrum (NMR–HSQC spectrum). Each peak in the spectrum is assigned to a backbone N–H pair of the protein aggregates. (c) The protein aggregates after H/D-exchange. Hydrogens forming hydrogen bonds in the cross-β-sheet are protected against deuterium replacement (positions 3, 4, and 5). (d) The exchange spectrum (NMR–HSQC spectrum). Remaining peaks in the spectrum correspond to backbone N–H pairs in the cross-β-sheet (positions 3, 4, and 5). (*See insert for color representation of the figure.*)

The ownership of these peaks is assigned with the help of two additional NMR spectra, HNCACB (Salzmann et al., 1999) and HNCOCANH (Bracken et al., 1997). Figure 7.4b shows the reference spectrum with all six assigned peaks.

H/D-Exchange The H/D-exchange experiment can identify backbone amide hydrogens that are protected against deuterium exchange. To perform the experiment, a sample that contains the same amount of protein aggregates as the reference sample is sedimented and resuspended in an exchange buffer, which uses D_2O as the solvent. Since the backbone amide hydrogens within the cross-β-sheet form hydrogen bonds, they are protected against deuterium exchange in the exchange buffer; whereas the amide hydrogens outside the cross-β-sheet are exposed and gradually replaced by deuterium (Hoshino et al., 2002). Figure 7.4c shows the protein aggregates after H/D-exchange, and hydrogens forming hydrogen bonds within cross-β-sheet are protected against deuterium exchange (positions 3, 4, and 5).

Exchange Spectrum After the sample incubates in the exchange buffer for a certain amount of time (e.g., 1 h), it is dehydrated and then dissolved in the DMSO-d6/TFA-d mixture as the reference sample. Because the DMSO is an aprotic solvent that lacks exchangeable hydrogen or deuterium, the deuterium exchange status of protein aggregates could be maximally preserved after they are dissolved in it (Zhang et al., 1995). However, because DMSO is also a hygroscopic solvent, it can absorb water from air and potentially cause deuterium-to-hydrogen exchange in the dissolved protein aggregates. To avoid this problem, a trace amount of TFA is added to adjust thepH of the DMSO solution to approximately 5.0. It was found that at this pH, the H/D-exchange in DMSO is maximally suppressed (Zhang et al., 1995). After the sample that has been exchanged is dissolved in the DMSO-d6/TFA-d mixture, an HSQC exchange spectrum is collected on it (Figure 7.4d). For hydrogens that are replaced by deuterium, their peaks disappear from the exchange spectrum. A comparison of Figure 7.4b and d shows that the remaining peaks in the exchange spectrum correspond to backbone N–H pairs at positions 3, 4, and 5.

Calculate the H/D-Exchange Rates Although a single exchange spectrum provides a qualitative evaluation of which peaks disappear from the spectrum, this evaluation may not be accurate. Because the initial peak intensities in the reference spectrum are usually different, a weak peak that experiences a minor deuterium exchange may disappear in the exchange spectrum, whereas a strong peak that experiences a major deuterium exchange may remain. To obtain a quantitative evaluation, we should calculate the

H/D-exchange rates for all backbone amide hydrogens of the protein. To do so, we need to repeat the H/D-exchange step, let protein aggregates incubate in the exchange buffer for different amounts of time (e.g., 0.1 h, 10 h, 100 h, and up to weeks), and collect their exchange spectra. For every backbone N–H pair, we record its peak intensity in every spectrum and plot its intensity change versus H/D-exchange time. The plot is then fitted to the exponential function to find the H/D-exchange rate (k_{ex}) of the backbone amide hydrogen:

$$I(t) = I(0)\exp(-k_{ex}t)$$

where t is the H/D-exchange time and $I(0)$ and $I(t)$ are peak intensities at time zero and time t, respectively. A slow exchange rate (large k_{ex}) suggests that the backbone amide hydrogen is located in a highly protected region, such as the cross-β-sheet structure; whereas a fast exchange rate (small k_{ex}) indicates the hydrogen in random coil or other secondary structures (Vilar et al., 2012).

7.3.2 Example: Structural Study of BMP2(13–74) Inclusion Bodies

Next we look at an example of use of the H/D exchange method to study the inclusion body structure of the fragment residue 13–74 of the human bone morphogenetic protein-2 (BMP2) (Scheufler et al., 1999; Allendorph et al., 2006; Wang et al., 2008). During recombinant protein production, BMP2(13–74) forms solely inclusion bodies without any soluble fraction. The purified inclusion bodies have an amorphous appearance (Figure 7.5a top) and they can bind to Thioflavin T and Congo Red. X-ray diffraction of these inclusion bodies shows two reflections, at 4.7 and 10 Å (Figure 7.5a middle), and production of the protein in *Escherichia coli* for 12 h yields inclusion bodies with fibril-like entities (Figure 7.5a bottom). These data suggest that BMP2(13–74) inclusion bodies may contain cross-β-sheet structure similar to amyloid fibrils.

To examine the residue-specific structure of BMP2(13–74) inclusion bodies, the protein was produced in ^{15}N–^{13}C isotopic labeled media. After protein production, inclusion bodies were formed and purified from hosting cells. One sample of inclusion bodies was dehydrated and then dissolved in 500 μL of DMSO-d6 (with 0.05% TFA-d) to a final concentration of 200 μM, followed by the NMR measurement to collect its reference spectrum (Figure 7.5b left). Each peak in the reference spectrum was assigned to a backbone N–H pair of the protein with the help of NMR HNCOCANH and HNCACB spectra. It should be noted that the proline residue lacks the backbone N–H pair, and therefore has no corresponding peak in the spectrum. The rest inclusion body samples were sedimented with centrifugation and resuspended in a D_2O exchange buffer. Every time after an H/D-exchange took

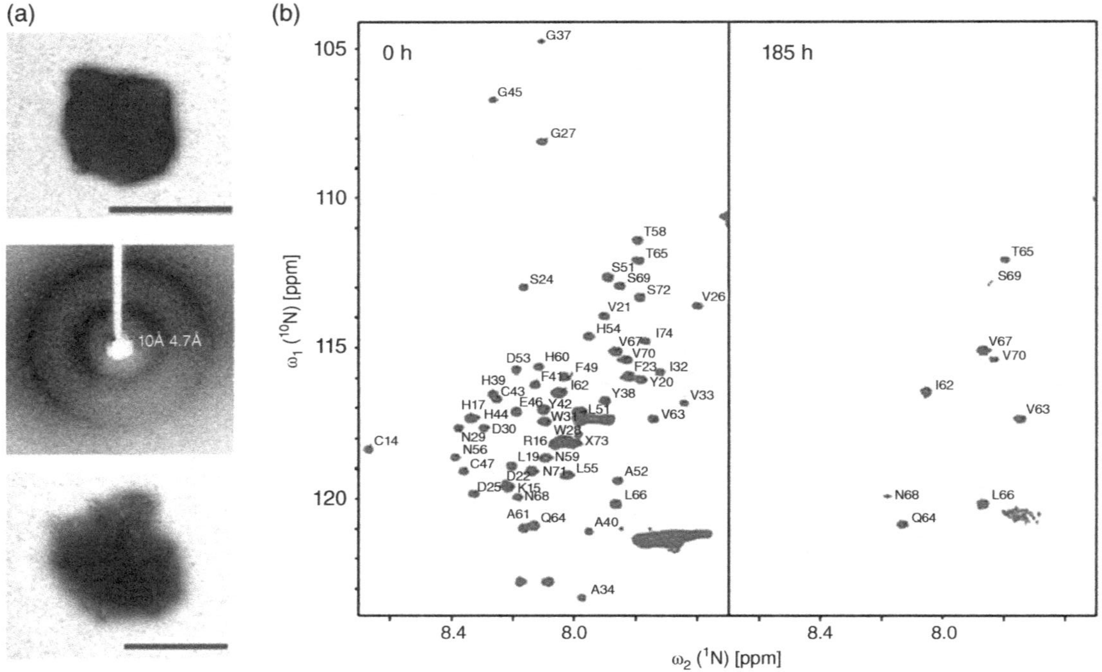

FIGURE 7.5 (a) Top: electron micrograph of a BMP2(13–74) inclusion body; middle: x-ray diffraction of BMP2(13–74) inclusion bodies; bottom: electron micrograph of a BMP2(13–74) inclusion body after protein production for 12 h; (b) NMR–HSQC spectra of BMP2(13–74) inclusion bodies before H/D-exchange (left) and after H/D-exchange for 185 h (right);

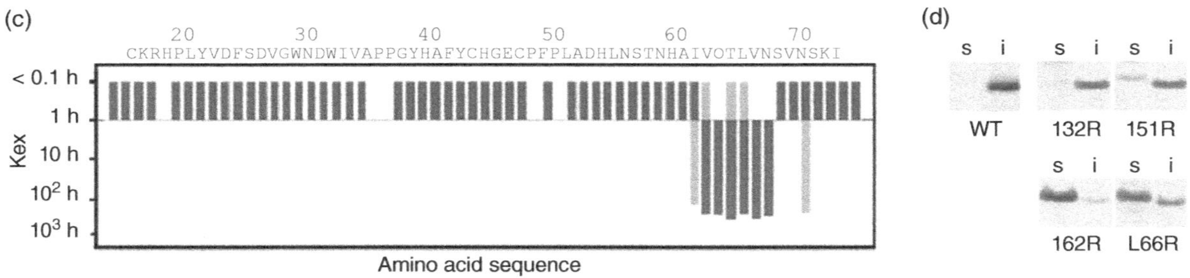

FIGURE 7.5 (*Continued*) (c) H/D-exchange rates k_{ex} of BMP2(13–74) inclusion bodies (dark gray bars) plotted versus the protein sequence. Due to the inhomogeneity of inclusion bodies, light gray bars represent the behavior of a minor population of inclusion bodies; (d) SDS gels of soluble (s) and insoluble (i) fractions of lysate of cells producing wild-type (WT) BMP2(13–74) and its mutants (I32R, I51R, I62R, L66R). [Reproduced from PLOS Biology (eISSN-1545-7885; ISSN-1544-9173).]

place for a certain amount of time (e.g., 0.1 h, 1 h, 10 h, 100 h, and up to weeks), one sample of inclusion bodies was taken out from the exchange buffer, dehydrated, and then dissolved in the DMSO-d6/TFA-d mixture to collect its exchange spectrum. Figure 7.5b (right) shows the spectrum of BMP2(13–74) inclusion bodies that were H/D-exchanged for 185 h.

For every backbone N–H pair of BMP2(13–74), its intensities in the reference spectrum and all H/D-exchange spectra were recorded and plotted versus the exchange time to calculate its H/D-exchange rate, k_{ex}. As seen in Figure 7.5c, residues 62 to 67 of BMP2(13–74) have very slow exchange rates between 10^2 and 10^3 h (dark gray bars), whereas the rest of the protein has much faster exchange rates, less than 1 h. The exchange rates suggest that residues 62 to 67 are protected against deuterium exchange and form the cross-β-sheet structure in BMP2(13–74) inclusion bodies. It should be noted that because inclusion bodies are usually inhomogeneous protein aggregates, the light gray bars in Figure 7.5c represent the exchange behavior of some minority BMP2(13–74) inclusion bodies. To confirm the potential of residues around 62 to 67 to be in a cross-β-sheet structure, a peptide corresponding to residues 59 to 71 of BMP2(13–74) was synthesized and was found to form amyloid fibrils at physiological conditions. To verify that residues 62 to 67 are indeed the dominant component during inclusion body formation, aggregation-prone residues in BMP2(13–74) were mutated to the aggregation-interfering residue arginine (Rousseau et al., 2006). It was found that mutations within the sequence segment (I62R and L66R) reduced the formation of inclusion bodies and brought some proteins to a soluble fraction, whereas mutations outside the segment (I32R and L51R) did not affect the formation of inclusion bodies (Figure 7.5d). Therefore, using solution-state NMR and the H/D-exchange method, we identified the fact that residues 62 to 67 of BMP2(13–74) are involved in the formation of cross-β-sheets in inclusion bodies.

7.4 STRUCTURAL STUDY OF INCLUSION BODIES WITH SOLID-STATE NMR SPECTROSCOPY

Solid-state NMR spectroscopy is a very powerful method to use to determine the molecular structure of protein aggregates because it can measure their structural restraints directly without the requirement of crystallization or molecular tumbling of the sample. In recent years, this method has been applied successfully to the structural determination of many amyloid fibrils: for example, HET-s(218–289) (Wasmer et al., 2008a; Van Melckebeke et al., 2010) and amyloid-β (Petkova et al., 2002; Paravastu et al., 2008, 2009). Although no three-dimensional structure of inclusion bodies has been determined yet with this method, it has provided valuable structural information on inclusion bodies.

7.4.1 Structural Study of HET-s(218–289) Inclusion Bodies

As introduced in Section 7.2.2, HET-s(218–289) is the core of a prion protein in filamentous fungus *Podospora anserine* (Balguerie et al., 2003). It can form amyloid fibrils under physiological conditions as well as inclusion bodies during recombinant protein production (Sabate et al., 2009). The structural study of HET-s(218–289) inclusion bodies involves two steps: (1) determine the structure of HET-s(218–289) amyloid fibrils (Wasmer et al., 2008a), and (2) compare the spectra of inclusion bodies and amyloid fibrils of HET-s(218–289) (Wasmer et al., 2009).

Determining the Structure of HET-s(218–289) Amyloid Fibrils To determine the structure of HET-s(218–289) amyloid fibrils, solid-state NMR spectra that selectively reflect their inter- or intramolecular orders were recorded. Internuclear distance restraints between pairs of protons or carbon atoms were derived from ^{13}C-detected proton-spin diffusion CHHC and NHHC experiments (Lange et al., 2002, 2005) or from proton-driven ^{13}C spin diffusion (PDSD) experiments (Szeverenyi et al., 1982), respectively. A sufficient total number of distance restraints could be extracted for the 72 amino acid residues of HET-s(218–289); initially, 134 unambiguous restraints were used (Wasmer et al., 2008a), and finally, 2631 automatically assigned, ambiguous restraints could be extracted from all spectra, comprising both proton–proton and carbon–carbon distances. Using all available experimental restraints, the structure of HET-s(218–289) amyloid fibrils was determined to have an average pairwise RMSD of 1.54 Å for the core region (residues 226–246 and 262–282) (Van Melckebeke et al., 2010) (Figure 7.3).

Comparing Spectra of Inclusion Bodies and Amyloid Fibrils of HET-s (218–289) To study the structure of HET-s(218–289) inclusion bodies, two inclusion body samples prepared differently were investigated. The first sample, called raw inclusion bodies, was obtained merely by washing the insoluble fraction of lysed cells in pure water; the second sample, purified inclusion bodies, was obtained by further washing the raw inclusion bodies with 8 M urea followed by 2% Triton X-100 (Wasmer et al., 2009).

For both inclusion body samples, two-dimensional ^{13}C–^{13}C proton-driven spin-diffusion fingerprint spectra (Szeverenyi et al., 1982) were recorded and compared with that of HET-s(218–289) amyloid fibrils. The spectrum of the raw inclusion bodies reproduced all peaks visible for amyloid fibrils; and as the sample contains the entire insoluble cellular fraction, additional resonance peaks not belonging to HET-s(218–289) appear in several regions of the spectrum. These additional peaks were tentatively assigned to phospholipids from the *Escherichia coli* membrane, as well as to other proteins and possibly, RNA. For the spectrum of purified inclusion bodies, all peaks of HET-s(218–289) amyloid fibrils were preserved, while those additional resonance peaks in the raw sample were considerably reduced.

Considering the strong dependence of NMR chemical shifts on protein conformation, the identical chemical shifts of HET-s(218–289) in inclusion bodies and in amyloid fibrils suggest that their molecular structures have to be virtually identical (Figure 7.6). The NMR line widths of amyloid fibrils and inclusion bodies are also indistinguishable, as judged from isolated signals in the two-dimensional spectra (Figure 7.6, inserts), indicating a highly defined local molecular structure of HET-s(218–289) inclusion bodies (Siemer et al., 2005). In addition to the solid-state NMR characterization, a solution-state NMR experiment with H/D-exchange was performed. The observed exchange pattern of the purified inclusion bodies closely resembles that of the amyloid fibrils of HET-s(218–289); therefore, it further verifies that their structures are virtually identical at the atomic level. In support of the NMR finding, a very careful inspection of electron micrographs also gave evidence of fibrillar structures in HET-s(218–289) inclusion bodies (Wasmer et al., 2009).

The result of solid-state NMR shows that HET-s(218–289) can form a highly ordered structure resembling amyloid fibrils in the crowded milieu of an *E. coli* cell and in the presence of folding modulators, chaperones, and high concentrations of other proteins. In addition, it shows that inclusion bodies can be highly ordered, in contrast to being "amorphous," as could be inferred from their appearance in electron micrographs. Whether HET-s(218–289) represents an exception or whether other proteins can also display a highly ordered structure resembling amyloid fibrils structure in inclusion bodies remains to be examined further.

7.4.2 Structural Study of FHA2 Inclusion Bodies

Many studies have shown that for some proteins, their native functions are partially preserved after they form inclusion bodies (Ami et al., 2005; García-Fruitós et al., 2005, 2007; Doglia et al., 2008; Lotti, 2011; Sans et al., 2012), suggesting that native structures of these proteins may remain in inclusion bodies besides the cross-β-sheet structure, or a fraction of inclusion bodies may consist of natively folded proteins (Cano-Garrido et al., 2012). The study of FHA2 inclusion bodies confirms the existence of native-like structural elements in inclusion bodies (Curtis-Fisk et al., 2008). FHA2 is the 185-residue functional domain of the HA2 subunit of the influenza virus

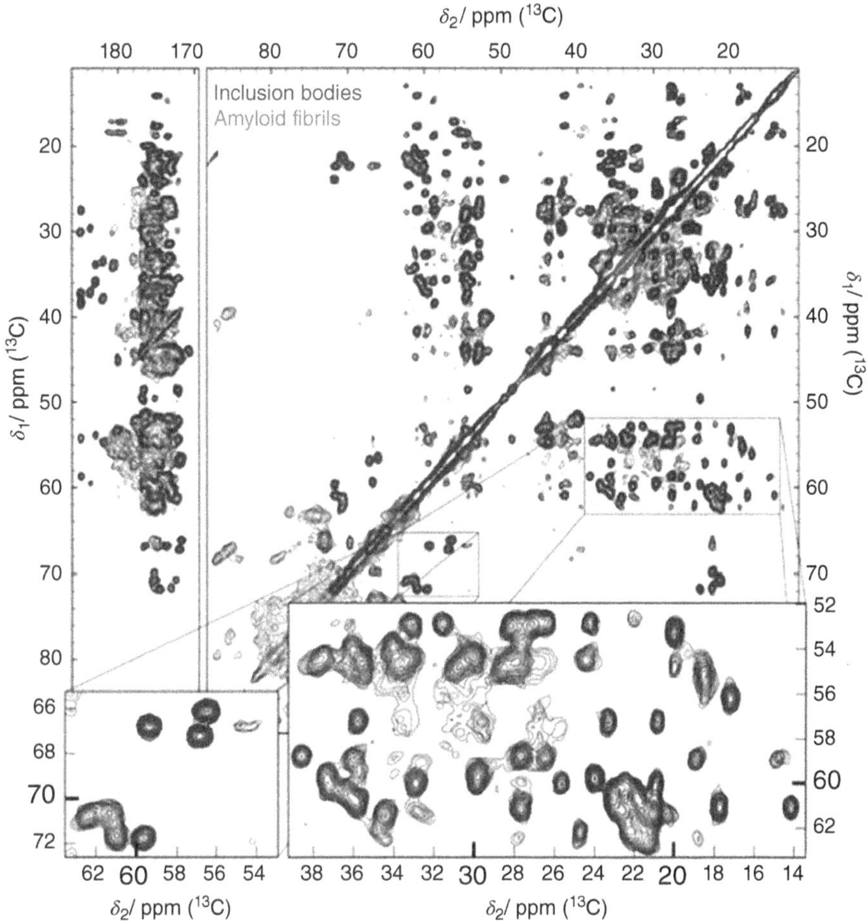

FIGURE 7.6 ^{13}C–^{13}C solid-state-NMR correlation spectrum (proton-driven spin diffusion with a mixing time of 50 ms) of purified HET-s(218–289) inclusion bodies (blue contours) compared with the same type of spectrum of HET-s(218–289) amyloid fibrils. All resonance peaks identified for the amyloid fibrils were also observed in the spectrum of purified inclusion bodies. The insets demonstrate that the linewidths of the two spectra are virtually identical, with no significant chemical-shift changes. (Reproduced with permission from Wiley-VCH Verlag GmbH.) (*See insert for color representation of the figure.*)

hemagglutin in protein (Curtis-Fisk et al., 2007). With ^{15}N–^{13}C isotopic labeling of special sequence pairs of FHA2, the rotational-echo double-resonance spectra (REDOR) (Gullionand Schaefer, 2011) were measured on FHA2 inclusion bodies with solid-state NMR. The REDOR spectra show that the backbone ^{13}CO chemical shift of residue Gly1, Gly4, Ala7, and Leu98 of FHA2 inclusion bodies resembles the chemical shift of an α-helix rather than a β-strand. Since these four residues locate in α-helical structures in the native FHA2, solid-state NMR studies suggest that the native structure of a protein may be present in inclusion bodies. However, to determine if the protein is partially folded in inclusion bodies, or natively folded but trapped in inclusion bodies, will require further investigation.

7.5 SUMMARY

Recent studies have shown that bacterial inclusion bodies have amyloid-like properties and/or native-like properties of the protein. They may contain cross-β-sheet conformation as well as partial native conformation. However, no high-resolution molecular structure of inclusion bodies is available yet because the non-crystalline nature and inhomogeneity of inclusion bodies present significant challenge to their direct structural determination. Current residue-specific structural information of inclusion bodies has been obtained through two indirect methods as described in this chapter. The solution-state NMR and H/D-exchange method has been used to identify the protein residues that are involved in the cross-β-sheet structure of inclusion bodies. The solid-state NMR method has been used to measure chemical shifts of residues in inclusion bodies, and compare them with chemical shifts of known structures to obtain structural information about inclusion bodies. Although these methods provide valuable insights of inclusion body structure, direct high-resolution structural characterization of inclusion bodies are still of high interest.

REFERENCES

Alexandrescu AT (2001) An NMR-based quenched hydrogen exchange investigation of model amyloid fibrils formed by cold shock protein A. Pac Symp Biocomput: 67–78.

Allendorph GP, Vale WW, Choe S (2006) Structure of the ternary signaling complex of a TGF-beta superfamily member. Proc Natl Acad Sci U S A 103:7643–7648.

Ami D, Natalello A, Gatti-Lafranconi P, Lotti M, Doglia SM (2005) Kinetics of inclusion body formation studied in intact cells by FT-IR spectroscopy. FEBS Lett 579:3433–3436.

Balguerie A, Dos Reis S, Ritter C, Chaignepain S, Coulary-Salin B, et al. (2003) Domain organization and structure–function relationship of the HET-s prion protein of *Podospora anserina*. EMBO J 22:2071–2081.

Bracken C, Palmer AG, 3rd, Cavanagh J (1997) (H)N(COCA)NH and HN(COCA) NH experiments for ^{1}H–^{15}N backbone assignments in $^{13}C/^{15}N$-labeled proteins. J Biomol NMR 9:94–100.

Cano-Garrido O, Rodriguez-Carmona E, Diez-Gil C, Vazquez E, Elizondo E, et al. (2012) Supramolecular organization of protein-releasing functional amyloids solved in bacterial inclusion bodies. Acta Biomater 9:6134–6142.

Carrió MM, Cubarsi R, Villaverde A (2000) Fine architecture of bacterial inclusion bodies. FEBS Lett 471:7–11.

Carrió M, González-Montalbán N, Vera A, Villaverde A, Ventura S (2005) Amyloid-like properties of bacterial inclusion bodies. J MolBiol 347:1025–1037.

Curtis-Fisk J, Preston C, Zheng Z, Worden RM, Weliky DP (2007) Solid-state NMR structural measurements on the membrane-associated influenza fusion protein ectodomain. J Am Chem Soc 129:11320–11321.

Curtis-Fisk J, Spencer RM, Weliky DP (2008) Native conformation at specific residues in recombinant inclusion body protein in whole cells determined with solid-state NMR spectroscopy. J Am Chem Soc 130:12568–12569.

de Groot NS, Castillo V, Grana-Montes R, Ventura S (2012) AGGRESCAN: method, application, and perspectives for drug design. Methods MolBiol 819:199–220.

Dobson CM (2003) Protein folding and misfolding. Nature 426:884–890.

Doglia SM, Ami D, Natalello A, Gatti-Lafranconi P, Lotti M (2008) Fourier transform infrared spectroscopy analysis of the conformational quality of recombinant proteins within inclusion bodies. Biotechnol J 3:193–201.

Fernandez-Escamilla AM, Rousseau F, Schymkowitz J, Serrano L (2004) Prediction of sequence-dependent and mutational effects on the aggregation of peptides and proteins. Nat Biotechnol 22:1302–1306.

García-Fruitós E, González-Montalbán N, Morell M, Vera A, Ferraz RM, et al. (2005) Aggregation as bacterial inclusion bodies does not imply inactivation of enzymes and fluorescent proteins. Microb Cell Fact 4:27.

García-Fruitós E, Aris A, Villaverde A (2007) Localization of functional polypeptides in bacterial inclusion bodies. Appl Environ Microbiol 73:289–294.

García-Fruitós E, Seras-Franzoso J, Vazquez E, Villaverde A (2010) Tunable geometry of bacterial inclusion bodies as substrate materials for tissue engineering. Nanotechnology 21:205101.

García-Fruitós E, Vazquez E, Diez-Gil C, Corchero JL, Seras-Franzoso J, et al. (2012) Bacterial inclusion bodies: making gold from waste. Trends Biotechnol 30:65–70.

Gatti-Lafranconi P, Natalello A, Ami D, Doglia SM, Lotti M (2011) Concepts and tools to exploit the potential of bacterial inclusion bodies in protein science and biotechnology. FEBS J 278:2408–2418.

Greenwald J, Riek R (2010) Biology of amyloid: structure, function, and regulation. Structure 18:1244–1260.

Gullion T, Schaefer J (2011) Rotational-echo double-resonance NMR. J MagnReson 213:413–417.

Hamodrakas SJ (2011) Protein aggregation and amyloid fibril formation prediction software from primary sequence: towards controlling the formation of bacterial inclusion bodies. FEBS J 278:2428–2435.

Hirota-Nakaoka N, Hasegawa K, Naiki H, Goto Y (2003) Dissolution of beta2-microglobulin amyloid fibrils by dimethylsulfoxide. J Biochem 134:159–164.

Hoshino M, Katou H, Hagihara Y, Hasegawa K, Naiki H, et al. (2002) Mapping the core of the beta(2)-microglobulin amyloid fibril by H/D exchange. Nat Struct Biol 9:332–336.

Ivanova MI, Sawaya MR, Gingery M, Attinger A, Eisenberg D (2004) An amyloid-forming segment of beta2-microglobulin suggests a molecular model for the fibril. Proc Natl Acad Sci U S A 101:10584–10589.

Ivanova MI, Sievers SA, Sawaya MR, Wall JS, Eisenberg D (2009) Molecular basis for insulin fibril assembly. Proc Natl Acad Sci U S A 106:18990–18995.

Kelly JW (2005) Attacking amyloid. N Engl J Med 352:722–723.

Klunk WE, Jacob RF, Mason RP (1999) Quantifying amyloid by Congo Red spectral shift assay. Methods Enzymol 309:285–305.

Lange A, Luca S, Baldus M (2002) Structural constraints from proton-mediated rare-spin correlation spectroscopy in rotating solids. J Am Chem Soc 124:9704–9705.

Lange A, Becker S, Seidel K, Giller K, Pongs O, et al. (2005) A concept for rapid protein-structure determination by solid-state NMR spectroscopy. Angew Chem Int Ed Engl 44:2089–2092.

Lee YH, Goto Y (2012) Kinetic intermediates of amyloid fibrillation studied by hydrogen exchange methods with nuclear magnetic resonance. Biochim Biophys Acta 1824:1307–1323.

LeVine H, 3rd (1999) Quantification of beta-sheet amyloid fibril structures with Thioflavin T. Methods Enzymol 309:274–284.

Lotti M (2011) Bacterial inclusion bodies as active and dynamic protein ensembles. FEBS J 278:2407.

Luhrs T, Ritter C, Adrian M, Riek-Loher D, Bohrmann B, et al. (2005) 3D structure of Alzheimer's amyloid-beta(1–42) fibrils. Proc Natl Acad Sci U S A 102:17342–17347.

Maji SK, Schubert D, Rivier C, Lee S, Rivier JE, et al. (2008) Amyloid as a depot for the formulation of long-acting drugs. PLoS Biol 6:e17.

Maji SK, Perrin MH, Sawaya MR, Jessberger S, Vadodaria K, et al. (2009a) Functional amyloids as natural storage of peptide hormones in pituitary secretory granules. Science 325:328–332.

Maji SK, Wang L, Greenwald J, Riek R (2009b) Structure–activity relationship of amyloid fibrils. FEBS Lett 583:2610–2617.

Morell M, Bravo R, Espargaro A, Sisquella X, Aviles FX, et al. (2008) Inclusion bodies: specificity in their aggregation process and amyloid-like structure. Biochim Biophys Acta 1783:1815–1825.

Nelson R, Sawaya MR, Balbirnie M, Madsen AO, Riekel C, et al. (2005) Structure of the cross-beta spine of amyloid-like fibrils. Nature 435:773–778.

Paravastu AK, Leapman RD, Yau WM, Tycko R (2008) Molecular structural basis for polymorphism in Alzheimer's beta-amyloid fibrils. Proc Natl Acad Sci U S A 105:18349–18354.

Paravastu AK, Qahwash I, Leapman RD, Meredith SC, Tycko R (2009) Seeded growth of beta-amyloid fibrils from Alzheimer's brain-derived fibrils produces a distinct fibril structure. Proc Natl Acad Sci U S A 106:7443–7448.

Petkova AT, Ishii Y, Balbach JJ, Antzutkin ON, Leapman RD, et al. (2002) A structural model for Alzheimer's beta-amyloid fibrils based on experimental constraints from solid state NMR. Proc Natl Acad Sci U S A 99:16742–16747.

Ritter C, Maddelein ML, Siemer AB, Luhrs T, Ernst M, et al. (2005) Correlation of structural elements and infectivity of the HET-s prion. Nature 435:844–848.

Rousseau F, Serrano L, Schymkowitz JW (2006) How evolutionary pressure against protein aggregation shaped chaperone specificity. J Mol Biol 355:1037–1047.

Sabate R, Espargaro A, Saupe SJ, Ventura S (2009) Characterization of the amyloid bacterial inclusion bodies of the HET-s fungal prion. Microb Cell Fact 8:56.

Salzmann M, Wider G, Pervushin K, Senn H, Wuthrich K (1999) TROSY-type triple-resonance experiments for sequential NMR assignments of large proteins. J Am Chem Soc 121:844–848.

Sans C, Garcia-Fruitos E, Ferraz RM, Gonzalez-Montalban N, Rinas U, et al. (2012) Inclusion bodies of fuculose-1-phosphate aldolase as stable and reusable biocatalysts. Biotechnol Prog 28:421–427.

Sawaya MR, Sambashivan S, Nelson R, Ivanova MI, Sievers SA, et al. (2007) Atomic structures of amyloid cross-beta spines reveal varied steric zippers. Nature 447: 453–457.

Scheufler C, Sebald W, Hulsmeyer M (1999) Crystal structure of human bone morphogenetic protein-2 at 2.7Å resolution. J Mol Biol 287:103–115.

Selkoe DJ (2003) Folding proteins in fatal ways. Nature 426:900–904.

Seras-Franzoso J, Diez-Gil C, Vazquez E, Garcia-Fruitos E, Cubarsi R, et al. (2012) Bioadhesiveness and efficient mechanotransduction stimuli synergistically provided by bacterial inclusion bodies as scaffolds for tissue engineering. Nanomedicine (Lond) 7:79–93.

Siemer AB, Ritter C, Ernst M, Riek R, Meier BH (2005) High-resolution solid-state NMR spectroscopy of the prion protein HET-s in its amyloid conformation. Angew Chem Int Ed Engl 44:2441–2444.

Stromer T, Serpell LC (2005) Structure and morphology of the Alzheimer's amyloid fibril. Microsc Res Tech 67:210–217.

Sunde M, Blake C (1997) The structure of amyloid fibrils by electron microscopy and X-ray diffraction. Adv Protein Chem 50:123–159.

Sunde M, Serpell LC, Bartlam M, Fraser PE, Pepys MB, et al. (1997) Common core structure of amyloid fibrils by synchrotron x-ray diffraction. J Mol Biol 273: 729–739.

Szeverenyi NM, Sullivan MJ, Maciel GE (1982) Observation of spin exchange by two-dimensional Fourier transform ^{13}C cross polarization-magic-angle spinning. J Magn Reson 47:462–475.

Tartaglia GG, Vendruscolo M (2008) The Zyggregator method for predicting protein aggregation propensities. Chem Soc Rev 37:1395–1401.

Thompson MJ, Sievers SA, Karanicolas J, Ivanova MI, Baker D, et al. (2006) The 3D profile method for identifying fibril-forming segments of proteins. Proc Natl Acad Sci U S A 103:4074–4078.

Van Melckebeke H, Wasmer C, Lange A, Ab E, Loquet A, et al. (2010) Atomic-resolution three-dimensional structure of HET-s(218–289) amyloid fibrils by solid-state NMR spectroscopy. J Am Chem Soc 132:13765–13775.

Vazquez E, Corchero JL, Burgueno JF, Seras-Franzoso J, Kosoy A, et al. (2012) Functional inclusion bodies produced in bacteria as naturally occurring nanopills for advanced cell therapies. Adv Mater 24:1742–1747.

Ventura S (2005) Sequence determinants of protein aggregation: tools to increase protein solubility. Microb Cell Fact 4:11.

Ventura S, Villaverde A (2006) Protein quality in bacterial inclusion bodies. Trends Biotechnol 24:179–185.

Vilar M, Chou HT, Luhrs T, Maji SK, Riek-Loher D, et al. (2008) The fold of alpha-synuclein fibrils. Proc Natl Acad Sci U S A 105:8637–8642.

Vilar M, Wang L, Riek R (2012) Structural studies of amyloids by quenched hydrogen–deuterium exchange by NMR. Methods Mol Biol 849:185–198.

Villaverde A (2012) Bacterial inclusion bodies: an emerging platform for drug delivery and cell therapy. Nanomedicine (Lond) 7:1277–1279.

Villaverde A, Garcia-Fruitos E, Rinas U, Seras-Franzoso J, Kosoy A, et al. (2012) Packaging protein drugs as bacterial inclusion bodies for therapeutic applications. Microb Cell Fact 11:76.

Wang L, Maji SK, Sawaya MR, Eisenberg D, Riek R (2008) Bacterial inclusion bodies contain amyloid-like structure. PLoS Biol 6:e195.

Wang L, Schubert D, Sawaya MR, Eisenberg D, Riek R (2010) Multidimensional structure–activity relationship of a protein in its aggregated states. Angew Chem Int Ed Engl 49:3904–3908.

Wasmer C, Lange A, Van Melckebeke H, Siemer AB, Riek R, et al. (2008a) Amyloid fibrils of the HET-s(218–289) prion form a beta solenoid with a triangular hydrophobic core. Science 319:1523–1526.

Wasmer C, Soragni A, Sabate R, Lange A, Riek R, et al. (2008b) Infectious and non-infectious amyloids of the HET-s(218–289) prion have different NMR spectra. Angew Chem Int Ed Engl 47:5839–5841.

Wasmer C, Benkemoun L, Sabate R, Steinmetz MO, Coulary-Salin B, et al. (2009) Solid-state NMR spectroscopy reveals that *E. coli* inclusion bodies of HET-s(218–289) are amyloids. Angew Chem Int Ed Engl 48:4858–4860.

Wiltzius JJ, Sievers SA, Sawaya MR, Cascio D, Popov D, et al. (2008) Atomic structure of the cross-beta spine of islet amyloid polypeptide (amylin). Protein Sci 17:1467–1474.

Zhang YZ, Paterson Y, Roder H (1995) Rapid amide proton exchange rates in peptides and proteins measured by solvent quenching and two-dimensional NMR. Protein Sci 4:804–814.

8

BIOMEDICAL APPLICATIONS OF BACTERIAL INCLUSION BODIES

IMMA RATERA, SPELA PETERNEL, JOAQUIN SERAS-FRANZOSO, OLIVIA CANO-GARRIDO, ELENA GARCÍA-FRUITÓS, RAFAEL CUBARSÍ, ESTHER VAZQUEZ, JOSÉ LUIS CORCHERO, ESCARLATA RODRÍGUEZ-CARMONA, JAUME VECIANA, AND ANTONIO VILLAVERDE

8.1 BIOLOGY OF IB PROTEINS

The production of recombinant proteins is a major technological platform in biotechnology, pharmacology, and associated disciplines, and a convenient source of polypeptides for the biocatalysis, food, and pharmaceutical industries. Proteins of industrial and medical interest cannot be purified from natural sources at working amounts through cost-effective procedures. Prokaryotic organisms, and especially the gram-negative bacterium *Escherichia coli*, have been adapted during three decades as cellular factories for protein production (Ferrer-Miralles et al., 2009; Makino et al., 2011). Alternative bacterial species, but also yeasts, fungi, algae, insect cells, and mammalian cells, as well as plants and animals, have been engineered to express foreign genes for the controlled biological synthesis of encoded proteins (Corchero et al., 2012). Using these biological platforms, enzymes, hormones, growth factors, cytoquines, antigens, immunogens, and a diversity of supramolecular complexes (including virus-like particles and other types of protein nanoparticles) have been obtained for characterization or to be

Protein Aggregation in Bacteria: Functional and Structural Properties of Inclusion Bodies in Bacterial Cells, First Edition. Edited by Silvia Maria Doglia and Marina Lotti.

used as tools in catalysis or biomedicine (Rodriguez-Carmona and Villaverde, 2010). Protein production processes are aimed at the manufacture of stable and soluble protein species (Baneyx and Mujacic, 2004; González-Montalbán et al., 2007). Therefore, proteolytic instability and aggregation are primary obstacles for protein production that had severely challenged the economic viability and feasibility of many processes. This is because many protein species are highly protease-sensitive and/or tend to aggregate as insoluble deposits upon biosynthesis in producing cells (Sørensen and Mortensen, 2005a,b).

In *E. coli*, particularly, the formation of large protein aggregates termed *inclusion bodies* (IBs) is very common (Villaverde and Carrió, 2003). These are spherical or pseudospherical proteinaceous clusters resulting from the aggregation of the recombinant protein, which also contain bacterial proteins and other macromolecules as minor contaminants (Neubauer et al., 2006). Similar particulate protein aggregates are observed in other cell platforms (e.g., aggresomes in mammalian or insect cells), but they have been in general less well studied than IBs (Kopito, 2000; Markossian and Kurganov, 2004; Martinez-Alonso et al., 2009a).

Bacterial IBs are highly porous, mechanically stable proteinaceous particles in a size range between the nano and micro scales (Bowden et al. 1991; Peternel et al., 2008a; García-Fruitós et al., 2009). Being refractile under optical microscope observation and electrodense by transmission electron microscopy, they tend to appear in the cytoplasm at the bacterial poles (Rokney et al., 2009) or in the periplasmic space in the case of secretory proteins (Miot and Betton, 2004). Although it had been believed that IBs are formed uniquely by unfolded or highly misfolded protein versions, and therefore are biologically inert (Baneyx and Mujacic, 2004), it was later determined that a significant fraction of the forming protein is functional and properly folded, adopting native or native-like secondary structure and exhibiting biological activity (García-Fruitós et al., 2005; Jevsevar et al., 2005; Ventura and Villaverde, 2006; González-Montalbán et al., 2007). This observation has supported the exploration of IBs formed by industrial enzymes as naturally immobilized biocatalysts with high operational stability (Nahálka and Nidetzky, 2007; Nahálka, 2008; Nahálka et al. 2008a,b; Nahálka et al., 2009; García-Fruitós and Villaverde, 2010; Sans et al., 2012), which indicates the high mechanical stability of the nanostructured particles, further confirmed by independent reports (Margreiter et al., 2008; García-Fruitós et al., 2009). On the other hand, IBs show marked amyloidal properties (Carrió et al., 2005; de Groot et al., 2009; Sabate et al., 2009; García-Fruitós et al., 2011; Espargaro et al., 2012; Villar-Piqué et al., 2012), as a fraction of the IB material is composed of proteinase K-resistant protein clusters, at least some of them organized as fibrils (Morell et al., 2008). Such a partitioning in the conformational status of IB protein species might be responsible for the unexpected potential of IBs in nanomedicine, as discussed below.

8.2 IB PROTEIN QUALITY

IBs and their forming proteins are not excluded from a cell's quality control, which surveys the conformational status of cellular polypeptides (González-Montalbán et al., 2007). Then, deficiencies in chaperones and proteases or the overexpression of these components along with the recombinant protein have a strong impact on the amount of IB protein and on its conformational and functional quality. Although co-production of folding modulators has largely been explored as a way to minimize aggregation and IB formation (Sørensen and Mortensen, 2005a; de Marco, 2007; de Marco et al., 2007; Kolaj et al., 2009), this approach often shows paradoxical effects, and it results in undesired events or even in a negative impact on the protein yield (García-Fruitós et al., 2007; Martinez-Alonso et al., 2007, 2009b, 2010; Platas et al., 2011), usually by enhancing degradation of the target protein (García-Fruitós et al., 2007; Martinez-Alonso et al., 2009a). In fact, chaperones (such as DnaK) and proteases (such as ClpP) seem to act coordinately on the IB surface (Mogk et al., 2003; Schlieker et al., 2004; Carrió and Villaverde, 2005), as a specific machinery to promote *in vivo* disaggregation for refolding attempts or, eventually, for in situ proteolysis (Carrió and Villaverde, 2003).

Interestingly, the conformational quality and physical properties of bacterial protein aggregates can be modulated through parameters such as culture conditions (e.g., growth temperature, harvesting time, dose of isopropyl β-D-1-thiogalactopyranoside or other chemical inducers of gene expression) and by selecting the genetic background of the producing strain (Table 8.1). In this context, the use of *E. coli* strains deficient in chaperones or proteases cannot only result into differently sized and shaped IBs, but it also serves to tune the conformational quality of the proteins forming the aggregates (Jevsevar et al., 2005; Vera et al., 2007; Peternel et al., 2008a) and the physicochemical properties of the whole particles (de Groot and Ventura, 2006; García-Fruitós et al., 2007, 2009). How functional proteins can occur together with amyloidal organization is not yet clear. However, this coexistence suggests the occurrence of heterogeneously folded protein species within IBs, a possibility that is fully supported by several observations, as discussed below.

TABLE 8.1 Main Parameters that Determine Final Size, Conformational Quality, and Physicochemical Properties of IBs[a]

	Size	Conformational Quality	Physicochemical Properties
Growth temperature	García-Fruitós et al., 2009	Jevsevar et al., 2005; Vera et al., 2007; de Groot and Ventura, 2006; Peternel, 2008b	ND
Harvesting time	García-Fruitós et al., 2007, 2009; Petersson et al., 2004	García-Fruitós et al., 2007	Carrió et al., 2000
Inducer concentration	Petersson et al., 2004	ND	ND
Genetic background	García-Fruitós et al., 2007, 2009	García-Fruitós et al., 2007, 2009	García-Fruitós et al., 2009, 2010; Diez-Gil et al., 2010

[a]ND, not described.

8.3 IB ARCHITECTURE

Scanning electron microscopy has provided some clues about the structure of IBs, revealing a wrinkled, porous structure (Figure 8.1a). In addition, independent studies indicated that IB proteins are released from IBs under nondenaturing conditions and that IBs are soluble in mild detergents at neutral pH (Jevsevar et al., 2005), especially when they are formed below 37°C and enriched with properly folded proteins. When IBs are suspended in mild detergent (e.g., 0.2% *N*-lauroylsarcosine or 0.1% deoxycholate), a fraction of properly folded proteins is washed away, and the inner structure, composed of small protoaggregates embedded in a cotton-like matrix, is exposed (Figure 8.1b) (Peternel et al., 2008a). However, when IBs are transferred to low-pH buffer, they irreversibly contract and became insoluble in mild detergents (Peternel et al., 2008a). Based on these results, a model for IB formation as well as their architecture structure was proposed, considering the distinct conformational status of IB protein subpopulations (Peternel and Komel, 2011). Such a heterogeneous composition is supported by the different sensitivity of IB protein populations to *in vitro* digestion with trypsin and proteinase K (Cano-Garrido et al., 2013). Protein species fully resistant to proteinase K seem to correspond to amyloidal structures within IBs, and these fibers might be the basis of the scaffolding matrix of IBs.

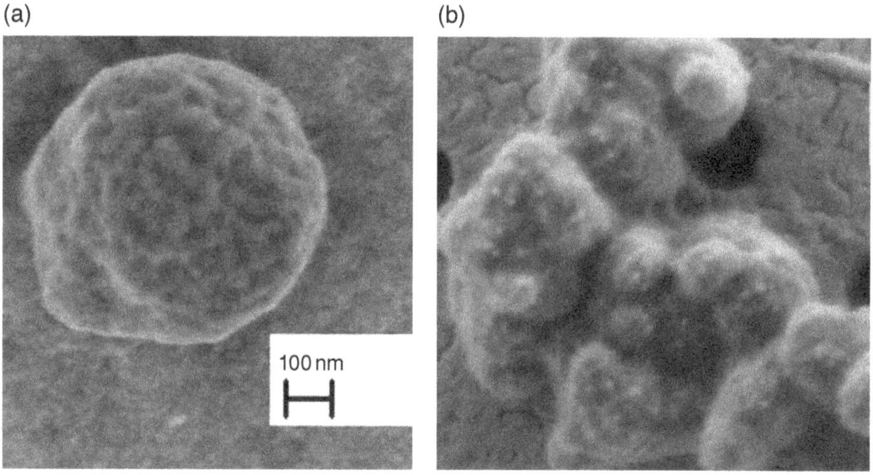

FIGURE 8.1 Scanning electron microscope images of (a) freshly purified IBs and (b) after detergent-mediated extraction of functional proteins.

As a bacterial cell is not compartmentalized, recombinant proteins are synthesized simultaneously on multiple locations in the cytoplasm, and many different transitional folding states of the protein are present simultaneously. In recombinant bacteria, some recombinant polypeptide chains remain unfolded or misfolded and start to aggregate into small and still soluble proto-aggregates (Schroedel and de Marco, 2005), acting as nucleation seeds. Folded or partially folded protein species might interact with these nuclei through cross-molecular stereospecific interactions, while cellular proteins are mostly excluded from these sequence-specific events (Speed et al., 1996; Carrió et al., 2005). Contaminant cell proteins present in IBs, such as folding modulators, could become entrapped because of their interaction with nascent polypeptides. At neutral pH, this network is loose, and folded proteins can easily be extracted (Makino et al., 2011). However, at an acidic pH the network of misfolded proteins irreversibly contracts and properly folded proteins remain trapped inside (Peternel et al., 2008a).

8.4 IBs AS BIOMATERIALS

Interestingly, IBs have recently been observed as emerging biomaterials with potential biomedical applications. This has been empowered by the fact that these protein clusters are biologically active and also mechanically stable. Once purified, IBs are regular particles with low polydispersity, usually spherical or ovoid, ranging from 50 to 1000 nm in diameter. IB size can be

TABLE 8.2 IB Features That Have an Impact in Their Use as Biomaterials

Determinant Parameters	IB Properties	Application
Genetic background	Geometry, stiffness, wettability zeta-potential, bioadhesiveness	Tissue engineering
Genetic background/ process variables	Size, specific activity, specific fluorescence	Biocatalysis
Process variables	Proteolytic stability	Cell therapy
Postproduction treatment	Density/porosity	Tissue engineering

tuned according to the particular protein species, genetic background of the bacteria, and production time (García-Fruitós et al., 2009), while other physicochemical properties of relevance in material sciences, such as stiffness, zeta potential, wettability, and morphology, can also be modulated by selective production in defined *E. coli* mutants (Table 8.2). In particular, deficiencies in the chaperones DnaK and ClpA or in the protease ClpP lead to anomalous quality control and altered protein deposition patterns, which significantly affect the nanoscale properties of IBs (García-Fruitós et al., 2012). To determine how the different genetics of IB bioproduction affects their elasticity, force spectroscopic atomic force microscopy (AFM) was used to measure Young's modulus of IBs in a statistical way. Young's modulus, a measure of the stiffness of elastic materials used to characterize elasticity, is defined as the ratio of the stress along an axis, which has units of pressure, over the strain along that axis, which is dimensionless, in the range of stress in which Hooke's law can be applied. Therefore, Young's modulus has units of pressure [usually megapascal (MPa)]. While wild-type IBs showed a monomodal elasticity distribution, IBs produced in specific phenotypes ($DnaK^-$ and $ClpA^-$) split into two elasticity populations, in agreement with wettability data. Wettability is the ability of a liquid, usually water, to maintain contact with a solid surface and it is important in the adherence of two materials. Adhesive forces between a liquid and a solid cause a liquid drop to spread across the surface. The same technique equipped with a closed-loop tracking system made it possible to obtain spatial distribution of the elasticity regimes over the various IBs and to develop stiffness maps in which each force versus indentation measurement was mapped over the target IB. Young's modulus was spread homogeneously over wild-type IBs, with mean values of 3.6 ± 0.56 MPa. On the contrary, for $DnaK^-$ and $ClpA^-$ IBs, two elasticity populations were observed, with the harder areas segregated from the softer areas. Specifically, they show two Young modulus distributions;

for the DnaK$^-$ IBs centered at 3.6 ± 0.48 and 7.7 ± 0.99 MPa and for the ClpA$^-$ IB centered at 5.01 ± 0.25 and 10.99 ± 0.30 MPa (César Díez-Gil et al., 2010). The data above indicated that the genetics of IB fabrication determines the coexistence of more rigid structures and softer ones, and that they seem to be localized on predetermined areas of individual IB particles. This fact confirmed the structurally diversity inside the protein aggregates, as discussed above.

8.5 PURIFICATION OF IBs

As biomaterials, IBs could eventually be used in biological interfaces, and purification protocols have been adapted to obtain these particles free from bacterial contaminants (Rodriguez-Carmona et al., 2011; Peternel, 2013). In general, IBs show mechanical and long-term stability and good tolerance to mechanical, chemical, and enzymatic cell disruption methods such as ultrasonication and high-pressure homogenization or physical procedures such as freeze/thaw cycles (Georgiou and Valax, 1999). However, IBs tend to co-sediment with bacterial cells and debris during downstream transport, and undesired contaminants such as lipids, lipopolysaccharides, nucleic acids, and remaining whole cells must be fully removed to assure the absence of cytotoxic effects upon exposure to mammalian cells. Moreover, downstream strategies must preserve conformational stability, mechanical properties, and the biological activity of the protein aggregates. In this context, Peternel and Komel compared (2010) three methods (enzymatic lysis and two mechanical methods, high-pressure homogenization and sonication) for the isolation of biologically active IBs. In this study, homogenization was most appropriate in terms of biological activity preservation of the proteins. On the other hand, a convenient combination of mechanical (sonication) and enzymatic (lysozyme) cell disruption methods followed by DNAse treatment and detergent washing steps has been adapted and optimized to provide preparations of cell-free IBs with viable cell contamination below 10^{-1} cfu/mL, while maintaining the particle size and the conformational status of the protein (Rodriguez-Carmona et al., 2010), summarized in Figure 8.2. This procedure seems to be suitable for most applications of IBs involving intimate contact with mammalian cells and organs, as discussed below.

8.6 IBs IN TISSUE ENGINEERING

Recently, numerous studies have supported functionalization and nanostructuring of surfaces for tissue engineering and regenerative medicine (Dalby, 2009). Among the approaches available, the generation of nanostructured and

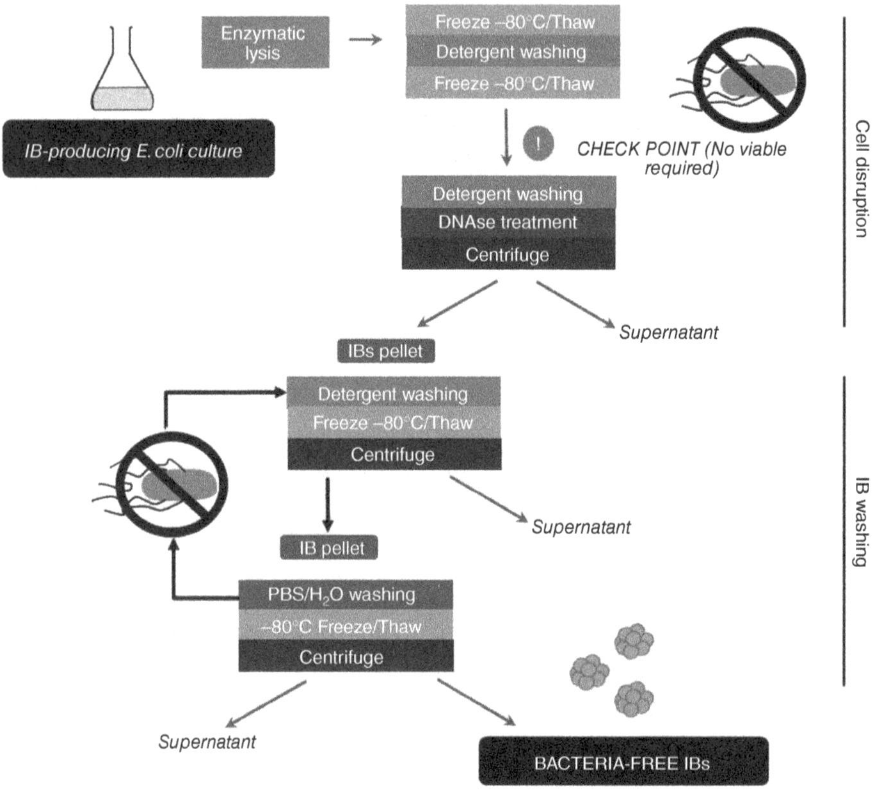

FIGURE 8.2 Schematic representation of IB purification to obtain bacteria-free IB preparations for biomedical applications. This process combines enzymatic digestion of the cell wall with soft mechanical disruption.

nanopatterned surfaces with either inorganic or organic materials is especially appealing (Harrison and Atala, 2007; Chien et al., 2009; Hu et al., 2010; Whited et al., 2011). As a contribution to this area and because of the tunable properties of IBs as described above, IBs formed by the green fluorescent protein (GFP) have been used successfully for a bottom-up approach to engineer surface topographies. These reports prove that IBs can be used as particulate biomaterials to stimulate surface colonization by mammalian cells without any sign of cytotoxicity (García-Fruitós et al., 2009, 2012; Diez-Gil et al., 2010). The fast colonization of IB-modified surfaces is supported by mechanical stimulation of mammalian cell proliferation on IB-decorated surfaces, as observed by enhanced phosphorylation of signal-regulated protein kinase and by the dramatic emission of filopodia–lamellipodia in the presence of IBs (Figure 8.3) (Seras-Franzoso et al., 2012a). Interestingly, it has also been shown that IBs are highly bioadhesive

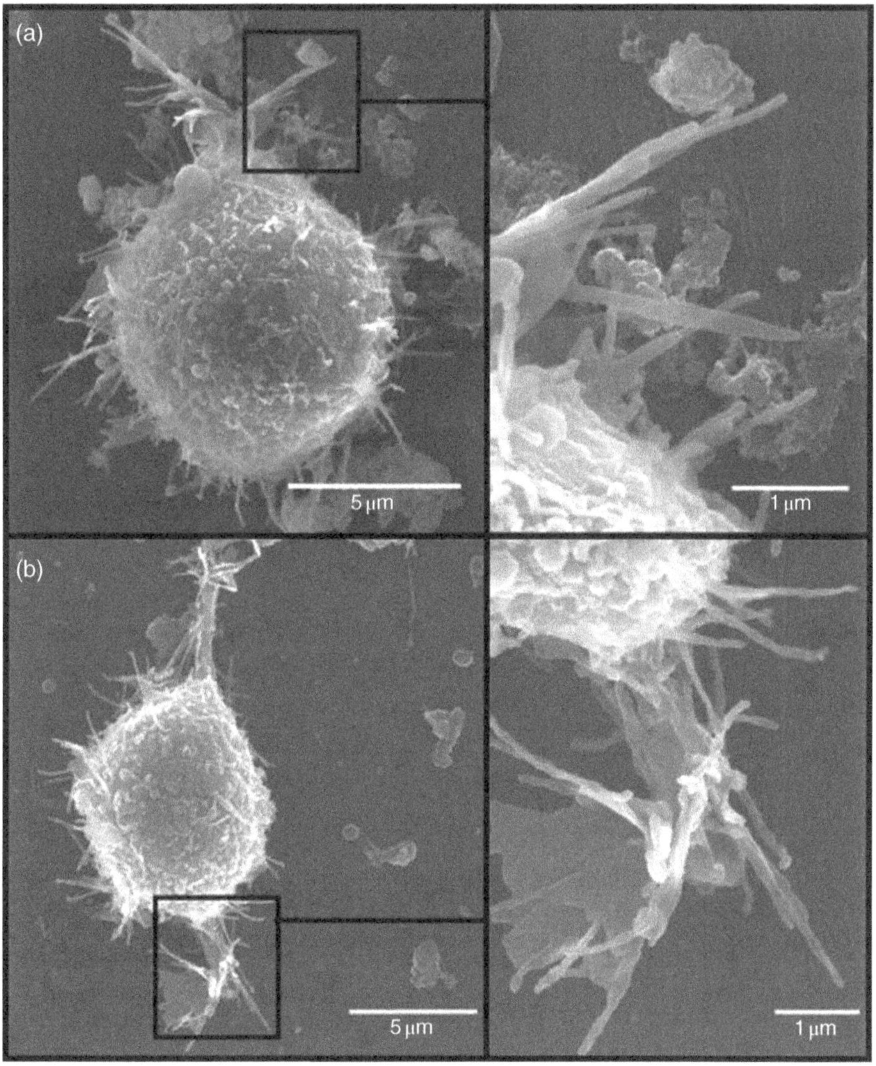

FIGURE 8.3 Scanning electron microscope images of PC12 cells cultured for 48 h on IB-based topographies, showing different individual cells (a and b). Inset frames provide magnifications with further detail of IB-induced filopodia.

materials and that mammalian cell expansion on IBs is supported synergistically by both enhanced adhesion and mechanical stimulation of cell division. The extent in which these events influence cell growth depends on the particular cell line, but it is also related to the genetic background of IB-producing bacteria, thus opening exciting possibilities for fine tailoring of the material (Seras-Franzoso et al., 2012a). In this regard, since IB properties can be tuned during their fabrication process (Peternel et al., 2008b; Martinez-Alonso et al., 2009b), these protein clusters can represent an alternative to the use of such adhesive proteins as fibronectin and vitronectin.

It is known that not only cell adhesion but also cell migration and differentiation can be controlled by engineering the cell–material interface (Ayala et al., 2011). Interestingly, the contact angle that is identified as optimal for surface colonization is within the range of angles exhibited by IBs and perfectly matches that of IB variants produced in $DnaK^-$ and $ClpA^-$ bacterial strains (Diez-Gil et al., 2010). This further supports the use of IBs as convenient substrate materials to favor mammalian cell colonization. Moreover, these biologically inert IBs have also been observed successfully as able to induce differentiation in hMSC cultures (Seras-Franzoso et al., 2013).

On the other hand, cell responses, such as positioning, morphological changes, and proliferation, are the result of complex chemical, topographical, and biological stimuli. In this context, cells respond positively to two-dimensional engineering of biological interfaces at the microscale using IBs. Data obtained with fibroblasts cultivated on IB-patterned supports demonstrate that these cells preferentially adhere to IB-covered areas and align and elongate according to IB patterns (Tatkiewicz et al., 2013). This result fills the gap between existing techniques which are based on the local modification of surface chemistry and techniques of modification of the topography at the nanoscale level (Malmstrom et al., 2011), and it also proves that IBs are interesting and useful biocompatible nanomaterials for biomedical applications.

8.7 SLOW DRUG RELEASE FROM BACTERIAL IBs USED AS NANOPILLS

IBs formed by proteins of therapeutic interest are able to deliver functional polypeptides to the cell cytoplasm and nucleus for a biological effect. Adsorbed to two-dimensional surfaces as functional materials (bioscaffolds), IBs formed by the antiapoptotic chaperone Hsp70 or the basic fibroblast growth factor (FGF-2) get embedded in top-growing cells, make their forming proteins available to them, and rescue or enhance cell viability, respectively (Seras-Franzoso et al., 2012b). Used in suspension as nanopills, IBs formed by these two particular polypeptides (and also by other enzymes or structural proteins) can penetrate the cellular and nuclear membranes in the absence of cell toxicity (Figure 8.4), promoting cell viability through a true protein replacement therapeutic effect at the cell level, for which protein release from IBs is suspected (Liovic et al., 2012; Vazquez et al., 2012).

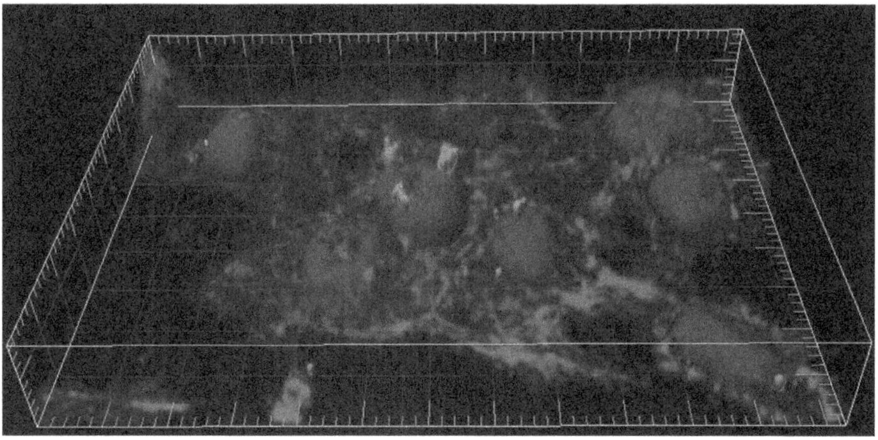

FIGURE 8.4 Confocal *xyz* stack of several sections generated with the Imaris 3D software (using the Isosurface module), showing penetration of GFP-based IBs into cultured HeLa cells. IBs are shown in green, nuclear DNA in blue, and the cell membranes in red. Experimental details can be found elsewhere (Vázquez et al., 2012). (*See insert for color representation of the figure.*)

Moreover, IBs are well tolerated in oral administration (Vazquez et al., 2012), which opens exciting opportunities for therapeutic uses of these protein particles (Villaverde, 2012; Villaverde et al., 2012). Again, functional protein release from therapeutic IBs seems to involve only a fraction of IB protein, as a mechanically stable scaffold is still observed in cell-internalized IBs. The cotton-like model of IBs seem's to account functionally and architectonically for the potential of these clusters as either bioscaffolds or nanopills, which might again be based on the heterogeneous conformational status of the IB protein. While amyloidal fibers and similar structures might confer mechanic stability, properly folded protein species are the functional part of the material that is released under convenient conditions. Interestingly, IBs seem to be natural mimetics of secretory granules of higher organisms, in which protein hormones are stored for secretion as repository amyloidal scaffolds (Maji et al., 2008, 2009; Mankar et al., 2011). The potential of bacterial IBs as functional materials in biomedicine is wide and highly promising, but it remains to be fully explored.

8.8 CONCLUSIONS

Bacterial IBs are naturally occurring proteinaceus particles that contain significant fractions of properly folded versions of recombinant proteins, conferring on them biological activity. Their mechanical stability could be based on amyloidal protein versions, which create a matrix-embedding functional species. The combination of mechanical stability and functionality, together with the potential of functional proteins to be released from IBs under nondenaturing conditions, has recently prompted researchers to explore them as functional materials for biomedical applications. Being fully biocompatible, they appear as interesting slow-delivery agents for protein replacement therapies at the organic level but also in regenerative medicine. Since essentially any polypeptide suitable to be produced in *E. coli* can be obtained as IBs, the potential of bacterial cells to generate therapeutic nanopills should be explored fully in the context of innovative medicines.

ACKNOWLEDGMENTS

We are indebted to the Cell Culture Unit of the Servei de Cultius Cel·lulars, Producció d'Anticossos i Citometria (SCAC), and Servei de Microscòpia, both at the Universitat Autònoma de Barcelona (UAB). We are also indebted to the Protein Production Platform (CIBER-BBN-UAB) for its helpful technical assistance (http://bbn.ciber-bbn.es/programas/plataformas/equipamiento). This study has been funded by MINECO (BFU2010-17450and CTQ2010-19501), AGAUR (2009SGR-00108 and 2009SGR-00516), and CIBER de Bioingeniería, Biomateriales y Nanomedicina (CIBER-BBN, Spain). CIBER-BBN is an initiative funded by the VI National R&D&I Plan *2008–2011*, Iniciativa Ingenio 2010, Consolider Program, CIBER Actions and financed by the Instituto de Salud Carlos III with assistance from the European Regional Development Fund. E.G.F. is supported by the Programa Personal de Técnico de Apoyo (Modalidad Infraestructuras científico-tecnológicas, MICINN). J.S.F. is a recipient of a PIF doctoral fellowship from UAB, O.C. of a FPU fellowship from MECD, and A.V. of an ICREA ACADEMIA award.

REFERENCES

Ayala R, Zhang C, Yang D, Hwang Y, Aung A, Shroff SS, Arce FT, Lal R, Arya G, Varghese S (2011) Engineering the cell-material interface for controlling stem cell adhesion, migration, and differentiation. Biomaterials 32:3700–3711.

Baneyx F, Mujacic M (2004) Recombinant protein folding and misfolding in *Escherichia coli*. Nat Biotechnol 22:1399–1408.

Bowden GA, Paredes AM, Georgiou G (1991) Structure and morphology of protein inclusion bodies in *Escherichia coli*. Biotechnology (N Y) 9:725–730.

Cano-Garrido O, Rodriguez-Carmona E, Diez-Gil C, Vázquez E, Elizondo E, Cubarsi R, Seras-Franzoso J, Corchero JL, Rinas U, Ratera I, Ventosa N, Veciana J,

Villaverde A, García-Fruitós E (2013) Supramolecular organization of protein-releasing functional amyloids solved in bacterial inclusion bodies. Acta Biomater 9:6134–6142.

Carrió MM, Villaverde A (2003) Role of molecular chaperones in inclusion body formation. FEBS Lett 537:215–221.

Carrió MM, Villaverde A (2005) Localization of chaperones DnaK and GroEL in bacterial inclusion bodies. J Bacteriol 187:3599–3601.

Carrió MM, Cubarsi R, Villaverde A (2000) Fine architecture of bacterial inclusion bodies. FEBS Lett 471:7–11.

Carrió M, González-Montalbán N, Vera A, Villaverde A, Ventura S (2005) Amyloid-like properties of bacterial inclusion bodies. J Mol Biol 347:1025–1037.

Chien HW, Chang TY, Tsai WB (2009) Spatial control of cellular adhesion using photo-crosslinked micropatterned polyelectrolyte multilayer films. Biomaterials 30:2209–2218.

Corchero JL, Gasser B, Resina D, Smith W, Parrilli E, Vázquez E, Abasolo I, Giuliani M, Jantti J, Ferrer P, Saloheimo M, Mattanovich D, Schwartz S Jr, Tutino L, Villaverde A (2012) Unconventional microbial systems for the cost-efficient production of high-quality protein therapeutics. Biotechnol Adv 31(2): 140–153.

Dalby MJ (2009) Nanostructured surfaces: cell engineering and cell biology. Nanomedirine (Lond) 4:247–248.

de Groot NS, Ventura S (2006) Effect of temperature on protein quality in bacterial inclusion bodies. FEBS Lett 580:6471–6476.

de Groot NS, Sabate R, Ventura S (2009) Amyloids in bacterial inclusion bodies. Trends Biochem Sci 34:408–416.

de Marco A (2007) Protocol for preparing proteins with improved solubility by co-expressing with molecular chaperones in *Escherichia coli*. Nat Protoc 2:2632–2639.

de Marco A, Deuerling E, Mogk A, Tomoyasu T, Bukau B (2007) Chaperone-based procedure to increase yields of soluble recombinant proteins produced in *E. coli*. BMC Biotechnol 7:32.

Diez-Gil C, Krabbenborg S, García-Fruitós E, Vázquez E, Rodriguez-Carmona E, Ratera I, Ventosa N, Seras-Franzoso J, Cano-Garrido O, Ferrer-Miralles N, Villaverde A, Veciana J (2010) The nanoscale properties of bacterial inclusion bodies and their effect on mammalian cell proliferation. Biomaterials 31: 5805–5812.

Espargaro A, Villar-Piqué A, Sabate R, Ventura S (2012) Yeast prions form infectious amyloid inclusion bodies in bacteria. Microb Cell Fact 11:89.

Ferrer-Miralles N, Domingo-Espin J, Corchero JL, Vázquez E, Villaverde A (2009) Microbial factories for recombinant pharmaceuticals. Microb Cell Fact 8:17.

García-Fruitós E, Villaverde A (2010) Friendly production of bacterial inclusion bodies. Korean J Chem Eng 27:385–389.

García-Fruitós E, González-Montalbán N, Morell M, Vera A, Ferraz RM, Aris A, Ventura S, Villaverde A (2005) Aggregation as bacterial inclusion bodies does not imply inactivation of enzymes and fluorescent proteins. Microb Cell Fact 4:27.

García-Fruitós E, Martinez-Alonso M, González-Montalbán N, Valli M, Mattanovich D, Villaverde A (2007) Divergent genetic control of protein solubility and conformational quality in *Escherichia coli*. J Mol Biol 374:195–205.

García-Fruitós E, Rodríguez-Carmona E, Díez-Gil C, Ferraz RM, Vázquez E, Corchero JL, Cano-Sarabia M, Ratera I, Ventosa N, Veciana J, Villaverde A (2009) Surface cell growth engineering assisted by a novel bacterial nanomaterial. Adv Mater 21:4249–4253.

García-Fruitós E, Seras-Franzoso J, Vázquez E, Villaverde A (2010) Tunable geometry of bacterial inclusion bodies as substrate materials for tissue engineering. Nanotechnology 21:205101.

García-Fruitós E, Sabate R, de Groot NS, Villaverde A, Ventura S (2011) Biological role of bacterial inclusion bodies: a model for amyloid aggregation. FEBS J 278:2419–2427.

García-Fruitós E, Vazquez E, Diez-Gil C, Corchero JL, Seras-Franzoso J, Ratera I, Veciana J, Villaverde A (2012) Bacterial inclusion bodies: making gold from waste. Trends Biotechnol 30:65–70.

Georgiou G, Valax P (1999) Isolating inclusion bodies from bacteria. Methods Enzymol 309:48–58.

González-Montalbán N, García-Fruitós E, Villaverde A (2007) Recombinant protein solubility—does more mean better? Nat Biotechnol 25:718–720.

Harrison BS, Atala A (2007) Carbon nanotube applications for tissue engineering. Biomaterials 28:344–353.

Hu Y, Cai K, Luo Z, Jandt KD (2010) Layer-by-layer assembly of beta-estradiol loaded mesoporous silica nanoparticles on titanium substrates and its implication for bone homeostasis. Adv Mater 22:4146–4150.

Jevsevar S, Gaberc-Porekar V, Fonda I, Podobnik B, Grdadolnik J, Menart V (2005) Production of nonclassical inclusion bodies from which correctly folded protein can be extracted. Biotechnol Prog 21:632–639.

Kolaj O, Spada S, Robin S, Wall JG (2009) Use of folding modulators to improve heterologous protein production in *Escherichia coli*. Microb Cell Fact 8:9.

Kopito RR (2000) Aggresomes, inclusion bodies and protein aggregation. Trends Cell Biol 10:524–530.

Liovic M, Ozir M, Bedina ZA, Peternel S, Komel R, Zupancic T (2012) Inclusion bodies as potential vehicles for recombinant protein delivery into epithelial cells. Microb Cell Fact 11:67.

Maji SK, Schubert D, Rivier C, Lee S, Rivier JE, Riek R (2008) Amyloid as a depot for the formulation of long-acting drugs. PLoS Biol 6:e17.

Maji SK, Perrin MH, Sawaya MR, Jessberger S, Vadodaria K, Rissman RA, Singru PS, Nilsson KP, Simon R, Schubert D, Eisenberg D, Rivier J, Sawchenko P,

Vale W, Riek R (2009) Functional amyloids as natural storage of peptide hormones in pituitary secretory granules. Science 325:328–332.

Makino T, Skretas G, Georgiou G (2011) Strain engineering for improved expression of recombinant proteins in bacteria. Microb Cell Fact 10:32.

Malmstrom J, Lovmand J, Kristensen S, Sundh M, Duch M, Sutherland DS (2011) Focal complex maturation and bridging on 200 nm vitronectin but not fibronectin patches reveal different mechanisms of focal adhesion formation. Nano Lett 11:2264–2271.

Mankar S, Anoop A, Sen S, Maji SK (2011) Nanomaterials: amyloids reflect their brighter side. Nano Rev 2.

Margreiter G, Messner P, Caldwell KD, Bayer K (2008) Size characterization of inclusion bodies by sedimentation field-flow fractionation. J Biotechnol 138: 67–73.

Markossian KA, Kurganov BI (2004) Protein folding, misfolding, and aggregation: formation of inclusion bodies and aggresomes. Biochemistry (Mosc) 69: 971–984.

Martinez-Alonso M, Vera A, Villaverde A (2007) Role of the chaperone DnaK in protein solubility and conformational quality in inclusion body-forming *Escherichia coli* cells. FEMS Microbiol Lett 273:187–195.

Martinez-Alonso M, Toledo-Rubio V, Noad R, Unzueta U, Ferrer-Miralles N, Roy P, Villaverde A (2009a) Re-hosting bacterial chaperones for high-quality protein production. Appl Environ Microbiol 75(24):7850–7854.

Martinez-Alonso M, González-Montalbán N, García-Fruitós E, Villaverde A (2009b) Learning about protein solubility from bacterial inclusion bodies. Microb Cell Fact 8:4.

Martinez-Alonso M, García-Fruitós E, Ferrer-Miralles N, Rinas U, Villaverde A (2010) Side effects of chaperone gene co-expression in recombinant protein production. Microb Cell Fact 9:64.

Miot M, Betton JM (2004) Protein quality control in the bacterial periplasm. Microb Cell Fact 3:4.

Mogk A, Schlieker C, Friedrich KL, Schonfeld HJ, Vierling E, Bukau B (2003) Refolding of substrates bound to small Hsps relies on a disaggregation reaction mediated most efficiently by ClpB/DnaK. J Biol Chem 278:31033–31042.

Morell M, Bravo R, Espargaro A, Sisquella X, Aviles FX, Fernandez-Busquets X, Ventura S (2008) Inclusion bodies: specificity in their aggregation process and amyloid-like structure. Biochim Biophys Acta 1783:1815–1825.

Nahálka J (2008) Physiological aggregation of maltodextrin phosphorylase from *Pyrococcus furiosus* and its application in a process of batch starch degradation to alpha-D-glucose-1-phosphate. J Ind Microbiol Biotechnol 35:219–223.

Nahálka J, Nidetzky B (2007) Fusion to a pull-down domain: a novel approach of producing *Trigonopsis variabilis* D-amino acid oxidase as insoluble enzyme aggregates. Biotechnol Bioeng 97:454–461.

Nahálka J, Dib I, Nidetzky B (2008a) Encapsulation of *Trigonopsis variabilis* D-amino acid oxidase and fast comparison of the operational stabilities of free and immobilized preparations of the enzyme. Biotechnol Bioeng 99: 251–260.

Nahálka J, Vikartovska A, Hrabarova E (2008b) A crosslinked inclusion body process for sialic acid synthesis. J Biotechnol 134:146–153.

Nahálka J, Mislovicova D, Kavcova H (2009) Targeting lectin activity into inclusion bodies for the characterisation of glycoproteins. Mol Biosyst 5:819–821.

Neubauer P, Fahnert B, Lilie H, Villaverde A (2006) Protein inclusion bodies in recombinant bacteria. In: Inclusions in Prokaryotes. Microbiology Monographs. JM Shively, ed. New York: Springer-Verlag. pp. 237–292.

Peternel S (2013) Bacterial cell disruption: a crucial step in protein production. Nat Biotechnol 30(2):250–254.

Peternel S, Komel R (2010) Isolation of biologically active nanomaterial (inclusion bodies) from bacterial cells. Microb Cell Fact 9:66.

Peternel S, Komel R (2011) Active protein aggregates produced in *Escherichia coli*. Int J Mol Sci 12:8275–8287.

Peternel S, Jevsevar S, Bele M, Gaberc-Porekar V, Menart V (2008a) New properties of inclusion bodies with implications for biotechnology. Biotechnol Appl Biochem 49:239–246.

Peternel S, Grdadolnik J, Gaberc-Porekar V, Komel R (2008b) Engineering inclusion bodies for non denaturing extraction of functional proteins. Microb Cell Fact 7:34.

Petersson L, Carrió MM, Vera A, Villaverde A (2004) The impact of dnaKJ overexpression on recombinant protein solubility results from antagonistic effects on the control of protein quality. Biotechnol Lett 26:595–601.

Platas G, Rodriguez-Carmona E, García-Fruitós E, Cano-Garrido O, Villaverde A (2011) Co-production of GroELS discriminates between intrinsic and thermally-induced recombinant protein aggregation during substrate quality control. Microb Cell Fact 10:79.

Rodriguez-Carmona E, Villaverde A (2010) Nanostructured bacterial materials for innovative medicines. Trends Microbiol 18:423–430.

Rodriguez-Carmona E, Cano-Garrido O, Seras-Franzoso J, Villaverde A, García-Fruitós E (2010) Isolation of cell-free bacterial inclusion bodies. Microb Cell Fact 9:71.

Rodriguez-Carmona E, Villaverde A, García-Fruitós E (2011) How to break recombinant bacteria: Does it matter? Bioeng Bugs 2:222–225.

Rokney A, Shagan M, Kessel M, Smith Y, Rosenshine I, Oppenheim AB (2009) *E. coli* transports aggregated proteins to the poles by a specific and energy-dependent process. J Mol Biol 392:589–601.

Sabate R, Espargaro A, Saupe SJ, Ventura S (2009) Characterization of the amyloid bacterial inclusion bodies of the HET-s fungal prion. Microb Cell Fact 8:56.

Sans C, García-Fruitós E, Ferraz RM, González-Montalbán N, Rinas U, Lopez-Santin J, Villaverde A, Alvaro G (2012) Inclusion bodies of fuculose-1-phosphate aldolase as stable and reusable biocatalysts. Biotechnol Prog 28:421–427.

Schlieker C, Tews I, Bukau B, Mogk A (2004) Solubilization of aggregated proteins by ClpB/DnaK relies on the continuous extraction of unfolded polypeptides. FEBS Lett 578:351–356.

Schrodel A, de Marco A (2005) Characterization of the aggregates formed during recombinant protein expression in bacteria. BMC Biochem 6:10.

Seras-Franzoso J, Diez-Gil C, Vázquez E, García-Fruitós E, Cubarsi R, Ratera I, Veciana J, Villaverde A (2012a) Bioadhesiveness and efficient mechanotransduction stimuli synergistically provided by bacterial inclusion bodies as scaffolds for tissue engineering. Nanomedicine (Lond) 7:79–93.

Seras-Franzoso J, Peebo K, Corchero JL, Tsimbouri PM, Unzueta U, Rinas U, Dalby M, Vázquez E, García-Fruitós E, Villaverde A (2012b) A nanostructured bacterial bio-scaffold for the sustained bottom-up delivery of protein drugs. Nanomedicine (Lond) 2012; 8:1587–1599.

Seras-Franzoso J, Tsimbouri PM, Burgess KV, Unzueta U, Garcia-Fruitos E, Vazquez E, Villaverde A, Dalby MJ (2013) Topographically targeted osteogenesis of mesenchymal stem cells stimulated by inclusion bodies decorating polycaprolactone surfaces. Nanomedicine. Apr 30 [Epub ahead of print].

Sørensen HP, Mortensen KK (2005a) Advanced genetic strategies for recombinant protein expression in *Escherichia coli*. J Biotechnol 115:113–128.

Sørensen HP, Mortensen KK (2005b) Soluble expression of recombinant proteins in the cytoplasm of *Escherichia coli*. Microb Cell Fact 4:1.

Speed MA, Wang DI, King J (1996) Specific aggregation of partially folded polypeptide chains: the molecular basis of inclusion body composition. Nat Biotechnol 14:1283–1287.

Tatkiewicz WI, Seras-Franzoso J, García-Fruitos E, Vazquez E, Ventosa N, Peebo K, Ratera I, Villaverde A, Veciana J (2013) 2D microscale engineering of novel protein-based nanomaterial for cell guidance. ACS Nano 7:4774–4784.

Vázquez E, Corchero JL, Burgueno JF, Seras-Franzoso J, Kosoy A, Bosser R, Mendoza R, Martinez-Lainez JM, Rinas U, Fernandez E, Ruiz-Avila L, García-Fruitós E, Villaverde A (2012) Functional inclusion bodies produced in bacteria as naturally occurring nanopills for advanced cell therapies. Adv Mater 24: 1742–1747.

Ventura S, Villaverde A (2006) Protein quality in bacterial inclusion bodies. Trends Biotechnol 24:179–185.

Vera A, González-Montalbán N, Aris A, Villaverde A (2007) The conformational quality of insoluble recombinant proteins is enhanced at low growth temperatures. Biotechnol Bioeng 96:1101–1106.

Villar-Piqué A, Espargaro A, Sabate R, de Groot NS, Ventura S (2012) Using bacterial inclusion bodies to screen for amyloid aggregation inhibitors. Microb Cell Fact 11:55.

Villaverde A (2012) Bacterial inclusion bodies: an emerging platform for drug delivery and cell therapy. Nanomedicine (Lond) 7:1277–1279.

Villaverde A, Carrió MM (2003) Protein aggregation in recombinant bacteria: biological role of inclusion bodies. Biotechnol Lett 25:1385–1395.

Villaverde A, García-Fruitós E, Rinas U, Seras-Franzoso J, Kosoy A, Corchero JL, Vazquez E (2012). Packaging protein drugs as bacterial inclusion bodies for therapeutic applications. Microb Cell Fact 11:76.

Whited BM, Whitney JR, Hofmann MC, Xu Y, Rylander MN (2011) Pre-osteoblast infiltration and differentiation in highly porous apatite-coated PLLA electrospun scaffolds. Biomaterials 32:2294–2304.

9

AGGREGATION OF RECOMBINANT PROTEINS: UNDERSTANDING BASIC ISSUES TO OVERCOME PRODUCTION BOTTLENECKS

Marina Lotti and Loredano Pollegioni

9.1 INTRODUCTION

Major contributions to our current knowledge regarding the molecular, physiological, and application issues related to the aggregation of recombinant proteins in inclusion bodies derive from a practical need: to identify bottlenecks that might be circumvented to improve the yield, solubility, and "quality" of the recombinant product. The knowledge derived from these studies is being applied thoroughly for biotechnological purposes, as summarized later in the chapter. The study of cell responses to protein misfolding and aggregation has also been boosted greatly by the evidence of similarities between aggregation in bacteria and the formation of amyloids in eukaryotic cells. This parallelism is described in Chapter 4 of the book and has stimulated the interest of the scientific community toward responses to protein aggregation in bacterial systems, which are much easier to handle than mammalian cells.

Protein Aggregation in Bacteria: Functional and Structural Properties of Inclusion Bodies in Bacterial Cells, First Edition. Edited by Silvia Maria Doglia and Marina Lotti.

9.2 HOW DO CELLS REACT TO THE OVERPRODUCTION OF A RECOMBINANT PROTEIN?

The overproduction of large amounts of recombinant proteins at non physiological rates is per se a stressful condition for recombinant cells in terms of an overload of the transcriptional, translational, and protein-folding machinery, as well as for the metabolic burden produced by the unbalanced use of material and energy resources. This elicits diverse and overlapping responses that are partly generic, in that they arise from the need to convey cellular resources toward the biosynthesis of a foreign product, and partly protein-specific, determined by the specific properties of the recombinant protein. To make the picture still more complex, aggregation has been recognized to induce specific reactions in the transcriptome, proteome, and lipid composition of the membranes of producing cells.

9.2.1 Physiological Responses to Protein Overproduction

As a matter of fact, when a bacterial cell is used as a cell factory, it is exposed to challenging conditions for a plethora of reasons, which although summarized into a few categories for the sake of simplicity are indeed complex and interconnected and therefore elicit overlapping reactions. Cultivation at high cell density, as is usually employed in the production of recombinant polypeptides, causes limitations in the nutrients and oxygen available, which may result in reduced growth ability and starvation. Moreover, the overproduction of a recombinant polypeptide diverts cell resources toward the synthesis of the foreign protein and of proteins involved in stress responses and folding helpers. Energy depletion and the shortage of particular molecules or building blocks usually slow the growth rate and might reshape metabolic fluxes (Hoffman and Rinas, 2004). To face the new requirements, cells can adjust their metabolic fluxes and also activate alternative metabolic pathways to overcome energy and material limitations. The occurrence of overlapping problems explains why often it is an advantage to fine-tune the growth of the culture and the expression of the recombinant product so they do not interfere with each other. A common strategy to deal with the issues mentioned is to uncouple them; that is, the culture is first grown to high cell density, and gene expression is induced in a second step.

9.2.2 Stress Response to Protein Overproduction

The expression of recombinant proteins is known to elicit a number of stress responses, some but not all of which are common to other stressful conditions. The presence at a high copy number of an expression vector can unbalance the cell metabolism because of the actions required to replicate and maintain the plasmid DNA in the cell and because of the chemicals used to induce expression (e.g., IPTG), which can also affect the cell physiology (Kosinski et al., 1992). Furthermore, a shortage of amino acylated tRNAs to support their own cell protein synthesis has been shown to trigger a stringent response that causes downregulation of the protein and amino acid biosynthesis machinery (Gallant, 1979; Potrykus, 2008). The DNA damage-induced SOS response is also reported to be activated, and although there is no agreement about how protein overproduction triggers this response, it is likely that high transcription rates of plasmid-encoded genes cause DNA suffering in cells (Wegrzyn and Wegrzyn, 2002). This is just to emphasize that even when the foreign protein is not toxic, its expression raises important physiological challenges to which bacterial cells react—in addition to reshaping their metabolic routes as far as possible—by activating their reaction potential toward physiological stress, which consists of different and somehow overlapping responses.

Indeed, during heterologous production, *Escherichia coli* cells have to protect themselves against the challenging physiological perturbation already outlined (starvation, stationary-phase stress, oxygen shortage, possibly even pH and oxidative stress) and against heavy accumulation in either the cytoplasm or the periplasm of misfolded or partly folded proteins, what is often described as *conformational stress* (Gasser et al., 2008). Thus, it is not surprising that bacteria activate all their molecular weapons involved in the *quality control system*, including chaperones and proteases (Baneyx and Mujacic, 2004). This mechanism shares similar features with the heat-shock response, evoked when the intracellular concentrations of misfolded proteins increases because of changes in temperature. In *E. coli*, the heat-shock response is triggered by binding to the RNA polymerase of alternative sigma factors, in particular σ^{32}, that regulate at the transcriptional level subgroups of stress-responsive proteins (Arsène et al., 2000): among them, genes coding for heat-shock proteins.

The intracellular effects of overproduction-induced stress have been tackled using complementary strategies, some of which analyzed variations in the expression of particular proteins (primarily chaperones), while others dealt with global changes in the transcriptome (via DNA microarrays) or in the proteome (Smith, 2007) or focused on stress-responsive genes only (Gill et al., 2000). Although the results of such studies can be biased by the use of different proteins or different cultivation conditions (i.e., culture medium, temperature, promoter of the expression vector, inducers), careful and comparative experimental work generated a solid ground of information and allowed for the definition of catalogs of genes and proteins upregulated during heterologous production. Studies on the transcriptional response analyzed global changes in gene expression with the aim of detecting changes due to both the metabolic and biosynthetic requirements for protein production and protein-specific effects, concerning in particular the solubility of the recombinant protein. Heterologous expression was found to induce global adjustments that reflect the biosynthetic and metabolic requirements for protein production (Gill et al., 2000; Smith, 2007) and the rapid onset (within a few minutes) of multiple stress responses with dramatic changes in the transcription of stress genes, which are part of different regulons (Gill et al., 2000), in particular of heat-shock genes (*ftsH, clP, lon, ompT, degP, groEL, aceA, ibpA*), of genes involved in the SOS response to DNA damage (recA, ℓon, IS5 transposase) and stationary-phase genes (*rpoS, ace A*). The major response to protein insolubility was identified in the high expression level of a subset of σ^{32}-regulated genes, markedly the components of the protein-folding machinery ClpB, DnaJ, DnaK, GroEL, GroES, GrpE, IbpA, IbpB (Smith, 2007). Accordingly, prolonged overexpression of σ^{32} during the synthesis of recombinant proteins promoted the solubility of the products (Cheng and Lee, 2010).

The identification of a limited set of genes responsive to protein misfolding opened the way for the development of strategies to improve the yield and conformational quality of recombinant proteins (see later). Consistent results derived from an analysis of the proteome of the host cells (Cheng and Lee, 2010). In particular, downregulation of enzymes of the steroid and amino acid synthesis was apparent, as well as of those belonging to the pentose phosphate pathway. Intriguingly, a shift was observed in gluconeogenesis from the malate dehydrogenase and phosphoenolpyruvate carboxylate pathway to the NADP-dependent malic enzyme (MaeB) under overexpression of the exogenous protein, which yielded to NADPH production for cell biosynthesis. Indeed, the enzyme *S*-ribosylhomocysteinase, which produces homocysteine, a metabolite that at high concentration blocks tRNA synthesis, was also downregulated. ClpB, HslU, GroEL chaperones, and the holding chaperones IbpA and IbpB were upregulated to different extents in response to overproduction of recombinant proteins, the specific increase depending on the host strain employed and on the overexpressed recombinant protein. As a general rule, the solubility of the recombinant protein showed a good degree of correlation with the expression of ClpB, HslU, IbpA, and IbpB.

Taken together, these studies emphasize a complex scenario in which cells activate different and partly overlapping reactions to cope with the effects of metabolic and conformational stress. While energy metabolism, SOS response, nutrient uptake, and the core of the heat-shock response undergo comparable changes, different recombinant proteins elicit specific reactions in cells that correlate with the properties of the protein expressed and share similarities with the heat-shock response (Gill et al., 2000; Smith, 2007;

Durrschmid et al., 2008). However, the nature and variability of the recombinant protein stress response suggests a more complex and regulated "heat-shock-like" mechanism (Harcum and Haddadin, 2006). Indeed, the heat-shock response is transient, fading away shortly after cells are released from stress, while increased synthesis rates of DnaK and GroEL chaperones and of the heat-shock protease Lon have been found to last for the entire length of overproduction. The small heat-shock proteins IbpA and IbpB are upregulated exclusively when proteins accumulate as inclusion bodies (IBs) and reduce the stress response, thus favoring growth (Lethanh et al., 2005).

9.2.3 Effects of Protein Overproduction on the Cell Membrane

Protein overproduction and aggregation also affect the cell membrane, which is emerging as an extremely sensitive sensor of aggregation-induced stress. This type of study was performed initially in eukaryotic cells to analyze the effect of amyloid formation or of exposure to substances that mimic heat-shock stress (Vigh et al., 2007). Along this line, a few studies on recombinant bacterial cells have shown that bacteria also react to protein overload and aggregation, with specific changes in the composition of the membrane involving both the protein and the lipid components (Ami et al., 2009; Villa et al., 2009). Protein aggregation was found to induce a decrease in the permeability and fluidity of the cell membrane. The presence in the lipid fraction of longer and/or more saturated acyl chains was detected consistently *in vivo* by infrared spectroscopy (Ami et al., 2009). Analysis of the cell proteome showed upregulation of the membrane protein OmpA, dihidrolipoyl dehydrogenase, and glycerophosphoryl-diester phosphodiesterase, the latter involved in lipid and antioxidant metabolism, and of superoxide dismutase, suggesting a transient oxidative stress. Even upregulation of IbpB, a marker of protein aggregation, makes sense in this context, since this small chaperone stabilizes the cell membrane (Tsvetkova et al., 2002). Independently on the recombinant protein, overexpression was found to modulate the abundance of some membrane proteins, with possible consequences to the traffic of nutrients and small solutes (Villa et al., 2009). Since changes in membrane proteins were not related to the solubility status of the overexpressed proteins, whereas changes in lipid were, it was suggested that the cell membrane reacts to protein production through two overlapping responses, in that the membrane proteome varies because of overproduction (generic reaction), while membrane lipids act as a second stress sensor responsive to aggregation.

Going more deeply into this issue, it was possible to show that the cell membrane also recognizes the complexity of the aggregates. As a matter of fact, protein aggregation within cells is far from being a two-state event, since the same protein, depending on the general conditions, can form aggregates of different complexity, which varies from soluble aggregates to insoluble aggregates to large protein deposits (Schroedel and de Marco, 2005). Based on this knowledge, Ami et al. (2009) studied whether aggregates of the same protein differing in their conformational state elicit different cell reactions. These studies were performed using fusions between glutathione-*S*-transferase (GST) and the green fluorescent protein (GFP), whose aggregation can be tuned by acting on the temperature of expression, and an aggregation detector, a vector in which expression of a reporter gene (β-galactosidase) is regulated by the IbpB promoter, activated by protein aggregation. This experimental setup made it possible to correlate changes in the membrane with the structure of GST–GFP aggregates. It was observed that IbpB-regulated expression of the reporter increased rapidly after induction but then decreased when proteins started getting entrapped in inclusion bodies. Modifications in membrane lipids followed the same trend, with the infrared signal of CH_2 components first increasing and then decreasing at the highest level of insoluble protein accumulation. Such temporal evolution of the response showed that misfolded proteins and soluble aggregates (but not the native protein or IBs) are responsible for membrane modification in the host cells.

9.3 STRUCTURE, COMPOSITION, AND MECHANISM OF DEPOSITION OF INCLUSION BODIES

Although under the microscope inclusion bodies appear like bullets of uncertain shape and variable composition, recent studies have unveiled unexpected specificity in the mechanism of aggregation and variability in structure and composition. Since the beginning of the recombinant protein era, "biotechnologists" reported that inclusion bodies are a valuable source of relatively pure recombinant polypeptides, and that in some cases aggregated proteins display the same biological activity that they have when soluble. Such (true) statements deserve to be treated in some detail, since they suggest that aggregation is much more than an erratic and unspecific event. For example, we may ask why inclusion bodies are made up primarily of a single protein and what is the role of other proteins eventually embedded therein. Since the exposition of hydrophobic patches in a partially folded protein promotes aggregation, it is reasonable to conclude that unfolded proteins stick together by means of nonspecific hydrophobic interactions. In this view, "pure" IBs would be formed because of the higher concentration of the recombinant protein with respect to all other newly synthesized polypeptides.

Alternatively, several lines of evidence suggest that the process of aggregation is also driven by the establishment of specific interactions that favor association between homologous proteins: *In vitro,* a mixture of protein-folding intermediates do not coaggregate with each other but only with themselves (Speed et al., 1996), and isolated IBs are able to seed aggregation of the soluble, partly folded counterpart in a dose-dependent manner (Carrió et al., 2005). Moreover, in the complex context of a cell environment, two different aggregation-prone polypeptides tagged with different fluorescent proteins and coexpressed were shown by FRET analysis to localize separately within aggregates (Morell et al., 2008).

If aggregation is a specific process, why are other proteins found within inclusion bodies? As a matter of fact, the presence within aggregates of DnaK, GroEL, IbpA, IbpB, and ClpB has been reported by several authors. This observation is consistent with the role of DnaK and GroEL in the folding of polypeptides and of DnaK and ClpB in the physiological process of disaggregation. Interestingly, electron microscopy demonstrated a different localization of the two chaperone types with GroEL within the core of the aggregate and with DnaK at its surface, which reflects their distinct role in protein folding and disaggregation of inclusion bodies already formed (Carrió and Villaverde, 2001). In similar cases, other proteins were also detected, possibly because they were interacting with DnaK when inclusion bodies were formed (Rinas et al., 2007). Small heat-shock proteins (IbpA and IbpB) are "holding chaperones", devoted to keeping proteins away from aggregating under severe stress by binding them at their surface. If they fail and aggregation occurs, it is not a surprise to find them entrapped in the inclusion bodies.

In vivo, the process of IB formation is reversible, as is shown by their disintegration upon the arrest of overall protein synthesis by chloramphenicol (Carrió and Villaverde, 2001). All these data converge in defining IBs as a dynamic ensemble of proteins that might be added or removed steadily consistent with the presence in the cells of a continuum of protein conformations that range from soluble to insoluble, with the equilibrium shifted in one or the other direction by the physiological conditions and the protein properties. That also means that IBs interact steadily with the quality control machinery, during both their deposition and their release. The mechanism by which proteins self-assemble in aggregates is still not understood completely; in particular, there have been discussions as to whether proteins first form small clusters that then associate to form bigger aggregates or whether IBs grow starting from a unique or a few nucleation sites (Kopito, 2000). Recently it was shown that recombinant asparaginase and human growth hormone (hGH) aggregates follow the first and second mechanisms, respectively. While asparaginase formed multiple nucleation seeds, resulting in small "loose" aggregates, hGH inclusion bodies grew by continuous deposition of proteins on a single nucleation seed.

Interestingly, the properties of the two types of aggregates were markedly different, with the hGH IBs more resistant to proteolytic attack and to chemical denaturation (Upadhyay et al., 2012). Novel insight into the events occurring in individual cells was provided by a study in which *in vivo* folding and aggregation of a β-galactosidase fused to the red fluorescent protein was analyzed at the single-cell level using cytofluorimetry, electron microscopy, and Western blotting (Zhao et al., 2012). Aggregation and solubilization of this protein are in dynamic equilibrium; therefore, the fusion can be present in cells both in soluble form and aggregated in IBs. An unexpected result of this work was that when analyzed at the single-cell level rather that in the whole culture, producing cells were found to exist in two very distinct states: Either they contained IBs or they contained soluble protein only. If after induction to allow expression of the recombinant protein, cells were cultivated in inducer-free medium, all of them shifted to a state producing mainly soluble protein. Such a cell-type transition was not accounted for by differences in the amount of misfolded protein since no changes were detected at the level of the recombinant protein. This observation is puzzling since it seems to introduce novel and still unknown players in the onset of aggregation.

Work reported above introduces a final issue that is being studied deeply and is of relevance in biotechnology: What is the structure of proteins within inclusion bodies? In a seminal work, Schrödel and de Marco (2005) analyzed by density gradient centrifugation the fate of recombinant GFP produced at different temperatures and in the presence or absence of the chaperone DnaK. They showed that the recombinant protein sedimented in discrete bands with the two extremes (from soluble to insoluble) intercalated by several intermediate states. Interestingly, the different aggregation states were shown to be endowed with different biological activity, from nonfluorescent in the samples completely aggregated, to fully active soluble protein and intermediate fluorescence in low-complexity aggregates.

Analysis of the structure of aggregated proteins is consistent with such a nonstatic view and accounts for the observation that inclusion bodies can be biologically active. Accordingly, the amount of residual structure might be modulated by acting on the expression conditions, a strategy today used largely to obtain a soluble protein or, whenever this is not possible, to produce active protein aggregates (see Chapter 10). Aggregates of proteins retaining high amounts of residual structure are also structurally "loose" and allow re-solubilization using mild experimental conditions (Peternel et al., 2008).

All this new information can be very roughly summarized in a few statements: (1) protein aggregation in bacteria is not an all-or-nothing process; (2) proteins can aggregate, while maintaining different degrees of structure; and (3) recombinant proteins can adopt a continuum of conformational forms. As a consequence, the concept of *protein quality* was developed, which goes well beyond a simple soluble or aggregated alternative. This concept is also of relevance in biotechnology since it questions whether solubility is always coincident with biological activity (García-Fruitós et al., 2012).

9.4 FROM KNOWLEDGE TO APPLICATION

Earlier we reported major information available to date about the molecular and physiological reactions evoked in bacterial cells by the metabolic and conformational stress induced by the production of recombinant proteins, the role of the quality control machinery, and current knowledge about mechanisms of aggregation and the structure of protein aggregates. We now illustrate how this information translates in the methodologies explored in the production of recombinant functional polypeptides, which involves yield, solubility, and activity.

High-level expression of a recombinant protein in *E. coli* is affected by a number of aspects, such as codon use, expression vector design, promoter strength, mRNA stability, translational initiation and termination, gene dosage, host design, and fermentation factors. Manipulation of expression conditions is therefore a major challenge in obtaining a high yield of protein production (Jana and Deb, 2005; Gustafsson et al., 2012). Below, specific attention will be paid to strategies aimed to obtain overproduced target polypeptides as soluble, folded proteins. Solubility is, in fact, still a major goal in biotechnology since it allows us to avoid steps related to the denaturation of inclusion bodies and refolding of the recombinant product.

9.4.1 Growth Conditions

As already noted, production of recombinant proteins induces a "metabolic burden," which corresponds to the amount of resources (raw material and energy) that are withdrawn from the host metabolism for maintenance and expression of the foreign DNA, while the formation of inclusion bodies takes place as a response to the accumulation of denatured protein. No direct link exists between metabolic burden and inclusion body formation; in fact, several recombinant proteins can be obtained in large quantities in soluble form. Nevertheless, the formation of soluble proteins *in vivo* can be controlled by a number of environmental factors affecting both the accumulation of denatured protein and the metabolic overload (Sørensen and Mortensen, 2005a,b). On the one hand, optimization of media composition (which is attained mostly by trial and error) has been considered, very carefully in particular by employing a factorial design approach (Romano et al., 2009; Volontè et al., 2010, 2011; Rosini et al., 2012). Recombinant *E. coli* cultures can be grown easily by batch cultivation, a condition showing limited control of growth during the process; cell densities and production levels are only moderate. On the other hand, in fed-batch cultivations, the concentration of energy sources and several other factors can be regulated, thus maximizing production levels in terms of target protein per biomass (Romano et al., 2009; Volontè et al., 2010). Intriguingly, real-time optimization of growth conditions is feasible in fed-batch cultivations by monitoring changes in intrinsic light scattering by flow cytometry related to the formation of inclusion bodies.

Various host strains have been developed to overcome the physiological and metabolic impact of heterologous production (Figure 9.1). For example, the *E. coli* mutant strains C41(DE3) and C43(DE3) allow overexpression of some globular and membrane proteins unable to be expressed efficiently at high levels in the parent BL21(DE3) strain. The *E. coli* Origami strains, in which the *trxB* and *gor* genes encoding two reductases have been disrupted, allow the formation of disulfide bonds in the cytoplasm favoring heterologous expression of disulfide bond–containing proteins. To limit the deposition of inclusion bodies, the temperature during expression can be reduced (Rosini et al., 2012), to decrease the rate of synthesis and to weaken protein–protein hydrophobic interactions (Figure 9.1). Moreover, lower temperatures are accompanied by a reduction in the heat-shock proteases, which are usually induced by overexpression. Indeed, the decreased amount of biomass obtained under this condition is paralleled by lower induction levels, which result in greater amounts of soluble protein because of the decrease in cellular protein concentration.

The folding of some proteins requires the presence of a specific cofactor (e.g., heme for hemoglobin or Mg^{2+} for metal-binding proteins), which can be added to the cultivation media, resulting in a significant increase in the yield of soluble protein (Sørensen and Mortensen, 2005b). Similarly, the solubility of subunits of hetero multimeric proteins can be favored by coexpression of the appropriate binding partner (Figure 9.1).

9.4.2 Coexpression with Chaperones and Folding Modulators

Several laboratories tried to optimize the process of heterologous protein production by coexpressing folding helpers (Figure 9.2). The overproduction of Hsp70 family members (DnaK/DnaJ/GrpE) made it possible to obtain a number of soluble proteins (for a review, see Schlieker et al., 2002; Kolaj et al., 2009; Martinez-Alonso et al., 2010). In these case studies, protein aggregation was prevented when Hsp70 proteins were coexpressed at two to three times its basal concentration, while higher DnaK/DnaJ/GrpE levels resulted in a reduced level of the target protein. It should be noted that improved solubility is not always related to an increase in protein quality

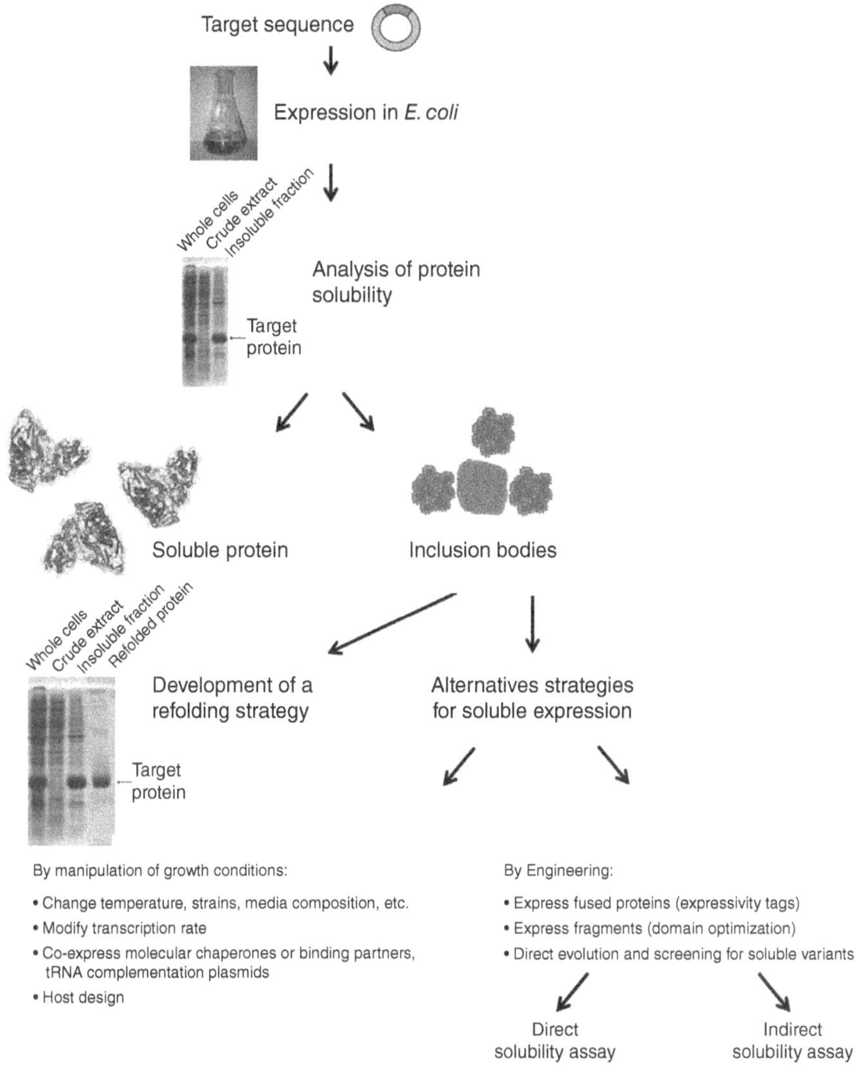

FIGURE 9.1 Downstream strategies employed to favor production of soluble proteins from recombinant *E. coli* cells. *In vivo* strategies aim at lowering the metabolic and conformational stress associated with recombinant expression (see the text for details).

since chaperones can increase the amount of soluble aggregates with variable specific activity. Overproduction of Hsp70 family members also positively affected the production of proteins targeted to the periplasm, apparently by increasing the solubility of the protein precursor prior to membrane translocation (Figure 9.2). Further, DnaK was shown to control the partition of active protein between soluble and insoluble fractions in cells producing IBs.

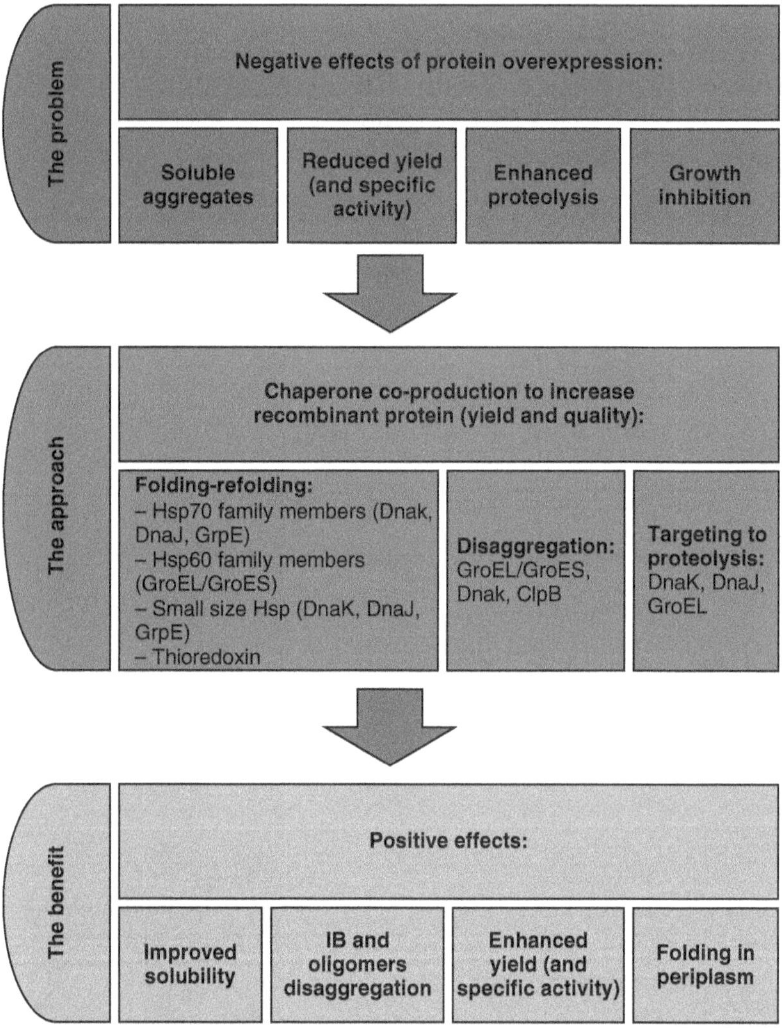

FIGURE 9.2 Co-production of individual or sets of chaperones represents a strategy aimed to increase the quality and solubility of recombinant proteins. Although a number of investigations reported on the positive effects of chaperone gene coexpression, other studies revealed that an excess of certain chaperones has negative effects on both protein yield and protein quality, related primarily to the role of chaperones in promoting proteolysis (see the text for details).

In the absence of DnaK, the activity and conformational nativeness of inclusion body proteins are impaired dramatically, whereas the soluble protein gains specific activity (González-Montalbán et al., 2006). DnaK acts as a negative regulator of the heat-shock response: Its enhancement above physiological levels can lead to the downregulation of other heat-shock proteins (Rodriguez et al., 2008). Indeed, DnaK is also involved in the degradation of aggregation-prone but functional proteins through their targeting to proteolysis (i.e., by Lon and ClpP) (Martinez-Alonso et al., 2009).

Overproduction in recombinant bacteria of GroEL/GroES chaperones makes it possible to increase the levels of a number of recombinant proteins otherwise deposited in inclusion bodies (see Table 1 in Kolaj et al., 2009), whereas it was not effective for some proteins with complex disulfide patterns or whose folding is limited by peptidyl-prolyl *cis-trans* isomerization, as well as for proteins destined to the periplasm. Coexpression of a set of chaperones (Hsp60, Hsp70, and/or the trigger factor) yielded interesting improvements, a few examples of which are described in the following. In the case of penicillin acylase, GroEL/GroES co-overproduction prevented intracellular proteolysis, while DnaJ/DnaK/GrpE led to improved solubilization and maturation of the protein precursor (Xu et al., 2005). The human basic fibroblast growth factor (hFGF-2) was coexpressed with chaperones of the DnaK and GroEL family, ClpB, and combinations thereof. Although none of these chaperones prevented the initial production of IBs, upon prolonged incubation they were able to mediate disaggregation of otherwise stable aggregates. The GroES/GroEL family members were efficient in IB dissolution but did not increase the amount of soluble recombinant growth factor since they increased proteolysis of hFGF-2 (Rinas et al., 2007). On the other hand, the cytoplasmatic trigger factor (TF) was demonstrated to be useful for only a few proteins.

Small heat-shock proteins (sHsp) bind nonnative proteins in an ATP-independent process: IbpA and IbpB protect misfolded proteins from irreversible aggregation and help the resolubilization of protein aggregates (Figure 9.2). The combined overproduction of four chaperone systems—GroEL/GroES, DnaK/DnaJ/GrpE, Clpb and IbpA/IbpB—was investigated in association with a number of recombinant proteins (de Marco et al., 2007). By using a two-step procedure to allow chaperone-mediated refolding of misfolded or aggregated proteins — based on blocking protein biosynthesis by the addition of an antibiotic to the culture medium after removal of IPTG — the co-production of IbpA/IbpB yielded an increase in the solubility of 20 of 23 target proteins tested.

In addition to conventional chaperones, various accessory proteins were used to improve the solubility of recombinant proteins. ClpB, a star-shaped hexameric protein that interacts with DnaK, shows the ability to disaggregate insoluble polypeptide aggregates (Figure 9.2); it was absolutely required to allow Hsp60- and Hsp70-mediated dissolution of macromolecular aggregates of hFGF-2 (Rinas et al., 2007). The size of aggregates was not a primary requirement affecting ClpB, which was, instead, related to the amount of nonnative secondary structure (particularly beta) in the denatured proteins forming the aggregates, as demonstrated *in vitro* using luciferase aggregates produced at different denaturant concentrations. Similarly, thioredoxin, a 12-kDa cytosolic protein with a Cys–X–X–Cys motif, allowed functional production and solubilization of proteins otherwise produced as inclusion bodies. To increase protein folding in the periplasm, Skp (a 17-kDa periplasmatic protein), DegP (a periplasmatic protease that shifts to function as a chaperone at low temperature), Dsb proteins (catalyzing disulfide bond formation, reduction, and isomerization), and FkpA and SurA (to promote peptidyl-prolyl *cis-trans* isomerization and folding) co-production was used (Schlapschy et al., 2006; Kolaj et al., 2009).

How can we rationalize this information? Coexpression with chaperones and folding helpers was an advantage in several cases (Figure 9.2) but it is still a trial-and-error approach, which can also be non influential or deleterious for the process. De Marco and de Marco (2004) demonstrated that when multiple vectors are employed to express different chaperones simultaneously, equilibrium is required in the host cell to optimize the co-transformation benefit. Results reported by different groups working with different polypeptides and in different conditions cannot readily be generalized and are still controversial. Coexpression of folding helpers can indeed be useful but requires that we determine which chaperone or which combination of folding modulators is better suited to the target protein. This information cannot be predicted; it must still be implemented experimentally.

Novel approaches based on reprogramming chaperone pathways or on transferring folding helpers to different cell contexts have also been explored. Starting from previous observation that co-production of DnaK/DnaJ in *E. coli* cells enhances solubility but also proteolytic degradation of the protein of interest by the Lon and ClpB proteases, both the target protein and the bacterial chaperones were expressed in eukaryotic cells, which lack the bacterial proteolytic system linked to DnaK/DnaJ. A strong improvement in yield, solubility, and proteolytic stability was obtained, which suggests that further investigation regarding the advantages of re-hosting chaperone systems is required (Martinez-Alonso et al., 2009).

9.4.3 Expressivity Tags

Expression vectors used to provide polypeptide sequences to improve protein solubility, or solubility detection (see below), have also been widely employed (Figure 9.1). A sequence improving basic expression levels independent of the target protein has not yet been identified, but a number of successful cases have been reported (for a review, see Sørensen and Mortensen, 2005a). Indeed, expressivity tags also allow purification of the fusion proteins by affinity chromatography, avoiding the need for a hexahistidine tag.

The most popular solubility partners are NusA, the highly soluble N-terminal fragment of the translation initiation factor IF2 (17.4 kDa), and barnase (Sørensen and Mortensen, 2005a). Similarly, when added to various target genes, the 21 nucleotides at the 5′-end of the *InfB* gene encoding for the initiation factor 2 yielded a higher expression of soluble recombinant proteins (Hansted et al., 2011). Maltose-binding protein has been used largely to improve protein solubility (and as a carrier protein for vaccination). The same protein, from the archea *Pyrococcus furiosus* (43 kDa), has also been used to enhance the thermostability of GFP (Huang et al., 2006). The soluble heme domain of cytochrome b_5 (98 amino acids) has also been used to solubilize a number of proteins from different sources (Mitra et al., 2005). This system provides the additional possibility of incorporating protease cleavage sites for various commonly used proteases, such as enterokinase and factor Xa. High-throughput screening strategies have been used to select for favorable fusion partners. More than 80% of the target proteins showed high levels of expression of soluble products with at least one of eight different fusion partners (including His-tag, intein, NusA, thioredoxin, MBP, calmodulin-binding protein, and GST) (Shi et al., 2002).

9.4.4 Protein Engineering as a Tool to Improve Protein Solubility

Domain Optimization A well-known approach used to overcome the inability to express a full-length protein in soluble form is the isolation of shorter constructs encompassing single- or multidomain protein fragments (Figure 9.1). In specific cases a rational elimination of regions that affect solubility and/or favor aggregation has been performed. As an example, the C-terminal region of flavin-containing monooxygenase (FMO, a membrane-associated enzyme) is highly hydrophobic, and it was proposed that it function as a membrane anchor in the endoplasmic reticulum. Removal of 26 residues from the C-terminus yielded a soluble, active, cytosolic recombinant FMO in *E. coli* cells (Krueger et al., 2006).

Domain optimization is not a trivial issue, in particular when a priori knowledge of the target protein is not available to guide construct design. Suggestions can be obtained both experimentally, by limited proteolysis of small amounts of the target protein and identification of stable fragments by mass spectrometry, and in silico, by multiple sequence alignments that can suggest domain location via evolutionarily conserved regions. When target proteins have no sequence or structural homologs, order and disorder prediction algorithms, such as DisEMBL, DISOPRED2, RONN, and IUPRED, may suggest which regions of a protein are folded (for a recent example, see Tallarita et al., 2012). Anyway, even in those cases where domain locations can be predicted with accuracy, various additional factors can prevent us from obtaining soluble proteins. These include protein toxicity, inhibitory mRNA secondary structure, rare codon usage, folding efficiency, and intramolecular stabilization, as well as requirement for interactions with chaperones, ligands, and protein partners (Yumerefendi et al., 2010).

Directed Evolution Another useful approach is provided by directed evolutionary strategies based on the generation of large libraries of random sub-full-length DNA molecules followed by high-throughput screening for rare clones expressing soluble proteins within the strong background of nonproductive constructs (Figure 9.1) (Savva et al., 2007).

For the generation of random genetic constructs, procedures based on exonuclease III truncation and DNA fragmentation by endonucleases or physical methods have been employed, as well as DNA shuffling technologies and error-prone PCR (Yumerefendi et al., 2010, 2011). Random library strategies show the advantage of (1) addressing empirically the unpredictable factors described above, (2) identifying domains, and (3) producing several variants similar to domains identified.

Screening procedures useful in identifying clones showing sufficient yields of soluble and stable proteins are classified into two major groups (Yumerefendi et al., 2010): (1) *direct* solubility assays based on physical separation of the target protein, and (2) *indirect* assays based on solubility reporters and small peptides. In all approaches, the assumption that a soluble protein is folded is made, the only exception being cases where an activity or function of the target protein is the output in the "direct" screen. Downstream biophysical analyses (i.e., circular dichroism spectroscopy, NMR, etc.) are used subsequently to confirm foldedness.

1. Concerning the *direct* assay of solubility, a systematic multiconstruct approach was carried out producing 10 different protein constructs for each targeted domain of about 400 human proteins, for a total of 5008 different constructs (Gräslund et al., 2008). Soluble expression in *E. coli* was obtained for about 50% of the target proteins, with a twofold increase in successful results as compared to more traditional approaches. The direct method of solubility assessment requires centrifugation of the cell lysate and is not adaptable for high-throughput screening procedures. Notwithstanding, lysate filtration coupled with protein tag detection has been used successfully for assessing the solubility of proteins; colony-based filtration approaches have been developed with the capacity to screen libraries of thousands of constructs (Cornvik et al., 2006; Dahlroth et al., 2006; Reich et al., 2006).
2. The most used library-compatible solubility screenings are based on *indirect* methods (Figure 9.1): They report solubility by phage display

> systems (Sieber et al., 1998) via C-terminally fused partners such as fluorescent proteins (Waldo, 2003; Heddle and Mazaleyrat, 2007) or complementation of antibiotic resistance proteins or enzymatic activities (Wigley et al., 2001; Liu et al., 2006; Dyson et al., 2008). In the fluorescence-based method, selection is based on the fluorescence associated with GFP, which is related directly to folding of the target recombinant protein. Poorly folded fusion proteins trap nonproductive folding intermediates of the fused GFP reporter, resulting in folding interference. For a list of successful expressions of recalcitrant proteins from different organisms by directed evolution and selection based on GFP fluorescence, see an article by Cabantous et al. (2008). A life-or-death method is based on the ability to render *E. coli* insensitive to the antibiotic trimethoprim (TMP) when the enzyme dihydrofolate reductase (DHFR) is expressed at an appropriate level (Liu et al., 2006). In this approach, the gene encoding an insoluble target protein is fused to the DHFR-encoding gene; the resulting chimeric protein forms IBs and inactivates DHFR, and the resulting cells are sensitive to TMP. Libraries of the target gene can be screened based on the observation that mutations increasing the solubility of the target protein will also increase the solubility of the fusion protein and DHFR will be partially active—the cells will be resistant to TMP. In the β-galactosidase complementation assay, an α-fragment of about 50 residues is fused to the C-terminus of the target protein. If the target protein is soluble, the β-galactosidase activity of a truncated lacZ form is restored by complementation in *trans* (Wigley et al., 2001). Additional screening methodologies are based on proteolytic selection by phage display (Sieber et al., 1998) or on cellular stress responses to misfolded proteins (Vostiar et al., 2004).

False positives are frequent in these solubility assays because of a "passenger solubilization" effect resulting from the fused reporter itself (Reich et al., 2006). This can be problematic when removal of large tags prior to application is mandatory. Indeed, small proteolytic or mistranslated fragments of the target protein (made apparent by SDS–PAGE or Western blot analyses) are often observed fused to the well-folded reporter. To minimize transient solubilization, a screening method based on small linear peptide tags was designed (Waugh, 2005): short peptides do not perturb solubility to a significant extent and are undetectable, presumably degraded, if not fused to a stabilizing domain. The split-GFP method uses the engineered self-complementing GFP fragments derived from the "superfolder" GFP: a 15-residue GFP fragment-strand 11 (S11, residues 216 to 228) that exhibits reduced folding interference from upstream fused polypeptides (Pedelacq et al., 2006) and a separately expressed GFP "1–10 detector" fragment (amino acids 1 to 215).

The S11 fragment is expressed at the C-terminus of the target protein, and GFP 1–10 is expressed from a different plasmid. If the target protein is stably expressed as a soluble protein, the S11 fragment is available for complementation by the independently expressed GFP 1–10 fragment. When the expression and folding of the recombinant protein is compromised, the resulting fluorescence signal reflects the proportion of the soluble fraction. This method allows screening for protein solubility both *in vitro* and *in vivo.*

The ESPRIT method employs a short C-terminal extension only, corresponding to the biotin acceptor peptide, a region used for protein immobilization on streptavidin-coated beads. This method differs significantly from the split-GFP method in the mechanism of solubility sensing, which is due to the post-translational biotinylation of the peptide *in vivo* by the cytoplasmic BirA enzyme. This procedure shows a high level of efficiency: thousands of constructs are analyzed simultaneously by streptavidin binding to robotically printed colony blots, and the signal intensity correlates to enrichment of putatively soluble constructs. By a combination of genetic truncation of the target gene (i.e., those encoding for *Helicobacter pylori* CagA, influenza polymerase PB2 subunit, or neurofibromatosis type 1 protein neurofibromin NF1) using an exonuclease III/Mung bean nuclease protocol and a high-throughput automated screen, nearly 30,000 constructs per run have been analyzed (Yumerefendi et al., 2010). Recently, the ORF-selector ESPRIT technology was presented, which allows gene fragmentation from both termini to access internal domains and that solved the resulting open reading frame problems by an unbiased, intein-based ORF selection yielding only frame DNA inserts (An et al., 2011).

9.4.5 What Have We Learned? Lessons from Directed Evolution

Surface Charge Distribution Directed evolution of a hyperthermophilic nucleotide diphosphate kinase resulted in a 90% soluble and active variant containing six point substitutions (Waldo et al., 1999). Two of these substitutions (i.e., E40K and G33D) gave more favorable charge distribution and hydrogen bonding at the monomer–monomer interface. An increase in the net charge of the N-terminal domain of the HypF protein also reduced the intrinsic propensity of the polypeptide chain to aggregate (Calloni et al., 2005).

Hydrophobicity and Folding Intermediates Directed evolution of serum paraoxonase 1 (PON1, a calcium-binding protein) produced soluble functional expression in *E. coli* of a variant containing eight point mutations (primarily substitutions of one hydrophobic residue for another) clustered in two regions of the protein sequence (Harel et al., 2004); the overall hydrophobicity of the PON1 variant was higher than that of the wild-type protein. Solution of the PON1 variant structure showed that six of the eight residues are part of a hydrophobic core of the protein; thus, they may increase its stability. Such stabilization is stronger for the apoprotein form of PON1 that accumulates in *E. coli* during overexpression under limited availability of the calcium ions (Harel et al., 2004). This result, as well as those of Calloni et al. (2005), identify the main determinant of active overexpression for a number of proteins as the thermodynamic and kinetic properties of an intermediate (e.g., an apoprotein, proprotein, monomeric state, or folding intermediate) rather than the stability and solubility of the final state.

Contribution of Single Residues A site-saturation mutagenesis approach applied to various solvent-exposed positions of ribonuclease provided hits to increase protein solubility in the absence of structural information (Trevino et al., 2007). This study showed that asparagine, glutamine, and threonine are good target residues because (1) they do not seem to contribute significantly to favor protein solubility, and (2) they are more (three-fold) solvent-exposed than are hydrophobic residues. Such residues can be replaced by acidic residues or serine.

Suggestions from Theoretical Analyses Up to now, the ability to predict successful expression of a target protein has been limited. As a general rule, the expression of soluble proteins is more successful when avoiding proteins with (1) a strong positive or negative charge; (2) a ratio of the hydrophobic/hydrophilic area under the curve of a plot reporting that the GC content versus sequence plot differs from 1 ± 0.5; (3) an isoelectric point below 6.5 or above 7.5; (4) high aliphatic content; (5) protein complexity above 1; (6) high β-sheet or coil content; and (7) low flexibility (Benita et al., 2006). Investigation of the expression of 547 human proteins in *E. coli* showed that proteins expressed successfully contained over-represented peptides that were rich in isoleucine, phenylalanine, aspartic acid, and glutamic acid compared with the average amino acid content of human proteins. Such peptides are hydrophilic and flexible (Benita et al., 2006).

9.5 ADVANTAGES OF USING INCLUSION BODIES

Isolated inclusion bodies are homogeneous pseudospherical particles, with limited size dispersion (diameter of 50 to 500 nm): Their size is related to the target protein, the genetic background of the host cell, and the harvest time. Indeed, many recombinant proteins in inclusion bodies maintain their biological activity (García-Fruitós et al., 2012). Accordingly, a number of enzymes (e.g., oxidases, reductases, kinases/phosphatases, aldolases, glucosidases) naturally immobilized in inclusion bodies have been exploited as industrial biocatalysts. The main advantages of *in vivo* immobilization are (García-Fruitós et al., 2012) (1) high yields of recombinant protein production; (2) low production cost (up to 20-fold lower than for soluble proteins); (3) long-term storage and incubation stability; (4) high tolerance to lyophilization; (5) efficient mass transfer through the inclusion body scaffold; (6) convenient immobilization in the solid phase; (7) low release of proteins to aqueous media during the first minutes of incubation; and (8) easy setup of the downstream process. The mechanical properties and operation stability can be increased further by entrapping or cross-linking the inclusion bodies in semipermeable gel matrices or microcapsules. These unique characteristics, and the ability of nonclassical inclusion bodies to release large amounts of functional polypeptides under nondenaturing conditions (i.e., to function as novel protein delivery systems), make inclusion body–based material platforms for innovative and promising bionanotechnology applications.

ACKNOWLEDGMENTS

The authors were supported by grants from Fondo di Ateneo per la Ricerca of the two universities. We acknowledge the help of Chiara Piatti in preparing Figure 9.1.

REFERENCES

Ami D, Natalello A, Schultz T, Gatti-Lafranconi P, Lotti M, Doglia SM, de Marco A (2009) Effects of recombinant protein misfolding and aggregation on bacterial membranes. Biochim Biophys Acta 1794:263–269.

An Y, Yumerefendi H, Mas PJ, Chesneau A, Hart DJ (2011) ORF-selector ESPRIT: a second generation library screen for soluble protein expression employing precise open reading frame selection. J Struct Biol 175(2):189–197.

Arsène F, Tomoyasu T, Bukau B (2000) The heat shock response of *Escherichia coli*. Int J Food Microbiol 55:3–9.

Baneyx F, Mujacic M (2004) Recombinant protein folding and misfolding in *Escherichia coli*. Nat Biotechnol 22(11):1399–1408.

Benita Y, Wise MJ, Lok MC, Humphery-Smith I, Oosting RS (2006) Analysis of high throughput protein expression in *Escherichia coli*. Mol Cell Proteomics 5(9): 1567–1580.

Cabantous S, Rogers Y, Terwilliger TC, Waldo GS (2008) New molecular reporters for rapid protein folding assays. PLoS One 3(6):e2387.

Calloni G, Zoffoli S, Stefani M, Dobson CM, Chiti F (2005) Investigating the effects of mutations on protein aggregation in the cell. J Biol Chem. 280(11):10607–10613.

Carrió M, Villaverde A (2001) Protein aggregation as bacterial inclusion bodies is reversible. FEBS Lett 489:29–33.

Carrió M, González-Montalbán N, Vera A, Villaverde A, Ventura S (2005) Amyloid-like properties of bacterial inclusion bodies, J. Mol. Biol. 347:1025–1037.

Cheng CH, Lee WC (2010) Protein solubility and differential proteomic profiling of recombinant *Escherichia coli* overexpressing double-tagged fusion proteins. Microb Cell Fact. 9:63.

Cornvik T, Dahlroth SL, Magnusdottir A, Flodin S, Engvall B, Lieu V, Ekberg M, Nordlund P (2006) An efficient and generic strategy for producing soluble human proteins and domains in *E. coli* by screening construct libraries. Proteins 65:266–273.

Dahlroth SL, Nordlund P, Cornvik T (2006) Colony filtration blotting for screening soluble expression in *Escherichia coli*. Nat Protoc 1:253–258.

de Marco A, de Marco V (2004) Bacteria co-transformed with recombinant proteins and chaperones cloned in independent plasmids are suitable for expression tuning. J Biotechnol 109(1–2):45–52.

de Marco A, Deuerling E, Mogk A, Tomoyasu T, Bukau B (2007). Chaperone-based procedure to increase yields of soluble recombinant proteins produced in *E. coli*. BMC Biotechnol 7:32.

Durrschmid K, Reischer H, Schmidt-Heck W, Hrebicek T, Guthke R, Rizzi A, Bayer K (2008) Monitoring of transcriptome and proteome profiles to investigate the cellular response of *E. coli* towards recombinant protein expression under defined chemostat conditions. J Biotechnol 135:34–44.

Dyson MR, Perera RL, Shadbolt SP, Biderman L, Bromek K, Murzina NV, McCafferty J (2008) Identification of soluble protein fragments by gene fragmentation and genetic selection Nucleic Acids Res 36:e51.

Gallant JA (1979) Stringent control in *E. coli*. Annu Rev Genet 13:393–415.

García-Fruitós E, Vázquez E, Díez-Gil C, Corchero JL, Seras-Franzoso J, Ratera I, Veciana J, Villaverde A (2012) Bacterial inclusion bodies: making gold from waste. Trends Biotechnol 30(2):65–70.

Gasser B, Saloheimo M, Rinas U, Dragosits M, Rodriguez-Carmona E, Baumann K, Giuliani M, Parrilli E, Branduardi P, Lang C, Porro D, Ferrer P, Tutino ML, Mattanovich D, Villaverde A (2008) Protein folding and conformational stress in microbial cells producing recombinant proteins: a host comparative overview. Microb Cell Fact 7:11.

Gill RT, Valdes JJ, Bentley WE (2000) A comparative study of global stress gene regulation in response to overexpression of recombinant proteins in *Escherichia coli*. Metab Eng 2:178–189.

González-Montalbán N, García-Fruitós E, Ventura S, Arís A, Villaverde A (2006) The chaperone DnaK controls the fractioning of functional protein between soluble and insoluble cell fractions in inclusion body-forming cells. Microb Cell Fact 5:26.

Gräslund S, Sagemark J, Berglund H, Dahlgren LG, Flores A, Hammarström M, Johansson I, Kotenyova T, Nilsson M, Nordlund P, Weigelt J (2008) The use of systematic N- and C-terminal deletions to promote production and structural studies of recombinant proteins. Protein Expr Purif 58(2):210–221.

Gustafsson C, Minshull J, Govindarajan S, Ness J, Villalobos A, Welch M (2012) Engineering genes for predictable protein expression. Protein Expr Purif 83(1):37–46.

Hansted JG, Pietikäinen L, Hög F, Sperling-Petersen HU, Mortensen KK (2011) Expressivity tag: a novel tool for increased expression in *Escherichia coli*. J Biotechnol 155(3):275–283.

Harcum SW, Haddadin FT (2006) Global transcriptome response of recombinant *Escherichia coli* to heat shock and dual heat-shock recombinant protein induction. J Ind Microbiol Biotechnol 33:801–814.

Harel M, Aharoni A, Gaidukov L, Brumshtein B, Khersonsky O, Meged R, Dvir H, Ravelli RB, McCarthy A, Toker L, Silman I, Sussman JL, Tawfik DS (2004) Structure and evolution of the serum paraoxonase family of detoxifying and anti-atherosclerotic enzymes. Nat Struct Mol Biol 11(5):412–419.

Heddle C, Mazaleyrat SL (2007) Development of a screening platform for directed evolution using the reef coral fluorescent protein ZsGreen as a solubility reporter. Protein Eng Des Sel 20:327–337.

Hoffmann F, Rinas U (2004) Stress induced by recombinant protein production in *Escherichia coli*. Adv Biochem Eng Biotechnol 89:73–92.

Huang H, Liu J, de Marco A (2006) Induced fit of passenger proteins fused to archaea maltose binding proteins. Biochem Biophys Res Commun 344(1):25–29.

Jana S, Deb JK (2005) Strategies for efficient production of heterologous proteins in *Escherichia coli*. Appl Microbiol Biotechnol 67(3):289–298.

Kolaj O, Spada S, Robin S, Wall JG (2009) Use of folding modulators to improve heterologous protein production in *Escherichia coli*. Microb Cell Fact 8:9.

Kopito RR (2000) Aggresomes, inclusion bodies and protein aggregation. Trends Cell Biol 10:524–530.

Kosinski MJ, Rinas U, Bailey JE (1992) Isopropyl–β-thiogalactopyranoside influences the metabolism of *Escherichia coli*. Appl Microbiol Biotechnol 36:782–784.

Krueger SK, Siddens LK, Henderson MC, VanDyke JE, Karplus PA, Pereira CB, Williams DE (2006) C-Terminal truncation of rabbit flavin-containing monooxygenase isoform 2 enhances solubility. Arch Biochem Biophys 450(2):149–156.

Lethanh H, Neubauer P, Hoffmann F (2005) The small heat-shock proteins IbpA and IbpB reduce the stress load of recombinant *Escherichia coli* and delay degradation of inclusion bodies. Microb Cell Fact 4:6.

Liu JW, Boucher Y, Stokes HW, Ollis DL (2006) Improving protein solubility: the use of the *Escherichia coli* dihydrofolate reductase gene as a fusion reporter. Protein Expr Purif 47(1):258–263.

Martínez-Alonso M, García-Fruitós E, Ferrer-Miralles N, Rinas U, Villaverde A (2010) Side effects of chaperone gene co-expression in recombinant protein production. Microb Cell Fact 9:64.

Martinez-Alonso M, Toledo-Rubio V, Noad R, Unzueta U, Ferrer-Miralles N, Roy P, Villaverde A (2009) Rehosting of bacterial chaperones for high-quality protein production. Appl Environ Microbiol 75(24):7850–7854.

Mitra A, Chakrabarti KS, Shahul Hameed MS, Srinivas KV, Senthil Kumar G, Sarma SP (2005) High level expression of peptides and proteins using cytochrome b5 as a fusion host. Protein Expr Purif 41(1):84–97.

Morell M, Bravo R, Espargaró A, Sisquella X, Avilés FX, Fernàndez-Busquets X, Ventura S (2008) Inclusion bodies: specificity in their aggregation process and amyloid-like structure. Biochim Biophys Acta 1783:1815–1825.

Pédelacq JD, Cabantous S, Tran T, Terwilliger TC, Waldo GS (2006) Engineering and characterization of a superfolder green fluorescent protein. Nat Biotechnol 24(1):79–88.

Peternel S, Grdadolnik J, Gaberc-Porekar V, Komel R (2008) Engineering inclusion bodies for non denaturing extraction of functional proteins. Microb Cell Fact 7:34.

Potrykus K, Cashel M (2008) (p)ppGpp: still magical? Annu Rev Microbiol 62:35–51.

Reich S, Puckey LH, Cheetham CL, Harris R, Ali AA, Bhattacharyya U, Maclagan K, Powell KA, Prodromou C, Pearl LH, Driscoll PC, Savva R (2006) Combinatorial domain hunting: an effective approach for the identification of soluble protein domains adaptable to high-throughput applications Protein Sci 15:2356–2365.

Rinas U, Hoffmann F, Betiku E, Estapé D, Marten S (2007) Inclusion body anatomy and functioning of chaperone-mediated in vivo inclusion body disassembly during high-level recombinant protein production in *Escherichia coli.* J Biotechnol 127(2):244–257.

Rodriguez F, Arsène-Ploetze F, Rist W, Rüdiger S, Schneider-Mergener J, Mayer MP, Bukau B (2008) Molecular basis for regulation of the heat shock transcription factor sigma32 by the DnaK and DnaJ chaperones. Mol Cell 32(3):347–358.

Rosini E, Nossa S, Valentino M, D'Arrigo P, Marinesco S, Pollegioni L (2012) Expression of rat diamine oxidase in *Escherichia coli.* J Mol Catal B82:115–120.

Romano D, Molla G, Pollegioni L, Marinelli F (2009) Optimization of human D-amino acid oxidase expression in *Escherichia coli.* Prot Expr Purif 68:72–78.

Savva R, C. Prodromou C, Driscoll PC (2007) DNA fragmentation based combinatorial approaches to soluble protein expression: II. Library expression, screening and scale-up. Drug Discov Today 12:939–947.

Schlapschy M, Grimm S, Skerra A (2006) A system for concomitant overexpression of four periplasmic folding catalysts to improve secretory protein production in *Escherichia coli.* Protein Eng Des Sel 19(8):385–390.

Schlieker C, Bukau B, Mogk A (2002) Prevention and reversion of protein aggregation by molecular chaperones in the *E. coli* cytosol: implications for their applicability in biotechnology. J Biotechnol 96(1):13–21.

Schrödel A, de Marco A (2005) Characterization of the aggregates formed during recombinant protein expression in bacteria. BMC Biochem 6:10.

Shi PY, Kung WM, Chen JC, Yeh CH, Wang AHJ, Wang TF (2002) Highthroughput screening of soluble recombinant proteins. Protein Sci 11:1714–1719.

Sieber V, Plückthun A, Schmid FX (1998) Selecting proteins with improved stability by a phage-based method. Nat Biotechnol 16(10):955–960.

Smith HE (2007) The transcriptional response of *Escherichia coli* to recombinant protein insolubility. J Struct Funct Genomics 8:27–35.

Sørensen HP, Mortensen KK (2005a) Advanced genetic strategies for recombinant protein expression in *Escherichia coli*. J Biotechnol 115(2):113–128.

Sørensen HP, Mortensen KK (2005b) Soluble expression of recombinant proteins in the cytoplasm of *Escherichia coli*. Microb Cell Fact 4(1):1.

Speed MA, Wang DI, King J (1996) Specific aggregation of partially folded polypeptide chains: the molecular basis of inclusion body composition. Nat Biotechnol 14:1283–1287.

Tallarita E, Pollegioni L, Servi S, Molla G (2012) Expression in *Escherichia coli* of the catalytic domain of human proline oxidase. Protein Expr Purif 82(2):345–351.

Trevino SR, Scholtz JM, Pace CN (2007) Amino acid contribution to protein solubility: Asp, Glu, and Ser contribute more favorably than the other hydrophilic amino acids in RNase Sa. J Mol Biol 366(2):449–460.

Tsvetkova NM, Horvath I, Török Z, Wolkers WF, Balogi Z, Shigapova N, Crowe LM, Tablin F, Vierling E, Crowe JH, Vigh L (2002) Small heat-shock proteins regulate membrane lipid polymorphism. Proc Natl Acad Soc U S A 99: 13504–13509.

Upadhyay AK, Murmu A, Singh A, Panda AK (2012) Kinetics of inclusion body formation and its correlation with the characteristics of protein aggregates in *Escherichia coli*. PlosONE 7(3):e33951.

Vigh L, Nakamoto H, Landry J, Gomez-Munoz A, Harwood JL, Horvath I (2007) Membrane regulation of the stress response from prokaryotic models to mammalian cells. Ann NY Acad Sci 1113:40–51.

Villa R, Lotti M, Gatti-Lafranconi P (2009) Components of the *E. coli* envelope are affected by and can react to protein over-production in the cytoplasm. Microb Cell Fact 8:32.

Volontè F, Pollegioni L, Molla G, Frattini L, Marinelli F, Piubelli L (2010) Production of recombinant cholesterol oxidase containing covalently bound FAD in *Escherichia coli*. BMC Biotechnol 21(10):33.

Volontè F, Pisanelli I, D'Arrigo P, Viani F, Molla G, Servi S, Pollegioni L (2011) Overexpression of a bacterial chymotrypsin: application for L-amino acid ester hydrolysis. Enz Microb Technol 49(6–7):560–566.

Vostiar I, Tkac J, Mandenius CF (2004) Off-line monitoring of bacterial stress response during recombinant protein production using an optical biosensor. J Biotechnol 111(2):191–201.

Waldo GS (2003) Genetic screens and directed evolution for protein solubility. Curr Opin Chem Bio 7:33–38.

Waldo GS, Standish BM, Berendzen J, Terwilliger TC (1999) Rapid protein-folding assay using green fluorescent protein. Nat Biotechnol 17(7):691–695.

Waugh DS (2005) Making the most of affinity tags. Trends Biotechnol 23:316–320.

Wegrzyn G, Wegrzyn A (2002) Stress responses and replication of plasmids in bacterial cells. Microb Cell Fact 1:2.

Wigley WC, Stidham RD, Smith NM, Hunt JF, Thomas PJ (2001) Protein solubility and folding monitored in vivo by structural complementation of a genetic marker protein. Nat Biotechnol 19(2):131–136.

Xu Y, Weng CL, Narayanan N, Hsieh MY, Anderson WA, Scharer JM, Moo-Young M, Chou CP (2005) Chaperone-mediated folding and maturation of the penicillin acylase precursor in the cytoplasm of *Escherichia coli*. Appl Environ Microbiol 71(10):6247–6253.

Yumerefendi H, Tarendeau F, Mas PJ, Hart DJ (2010) ESPRIT: an automated, library-based method for mapping and soluble expression of protein domains from challenging targets. J Struct Biol 172(1):66–74.

Yumerefendi H, Desravines DC, Hart DJ (2011). Library-based methods for identification of soluble expression constructs. Methods 55(1):38–43.

Zhao Y, Liu W-F, Liu C-C, Feng l-K, Sun L, Yan Y-B, Hang J-Y (2012) Two distinct states of *Escherichia coli* cells that overexpress recombinant heterogeneous β-galactosidase. J Biol Chem 287:9259–9268.

10

FUSION TO A PULL-DOWN MODULE: DESIGNING ENZYMES TO FORM BIOCATALYTICALLY ACTIVE INSOLUBLE AGGREGATES

BERND NIDETZKY

10.1 INTRODUCTION

Biocatalysis has recently received increased attention for process development in the chemical and allied industries (Nestl et al., 2011; Illanes et al., 2012; Huisman and Collier, 2013). A large number of catalytic transformations are carried out advantageously at present using enzymes instead of chemical catalysts, especially in situations where selectivity is a key issue (Patel, 2007). Biocatalysis promotes inherently safe and sustainable process development, thus contributing to fulfillment of green chemistry principles at large (Jimenez-Gonzalez and Constable, 2011; Sheldon, 2012). Process development in the chemical industries often involves a transition from a batch mode of operation used initially to a fully continuous process at the manufacturing scale (Rao et al., 2009). Therefore, because heterogeneous catalysis is ideal for processes to be performed continuously, there is a general trend in process chemistry that solid supported insoluble catalysts are favored over soluble catalysts (Liese and Hilterhaus, 2013). For that reason, insoluble enzymes are widely recognized as useful biocatalysts, and the majority of large-scale biocatalytic transformations

Protein Aggregation in Bacteria: Functional and Structural Properties of Inclusion Bodies in Bacterial Cells, First Edition. Edited by Silvia Maria Doglia and Marina Lotti.

in the fine chemicals industries are built on heterogeneous enzyme preparations (Buchholz et al., 2005; Cao, 2005). Among a variety of methods known for the preparation of an insoluble biocatalyst, enzyme immobilization on a solid support appears to be the one most widely used. During immobilization, the originally soluble enzyme is made insoluble by its attachment to the surface of a solid carrier.

The general principle of enzyme immobilization has been diversified into a large number of protocols and technologies, reflecting variations in the type of carrier used and the method employed to fix the enzyme to the carrier (Guisan, 2006; Torres-Salas et al., 2011). Encapsulation in organic or inorganic gels represents a special mode of carrier-dependent enzyme immobilization (Brady and Jordaan, 2009). Despite the clear advantages of solid-supported enzymes, perhaps the most important being gain in operational stability and flexible use in different continuous reactor configurations, there are also important drawbacks (Roessl et al., 2010). One is that enzymes are often partly inactivated as a result of their attachment to a solid carrier. Another is that the use of solid carriers usually introduces a large amount of noncatalytic mass into the bioreactor. The active enzyme may represent only a minor portion (<10 to 20%) of the total mass of the heterogeneous biocatalyst. Finally, some carriers constitute high-tech macroporous materials (Buchholz et al., 2005; Cao 2005; Hudson et al., 2008) and are therefore often more expensive than the enzymes bound to them. Regeneration of carriers for multiple rounds of immobilization to be performed on them might therefore be an economic necessity, but it is often impractical, especially in cases where covalent coupling chemistry was employed for attaching the enzymes to the solid surface. Therefore, if one succeeded in precipitating the target enzyme into a sizable insoluble aggregate that shows both good retention of the specific activity of the soluble enzyme and useful resistance to mechanical stresses, a powerful carrier-free, yet heterogeneous biocatalyst might be obtained, and problems with the use of solid carriers could be partially eliminated. In this chapter we build on prior work from this laboratory (Roessl et al., 2010) and summarize recent developments in the preparation and application of insoluble enzyme aggregates. We focus in particular on molecular approaches by protein design, where an aggregation-prone or self-assembling structural element, herein referred to as a *pull-down module* (PDM), is appended to the enzyme of interest. Recombinant production of the PDM–enzyme fusion (e.g., in *Escherichia. coli*) should thus yield an insoluble protein aggregate that can be recovered easily, processed further (e.g., cross-linked) if needed, and then used as an insoluble biocatalyst. Fusion to PDMs might thus become a generally useful tool in the preparation of carrier-free insoluble enzyme aggregates, offering the prospect of achieving control over functional and structural–morphological characteristics of the protein particles produced.

The well-known technology of cross-linked enzyme aggregates (CLEAs; Sheldon, 2011) is discussed here briefly for reference.

10.2 A SHORT GLANCE AT CLEAs

The technology of CLEA production was introduced several years ago (Cao et al., 2000; Sheldon, 2011). CLEA is now a registered trade name, and the underlying technology has been commercialized. As first demonstrated for penicillin G acylase and subsequently applied to a number of enzymes from different EC classes (e.g., hydrolases, oxioreductases, lyases), CLEAs are made by a general procedure that involves salt- or solvent-induced microaggregation of the enzyme or a desired combination of several enzymes, followed by chemical cross-linking to obtain a stable all-protein precipitate (Figure 10.1). Ammonium sulfate, *tert*-butyl alcohol, or poly(ethylene glycol) was used to promote aggregation of penicillin G acylase. The cross-linking reagent can also be varied according to the requirements of the particular

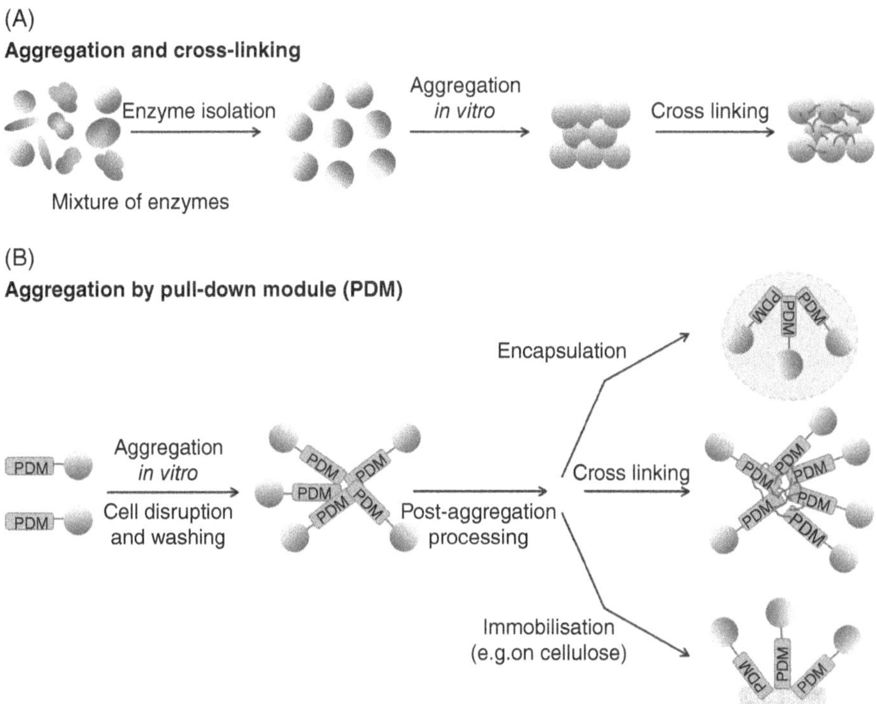

FIGURE 10.1 Preparation of insoluble enzyme aggregates using (A) CLEA technology or (B) aggregation by pull-down module. (*See insert for color representation of the figure.*)

enzyme used (Mateo et al., 2004; Kubáč et al., 2008). However, the bifunctional cross-linker glutaraldehyde was most often used. Co-aggregation of enzyme with a soluble polymer [e.g., poly(ethylenimine), bovine serum albumin] was applied to obtain robust catalyst preparations (Sheldon, 2011). Multienzyme aggregates are called combi-CLEAs, and several of them have been reported (Mateo et al., 2006) A "tripledecker" combi-CLEA consisting of oxynitrilase, nitrilase, and amidase was used in the preparation of chiral (*S*)-mandelic acid from benzaldehyde using a three-enzymatic cascade transformation (Sheldon, 2012). Several CLEAs were shown to have useful operational stability and to be recyclable for multiple rounds of conversion.

Various reactor configurations for the application of CLEAs in (semi)-continuous transformations have been considered. These include conventionally mixed (stirred or shaken) reactors operated at different scales and in batch mode, but also microchannel flow reactors and fluidized beds. Sheldon and co-workers applied a slurry reactor equipped with a filtration unit to achieve retention of CLEAs during continuous operation (Sheldon, 2011). Magnetic CLEAs might be applied in combination with fluidized beds, using magnetic field for retention of the insoluble enzyme (Sheldon, 2011, 2012).

The use of CLEAs has been widely adopted by the biocatalysis community. However, preparation of CLEAs has the inherent complication that the exact protocol of aggregation and cross-linking must be established for each enzyme anew, implying that a substantial effort has to be given to the optimization of conditions. Good retention of specific activity of the soluble enzyme in the CLEA preparation is clearly essential. In a recent review, Sheldon (2011) raises the question of determination of activity recovery in CLEAs. An important point is that activities of CLEAs and the corresponding soluble enzyme must be compared through measurements done under exactly the same conditions. It is undoubtedly useful to apply assay conditions similar to those expected in the real process. However, it should also be noted that soluble enzymes are sometimes extremely unstable under the conditions of organic synthesis (e.g., due to use of a nonconventional solvent) and might therefore constitute poor benchmarks for CLEAs. An alternative immobilizate may be more suitable in such cases. Structural properties of the CLEAs such as size, porosity, and resistance to mechanical stress are also important. The cross-linking reagent and its concentration relative to the protein concentration used are critical process variables to be optimized. Addition of BSA or poly(ethylenimine) during the cross-linking step sometimes serves the purpose of stabilizing the enzyme activity and providing additional amino groups for reaction with glutaraldehyde (Lopez-Gallego et al., 2005). CLEA size is affected by many process variables and is therefore not easy to control. Another, potentially troublesome aspect of CLEA preparation is that some purification of the crude protein will in most cases be required prior to its precipitation. In crude cell lysates, for example, it should be minimally necessary to remove nucleic acids prior to the protein precipitation. It has been argued on the contrary (Sheldon, 2011) that aggregating an enzyme into a CLEA constitutes "a combination of purification and immobilization in one operation." To achieve purification to a significant degree, however, the precipitation must be adequately selective for the enzyme(s) of interest.

Published work on CLEAs has focused mostly on biocatalytic applications in organic synthesis, often in nonconventional solvent systems. Detailed biochemical engineering analysis of the use of CLEAs in (continuous) processes has also been done—less often, however. Several authors have recognized the problem of weak resistance of CLEAs to mechanical stresses. Encapsulation of CLEAs, in a polymeric matrix prepared from poly(vinyl alcohol) (Wilson et al., 2004), called LentiKats, might present a possible solution. Entrapment into hierarchically ordered mesoporous silica is another approach for CLEA stabilization against mechanical stress.

10.3 PULL-DOWN MODULES AND THEIR USE IN INDUCING SELECTIVE PROTEIN AGGREGATION

A significant improvement in enzyme aggregate production could come from molecular design of the aggregation process such that it features certain elements of biological recognition and specificity. Not only would laborious optimization of conditions for enzyme precipitation no longer be necessary, but it should also be much easier to achieve control over retention of activity and final particle size during the formation of insoluble protein aggregates. Fusion to solubility enhancer modules is a well-known approach to improve the solubility of target proteins that are otherwise prone to aggregation (inclusion body formation) under conditions of recombinant protein production. Together with Jozef Nahálka, I have suggested exploitation of the reverse approach to the production of enzyme aggregates. This concept, called *fusion to a pull-down module* works as follows. A chimeric protein is generated that harbors the enzyme of interest fused to a very poorly soluble protein module, such as the cellulose-binding module of *Clostridium cellulovorans* (*Cc*CBM) (Nahálka and Nidetzky, 2007). Protein production in a heterologous host such as *Escherichia coli* is carried out under conditions that induce selective pull-down of the folded chimeric protein via intermolecular self-aggregation of the *Cc*CBM. The idea that enzymes trapped in insoluble aggregates can be catalytically active and therefore are not completely useless is certainly not new. Following early discoveries (for a review, see García-Fruitós et al., 2011), seminal studies of Villaverde and co-workers have demonstrated that retention of biological function in such aggregated proteins is actually quite common (García-Fruitós et al., 2007, 2010, 2011; González-Montalbán et al., 2007, Vera et al., 2007). Elegant structural analyses confirm the notion of retention of native-like structural features in the aggregated protein (Gatti-Lafranconi et al., 2011). However, in the context of this article it seems justified to distinguish insoluble enzyme aggregates derived through fusion to a PDM (Nahálka and Nidetzky, 2007) from "classical" inclusion bodies that may also show (high) activity (García-Fruitós et al., 2011). It could be interesting to apply methods of structural analysis of insoluble protein aggregates (e.g., Fourier transform infrared spectroscopy; Ventura and Villaverde, 2006; Gatti-Lafranconi et al., 2011) to a comparative characterization of different aggregate types of the same enzyme. From the work of this laboratory, D-amino acid oxidase from *Trigonopsis variabilis* (*Tv*DAO) could be an interesting candidate.

A first case for application of PDMs was provided using the dimeric flavoenzyme *Tv*DAO (Nahálka and Nidetzky, 2007). Upon fusion of *Cc*CBM to the N-terminus of the enzyme, *Tv*DAO produced in *E. coli* BL21 (DE3) was trapped almost completely into intracellular protein particles. Enzyme lacking *Cc*CBM was almost completely soluble under otherwise identical conditions for recombinant protein production, indicating that the PDM fulfilled the expected solubility switch function very efficiently. Aggregates of *Cc*CBM–*Tv*DAO displayed about 40% of the specific activity of the purified soluble oxidase. The high activity level in oxidase aggregates was interesting for two reasons. First, it demonstrated that the PDM showed minor interference with the folding and functioning of its fusion partner, despite the structural complexity of *Tv*DAO. Second, it suggested that the overall catalytic activity deposited in the aggregates was not strongly compromised by mass transfer effects, or, in other words, the insoluble enzyme was readily accessible to its substrates, D-amino acid and O_2. Using SDS-PAGE it was shown that the *Cc*CBM–*Tv*DAO preparation as isolated contained mainly the protein of interest. *Cc*CBM–*Tv*DAO displayed weak binding affinity to microcrystalline cellulose and was substantially more stable than the soluble oxidase under conditions of D-methionine conversion in a bubble-aerated bioreactor (Figure 10.2). Encapsulation of *Cc*CBM–*Tv*DAO contributed additional stability (about three-fold) to the insoluble enzyme aggregates under conditions of bubble aeration (Nahálka et al., 2008b). Semipermeable microcapsules were prepared from the polycation poly(methylene-*co*-guanidine) and the polyanions alginate and cellulose sulfate in the presence of $CaCl_2$ (Figure 10.3).

Nahálka and colleagues have demonstrated the scope of *Cc*CBM as a flexibly useful PDM. They describe catalytically active aggregates of a maltodextrin phosphorylase (*Pyrococcus furiosus*) (Nahálka, 2008), of a sialic acid aldolase (SSA; *E. coli* K-12) (Nahálka et al., 2008a), and of a polyphosphate kinase (PPK; *Silicibacter pomeroyi*) (Nahálka and Pätoprstý, 2009). Useful retention (≥ 83%) of the specific activity of the respective soluble enzyme was obtained in all cases. *Cc*CBM–SSA particles were further entrapped in

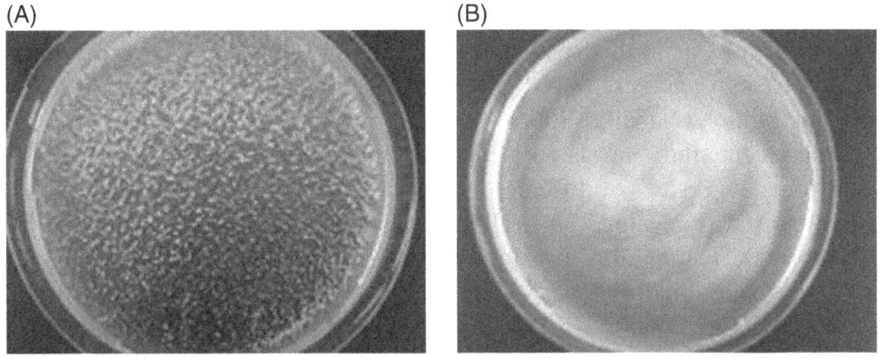

FIGURE 10.2 Aggregates of *Cc*CBM–*Tv*DAO bind to microcrystalline cellulose, resulting in flocculation of the cellulose. (A) Whatman CC 41 microcrystalline cellulose (10 mg/mL; 5 mL) is shown after addition of *Cc*CBM–*Tv*DAO (about 120 μg/mL); (B) the same cellulose preparation is shown in the absence of *Cc*CBM–*Tv*DAO. [From Nahálka and Nidetzky (2007).]

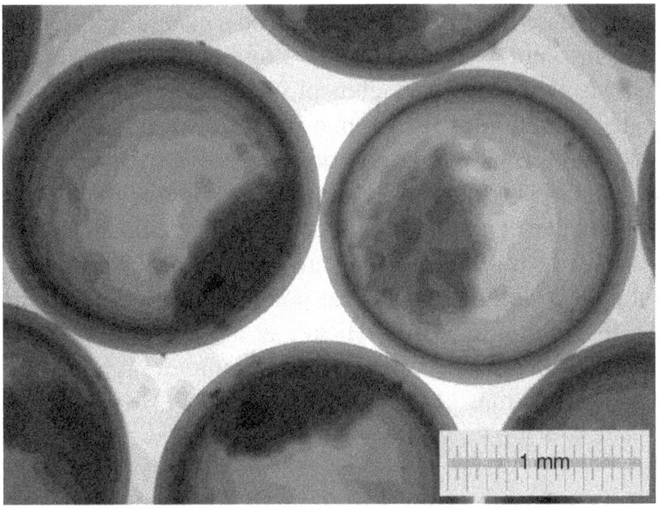

FIGURE 10.3 Encapsulated aggregates of *Cc*CBM–*Tv*DAO. The microcapsules are stable and recyclable biocatalysts in reactions performed under conditions of bubble aeration, which result in rapid inactivation of the soluble enzyme. [From Nahálka et al. (2008b).]

alginate hydrogels stabilized by cross-linking with glutaraldehyde (Nahálka et al., 2008a). The insoluble SSA catalyst was still very active (51%) and displayed high operational stability as well as useful recyclability in a repeated batch process for neuraminic acid synthesis. It could be freeze-dried and rehydrated with nearly full retention of activity. Work with *Cc*CBM–PPK demonstrates yet another important advantage of selective pull-down of the target enzyme, as in this case, contaminating phosphatase activity from the *E. coli* background is removed efficiently by simple washes of the insoluble protein fraction (Nahálka, 2008).

In a recent study unrelated to biocatalysis, *Cc*CBM was used to prepare functional aggregates of different lectins (Nahálka et al., 2009). A microtiter plate assay was developed in which the insoluble lectins should be used for glycoprotein recognition in a high-throughput format. *Cc*CBM fusions of two sialic acid–binding lectins were shown to precipitate sialylated proteins under the assay conditions, leading to a positive readout when multivalent binding by the lectin occurred (Nahálka et al., 2009).

More recently, Huang and colleagues suggested that green fluorescent protein (GFP) could be applied as a PDM. The authors fused the GFP to the phosphatase PhoC from *Enterobacter aerogenes* and obtained highly active enzyme trapped in insoluble aggregates. They applied flexible and rigid types of linkers to connect GFP to the phosphatase. Both linkers gave useful results (Huang et al., 2013).

In summary, fusion to a PDM appears to be a generally applicable method of producing insoluble, yet highly active aggregates of different enzymes. Using *Cc*CBM, noncovalent attachment of the *Cc*CBM-fusion to cellulose will facilitate separation of the protein particles from bulk liquid (Nahálka and Nidetzky, 2007). However, additional processing steps (encapsulation, cross-linking) may be needed to finally obtain an operationally robust biocatalyst (Nahálka et al., 2008b). Because aggregation under the control of a PDM is unlikely to enhance the intrinsic stability of the enzyme used (Nahálka et al., 2008a), there is a clear potential in applying (thermo)-stable catalysts (Littlechild et al., 2007; Hickey et al., 2007, 2009), provided that extreme process conditions (e.g., high temperature or high salt) do not cause unwanted disaggregation.

10.4 SELF-AGGREGATING TAGS AS PULL-DOWN MODULES

Elastin-like polypeptides (ELPs) contain the penta-peptide repeat Val–Pro–Gly–Xxx–Gly where Xxx is any amino acid (Meyer and Chilkoti, 1999). ELPs precipitate reversibly upon heating above the transition temperature. Numerous ELP fusions were developed with the aim of facilitating downstream processing of recombinant proteins through their selective precipitation from crude cell extracts of the host organism (Meyer et al., 2001; Banki and Wood 2005). The technology holds considerable promise for biocatalysis but has not been employed widely in the field (Fujita et al., 2009). One could imagine a general procedure in which an aggregate of ELP-enzyme is formed first and stabilized by an ensuing cross-linking step. Methods for cross-linking of elastin-like polypetides are proposed in the literature (Lim et al., 2007). An interesting alternative approach was proposed by Lin and co-workers, showing that fusion to amphipathic peptides capable of self-assembly during expression in *E. coli* resulted in the formation of biologically active aggregates of different enzymes, including lipase, xylosidase, and amadoriase (Wu et al., 2011a; Xing et al., 2011).

10.5 PULL-DOWN MODULES UNDERGOING ORDERED SELF-ASSEMBLY

Well known for lipids and nonnatural polymers, self-assembly into insoluble ordered structures is a feature recognized for a growing number of proteins (Sleytr et al., 2007). Using chimeric proteins in which the enzyme of interest is fused to a module capable of self-assembly, it could be possible to obtain insoluble enzyme preparations that have a well-defined, perhaps crystalline-like association state. Exploitation of self-assembly properties may allow one to achieve relatively precise control over the size and shape of the resulting enzyme particles. Although of interest, encapsulation through self-assembly of another protein or peptide entity (disjoined from the target enzyme) is not considered here.

Cell surface layers (S-layers) are common "protein membrane" structures of the cell envelope of archaea and bacteria. They have a lattice-like appearance and are formed through a self-assembly process. Isolated protein units of natural S-layers are capable of recrystallization into precisely defined monolayers in suspension and at different interfaces. These properties of S-layer proteins have promoted a wide range of (nano)-biotechnological applications (Sleytr et al., 2007). Although fusion to an S-layer protein has been explored for protein array development (Tang et al., 2008) and other analytical purposes (Sleytr et al., 2007), there are surprisingly few examples showing application of S-layer technologies to enzyme immobilization and biocatalyst development. However, a notable common feature of S-layer fusions was retention of the original function of the individual components of the chimeric protein, that is, the self-assembly properties and the respective biological activity (Sleytr et al., 2007).

A thermophilic laminarinase (LamA from *Pyrococcus furiosus*) fused to the S-layer protein SbpA from *Bacillus sphericus* was produced as an insoluble aggregate in *E. coli* that was unfortunately not characterized. However, after solubilization by chemical unfolding and renaturation, the SbpA–LamA fusion was immobilized successfully on different solid supports (Tschiggerl et al., 2008). Liposomes featuring a self-assembled fusion of an enzyme and an S-layer protein on the surface constitute an interesting approach to potentially useful biocatalysts. The easily tunable size of liposomes could provide flexibility in the design of an immobilized enzyme preparation. Using glucose-1-phosphate thymidyltransferase fused to the C-terminus of the S-layer protein SgsE from *Geobacillus stearothermophilus*, Schäfer et al. (2007) also demonstrated a possibility for recycling of the insoluble, liposome-supported catalyst.

It is worth noting that production of S-layer fusion proteins in *E. coli* and other hosts (*Saccharomycescerevisiae* or HeLa cells) generally appears to yield insoluble products resulting from intracellular self-assembly or aggregation (Blecha et al., 2005). These "inclusion bodies" might already be useful as carrier-free insoluble enzyme preparations. Nonamyloidogenic peptide tags might also be useful for fusion to enzymes. Used previously to generate all-protein nanoparticles, they might become useful in the preparation of insoluble ordered enzyme aggregates.

10.6 AUTO-ENCAPSULATION

The silaffin polypeptide from the diatom *Cylindrotheca fusiformis* induces precipitation of silica from monosilicic acid solution. The repeat unit of the silaffin polypeptide, the R5 peptide (H_2N–SSKKSGSYSGSKGSKRRIL–COOH), was utilized to immobilize butyrylcholinesterase. Highly active and stable biosilica spheres were obtained (Luckarift et al., 2004; Betancor and Luckarift, 2008). The addition of R5 peptide to a solution containing enzyme and silicic acid promoted biosilification at room temperature and under mild chemical reaction conditions (Naik et al., 2004). Using R5 fusions, phosphodiesterase and organophosphate hydrolase peptide were recently produced as immobilized enzyme preparations in *E. coli*. Each of the fusion proteins initiated silica polycondensation, and enzymatic activity (69 to 97%) was retained in the resulting silica spheres (Marner et al., 2009). An additional metal ion-chelating peptide $(His)_6$-tag enabled direct recovery of the "silified" insoluble catalyst from crude cell extract, as demonstrated for D-amino acid oxidase (Chien and Lee, 2007). *In vivo* immobilization of an enzyme in a biomineral has also been demonstrated (Poulsen et al., 2007).

10.7 EVEN SIMPLER ENZYMATICALLY ACTIVE INCLUSION BODIES DERIVED FROM RECOMBINANT PROTEINS AS SUCH

Work of Villaverde and colleagues (Sans et al., 2012) showed that variation in conditions used for inclusion body production are useful to modulate the properties of the resulting enzyme aggregates. Key properties include the specific enzyme activity, but also the size and other morphological characteristics of the protein particles. Insoluble preparations of an *E. coli* fuculose-1-phosphate aldolase were used directly as recyclable catalysts in aldol condensation reactions. Lentikats immobilizates of the inclusion bodies were also obtained and shown to be active for two months without leakage of activity. Despite these notable developments, the clear disadvantage of working with "native" inclusion bodies is that conditions for *in vivo* aggregation have to be developed anew for each enzyme, just as precipitation conditions for CLEA production have to be established through case-specific optimization.

10.8 CONCLUSIONS

Different methods of carrier-free immobilization were discussed in this chapter. Pros and cons related to biocatalyst preparation and process operation were pointed out. Among the methods available, CLEA technology is clearly the one most advanced and widely used. However, fusion proteins designed for aggregation using PDMs present a potentially useful alternative (Figure 10.1). They could be applied to improve the current protocols for production of CLEAs. Coaggregation of two or more enzymes harboring the same PDM should be possible in principle. Encapsulation or an other form of processing postaggregation may eventually be necessary in all cases to obtain mechanically stable biocatalysts. In addition to the procedures reported (Nahálka et al., 2008b), new protocols might be utilized (Wu et al., 2011b). The preparation of composites from protein aggregates and other materials (e.g., silica) could be an interesting option.

REFERENCES

Banki MR, Wood DW (2005) Intensin and affinity resin substitutes for protein purification and scale up. Microb Cell Fact 4:32.

Betancor L, Luckarift HR, (2008) Bioinspired enzyme encapsulation for biocatalysis. Trends Biotechnol 26:566–572.

Blecha A, Zarschler K, Sjollema KA, Veenhuis M, Rödel G (2005) Expression and cytosolic assembly of the S-layer fusion protein mSbsC-EGFP in eukaryotic cells. Microb Cell Fact 4:28.

Brady D, Jordaan J (2009) Advances in enzyme immobilisation. Biotechnol Lett 31:1639–1650.

Buchholz K, Kasche V, Bornscheuer UT (2005) Biocatalysts and Enzyme Technology. Weinheim, Germany: Wiley-VCH.

Cao L (2005) Carrier-Bound Immobilized Enzymes: Principles, Applications and Design. Weinheim, Germany: Wiley-VCH.

Cao L, van Rantwijk F, Sheldon RA (2000) Cross-linked enzyme aggregates: a simple and effective method for the immobilization of penicillin acylase. Org Lett 2:1361–1364.

Chien L-J, Lee C-K (2007) Biosilicification of dual-fusion enzyme immobilized on magnetic nanoparticle. Biotechnol Bioeng 100:223–230.

Fujita Y, Mie M, Kobatake E (2009) Construction of nanoscale protein particle using temperature-sensitive elastin-like peptide and polyaspartic acid chain. Biomaterials 30:3450–3457.

García-Fruitós E (2010) Inclusion bodies: a new concept. Microb Cell Fact 9:80.

García-Fruitós E, Aris A, Villaverde A (2007) Localization of functional polypeptides in bacterial inclusion bodies. Appl Environ Microbiol 73:289–294.

García-Fruitós E, Vázquez E, Diez-Gil C, Corchero JL, Seras-Franzoso J, Ratera I, Veciana J, Villaverde A (2011) Bacterial inclusion bodies: making gold from waste. Trends Biotechnol 30:65–70.

Gatti-Lanfranconi P, Natalello A, Ami D, Doglia SM, Lotti M (2011) Concepts and tools to exploit the potential of bacterial inclusion bodies in protein science and biotechnology. FEBS J 278:2408–2418.

González-Montalbán N, García-Fruitós E, Villaverde A (2007) Recombinant protein solubility—Does more mean better? Nat Biotechnol 23:718–720.

Guisan JM (Editor) (2006) Immobilization of enzymes and cells. Methods Biotechnol 22.

Hickey AM, Marle L, McCreedy T, Watts P, Greenway GM, Littlechild JA (2007) Immobilization of thermophilic enzymes in miniaturized flow reactors. Biochem Soc Trans 35:1621–1623.

Hickey AM, Ngamsom B, Wiles C, Greenway GM, Watts P, Littlechild JA (2009) A microreactor for the study of biotransformations by a cross-linked γ-lactamase enzyme. Biotechnol J 4:510–516.

Huang Z, Zhang C, Chen S, Ye F, Xing XH (2013) Active inclusion bodies of acid phosphatase PhoC: aggregation induced by GFP fusion and activities modulated by linker flexibility. Microb Cell Fact 12:25.

Hudson S, Cooney J, Magner E (2008) Proteins in mesoporous silicates. Angew Chem Int Ed Engl 47:8582–8594.

Huisman GW, Collier SW (2013) On the development of new biocatalytic processes for practical pharmaceutical synthesis. Curr Opin Chem Biol 17:284–292.

Illanes A, Cauerhff A, Wilson L, Castro GR (2012) Recent trends in biocatalysis engineering. Biores Technol 115:48–57.

Jimenez-Gonzalez C, Constable DJC (2011) Green Chemistry and Engineering: A Practical Design Approach. Hoboken, NJ: Wiley-VCH.

Kubáč D, Kaplan O, Elišáková V, Pátek M, Vejvoda V, Slámová K, Tóthová A, Lemaire M, Gallienne E, Lutz-Wahl S, Fischer L, Kuzma M, Pelantová H, van Pelt S, Bolte J, Křen V, Martínková L (2008) Biotransformation of nitriles to amides using soluble and immobilized nitrile hydratase from *Rhodococcus erythropolis* A4. J Mol Catal B50:107–113.

Liese A, Hilterhaus L (2013) Evaluation of enzymes for industrial application. Chem Soc Rev 42:6236–6249.

Lim DW, Nettles DL, Setton LA, Chilkoti A (2007) Rapid cross-linking of elastin-like polypeptides with hydroxymethylphosphines in aqueous solution. Biomacromol 8:1463–1470.

Littlechild JA, Guy J, Connelly S, Mallett L, Waddell S, Rye CA, Line K, Isupov M (2007) Natural methods of protein stabilization: thermostable biocatalysts. Biochem Soc Trans 35:1558–1563.

López-Gallego F, Betancor L, Hidalgo A, Alonso N, Fernández-Lafuente R, Guisán JM (2005) Co-aggregation of enzymes and polyethyleneimine: a simple method to prepare stable and immobilized derivatives of glutaryl acylase. Biomacromol 6:1839–1842.

Luckarift HR, Spain JC, Naik RR, Stone MO (2004) Enzyme immobilization in a biomimetic silica support. Nat Biotechnol 22:211–213.

Marner WD, Shaikh AS, Muller SJ, Keasling JD (2009) Enzyme immobilization via silaffin-mediated autoencapsulation in a biosilica support. Biotechnol Prog 25:417–423.

Mateo C, Palomo JM, van Langen LM, van Rantwijk F, Sheldon RA (2004) A new, mild cross-linking methodology to prepare cross-linked enzyme aggregates. Biotechnol Bioeng 86:273–276.

Mateo C, Chmura A, Rustler S, van Rantwijk F, Stolz A, Sheldon RA (2006) Synthesis of enantiomerically pure (*S*)-mandelic acid using an oxynitrilase–nitrilase bienzymatic cascade: a nitrilase surprisingly shows nitrile hydratase activity. Tetrahedron: Asymmetry 17:320–323.

Meyer DE, Chilkoti A (1999) Purification of recombinant proteins by fusion with thermally-responsive polypeptides. Nat Biotechnol 17:1112–1115.

Meyer DE, Trabbic-Carlson K, Chilkoti A (2001) Protein purification by fusion with an environmentally responsive elastin-like polypeptide: effect of polypeptide length on the purification of thioredoxin. Biotechnol Prog 17:720–728.

Nahálka J (2008) Physiological aggregation of maltodextrin phosphorylase from *Pyrococcus furiosus* and its application in a process of batch starch degradation to α-D-glucose-1-phosphate. J Ind Microbiol Biotechnol 35:219–223.

Nahálka J, Nidetzky B, (2007) Fusion to a pull-down domain: a novel approach of producing *Trigonopsis variabilis* D-amino acid oxidase as insoluble enzyme aggregates. Biotechnol Bioeng 97:454–461.

Nahálka J, Pätoprstý V (2009) Enzymatic synthesis of sialylation substrates powered by a novel polyphosphate kinase (PPK3). Org Biomol Chem 7:1778–1780.

Nahálka J, Vikartovská A, Hrabárová E (2008a) A cross-linked inclusion body process for sialic acid synthesis. J Biotechnol 134:146–153.

Nahálka J, Dib I, Nidetzky B (2008b) Encapsulation of *Trigonopsis variabilis* D-amino acid oxidase and fast comparison of the operational stabilities of free and immobilized preparations of the enzyme. Biotechnol Bioeng 99:251–260.

Nahálka J, Mislovičová D, Kavcová H (2009) Targeting lectin activity into inclusion bodies for the characterisation of glycoproteins. Mol Biosyst 5:819–821.

Naik RR, Tomczak MM, Luckarift HR, Spain JC, Stone MO (2004) Entrapment of enzymes and nanoparticles using biomimetically synthesized silica. Chem Commun1684–1685.

Nestl B, Nebel BA, Hauer B (2011) Recent progress in industrial biocatalysis. Curr Opin Chem Biol 15:187–193.

Patel RN (Editor) (2007) Biocatalysis in the Pharmaceutical and Biotechnology Industries. CRC Press: Boca Raton, FL.

Poulsen N, Berne C, Spain J, Kröger N (2007) Silica immobilization of an enzyme through genetic engineering of the diatom *Thalassiosira pseudonana*. Angew Chem 119:1875–1878.

Rao RN, Lütz S, Würges K, Minör D (2009) Continuous biocatalytic processes. Org Proc Res Dev 13:607–616.

Roessl U, Nahálka J, Nidetzky B (2010) Carrier-free immobilized enzymes for biocatalysis. Biotechnol Lett 32:341–350.

Sans C, Ferraz RM, González-Montalbán N, Rinas U, Lopez-Santin J, Villaverde A, Alvaro G (2012) Inclusion bodies of fuculose-1-phosphate aldolase as stable and reusable biocatalysts. Biotechnol Prog 28:421–427.

Schäffer C, Novotny R, Küpcü S, Zayni S, Scheberl A, Friedmann J, Sleytr UB, Messner P (2007) Novel biocatalysts based on S-layer self-assembly of *Geobacillus stearothermophilus* NRS 2004/3a: a nanobiotechnological approach. Small 3:1549–1559.

Sheldon RA (2011) Characteristic features and biotechnological applications of cross-linked enzyme aggregates (CLEAs). Appl Microbiol Biotechnol 92:467–477.

Sheldon RA (2012) Fundamentals of green chemistry: efficiency in reaction design. Chem Soc Rev 41:1437–1451.

Sleytr UB, Huber C, Ilk N, Pum D, Schuster B, Egelseer EM (2007) S-layers as a tool kit for nanobiotechnological applications. Microbiol Lett 267:131–144.

Tang J, Badelt-Lichtblau H, Ebner A, Preiner J, Kraxberger B, Gruber HJ, Sleytr UB, Ilk N, Hinterdorfer P (2008) Fabrication of highly ordered gold nanoparticle arrays templated by crystalline lattices of bacterial S-layer protein. ChemPhysChem 9:2317–2320.

Torres-Salas P, del Monte-Martinez A, Cutino-Avila B, Rodriguez-Colinas B, Alcalde M, Ballesteros AO, Plou FJ (2011) Immobilized biocatalysts: novel approaches and tools for binding enzymes to supports. Adv Mater 23:5275–5282.

Tschiggerl H, Breitwieser A, de Roo G, Verwoerd T, Schäffer C, Sleytr UB (2008) Exploitation of the S-layer self-assembly system for site directed immobilization of enzymes demonstrated for an extremophilic laminarinase from *Pyrococcus furiosus*. J Biotechnol 133:403–411.

Ventura S, Villaverde A (2006) Protein quality in bacterial inclusion bodies. Trends Biotechnol 24:179–185.

Vera A, Gonzalez-Montalban N, Aris A, Villaverde A (2007) The conformational quality of insoluble recombinant proteins is enhanced at low growth temperatues. Biotechnol Bioeng 96:1101–1106.

Wilson L, Illanes A, Pessela BCC, Abian O, Fernández-Lafuente R, Guisán JM (2004) Encapsulation of cross-linked penicillin G acylase aggregates in lentikats: evaluation of a novel biocatalyst in organic media. Biotechnol Bioeng 86, 5:558–562.

Wu W, Xing L, Zhou B, Lin Z (2011a) Active protein aggregates induced by terminally attached self-assembling peptide ELK16 in *Escherichia coli*. Microb Cell Fact 10:9.

Wu C, Bai S, Ansorge-Schumacher MB, Wang D (2011b) Nanoparticle cages for enzyme catalysis in organic media. Adv Mater 23:5694–5699.

Xing L, Wu W, Zhou B, Lin Z (2011) Streamlined protein expression and purification using cleavable self-aggregating tags. Microb Cell Fact 10:42.

INDEX

Page numbers followed by "*f*" and "*t*" indicate figures and tables.

Protein Aggregation in Bacteria: Functional and Structural Properties of Inclusion Bodies in Bacterial Cells, First Edition. Edited by Silvia Maria Doglia and Marina Lotti.